Introduction

This book has been written to support the OCR free standing unit, Additional Mathematics, but you may also use it independently as an introduction to mathematics beyond GCSE. It is expected that many of the students using this book could be working with little day-to-day teacher support. With this in mind the text has been writen in an interactive way and the answers are fuller than is often the case in books of this nature.

There are two symbols that may need some introduction.

? This denotes a 'discussion point'. These are prompts to help you to understand the theory that has been, or is about to be, introduced. Answers to these are also included.

! This 'warning sign' is alerting you, either to restrictions that need to be imposed or to possible pitfalls.

In addition the book includes a number of activities. These are often used to introduce a new concept, or to reinforce the examples in the text. Throughout the book the emphasis is on understanding the mathematics being used rather than merely being able to perform the calculations but the exercises do, nonetheless, provide plenty of scope for practising basic techniques.

It is hoped that students who use this book will develop a fascination for mathematics and will be inspired to study the subject at a higher level.

I would like to thank Roger Porkess in particular for the numerous suggestions that he has made, and also my students who have, often unwittingly, trialled many of the exercises.

Val Hanrahan

Additional Mathematics for OCR

Editor: Roger Porkess

Val Hanrahan

Acknowledgements

We are grateful to the following companies, institutions and individuals who have given permission to reproduce photographs in this book. Every effort has been made to trace and acknowledge ownership of copyright. The publishers will be glad to make suitable arrangements with any copyright holder it has not been possible to contact.

Corbis; pages 60, 73, 121, 124 and 213 (top); Associated Press, AP; page 169; National Portrait Gallery; London, page 213 (bottom).

Exam questions on pages 57 and 126 reproduced with the kind permission of OCR.

Artwork was drawn by Jeff Edwards and Josephine Blake

Map
The cartographic information which appears on page 59 in this publication was supplied by
© Automobile Association Developments Ltd 2003 All rights reserved Licence 399221. A02062
The map on page 155 was adapted with permission of Philip's, 11 Salusbury Road, London, NW6 6RG

Orders: please contact Bookpoint Ltd, 130 Milton Park, Abingdon, Oxon OX14 4SB. Telephone: (44) 01235 827720. Fax: (44) 01235 400454. Lines are open from 9.00 to 5.00, Monday to Saturday, with a 24-hour message-answering service. You can also order through our website *www.hodderheadline.co.uk*.

British Library Cataloguing in Publication Data
A catalogue record for this title is available from the British Library

ISBN: 978 0 340 86960 4

First Published 2004
Impression number 15
Year 2011

Paper used in this book are natural, renewable and recyclable products. They are made from wood grown in sustainable forests. The logging and manufacturing processes conform to the environmental regulations of the country of origin.

Typeset by Pantek Arts Ltd, Maidstone, Kent.
Printed in Great Britain for Hodder Education, an Hachette UK company,
338 Euston Road, London NW1 3BH by the MPG Books Group, Bodmin.

Contents

SECTION 1 ALGEBRA

SECTION 2 CO-ORDINATE GEOMETRY

SECTION 3 TRIGONOMETRY

SECTION 4 CALCULUS

SECTION 1
Algebra

1 Algebra I – review

The only thing that separates successful people from the ones who aren't is the willingness to work very, very hard.

Helen Gurley Brown (American Businesswoman)

❓ A large ice cream costs 40p more than a small one. Two large ice creams cost the same as three small ones. What is the cost of each size of ice cream?

This is an example of the type of question that you might find in a puzzle book or the puzzle section of a newspaper or magazine. How would you set about tackling it?

❓ You may think that the following question appears to be very similar. What happens when you try to solve it?

A large ice cream costs 40p more than a small one. Five small ice creams plus three large ones cost 80p less than three small ice creams plus five large ones. What is the cost of each size of ice cream?

Linear expressions

When you are asked to *simplify* an algebraic expression you need to write it in its most compact form. This will involve techniques such as collecting like terms, removing brackets, factorising and finding a common denominator (if the equation includes fractions).

EXAMPLE 1.1

Simplify this expression.

$$3a + 4b - 2c + a - 3b - c$$

SOLUTION

$$\begin{aligned}\text{Expression} &= 3a + a + 4b - 3b - 2c - c \quad \text{collecting like terms}\\ &= 4a + b - 3c\end{aligned}$$

EXAMPLE 1.2

Simplify this expression.

$$2(3x - 4y) - 3(x + 2y)$$

Notice that $(-3)(2y) = -6y$.

SOLUTION

$$\begin{aligned}\text{Expression} &= 6x - 8y - 3x - 6y \quad \text{removing the brackets}\\ &= 3x - 14y\end{aligned}$$

EXAMPLE 1.3

Simplify this expression.

$$3x^2yz \times 2xy^3$$

SOLUTION

$$\begin{aligned}\text{Expression} &= (3 \times 2) \times (x^2 \times x) \times (y \times y^3) \times z \quad \text{collecting like terms}\\ &= 6x^3y^4z\end{aligned}$$

EXAMPLE 1.4

Simplify this expression.

$$\frac{6a^2b^3c}{3ab^4c^3}$$

SOLUTION

Look where the higher powers of a, b and c occur. You may find that it helps if you split them up like this.

$$\frac{6a^2b^3c}{3ab^4c^3} = \frac{\cancel{3} \times 2 \times \cancel{a} \times a \times \cancel{b^3} \times \cancel{c}}{\cancel{3} \times \cancel{a} \times \cancel{b^3} \times b \times \cancel{c} \times c^2}$$

You can then cancel as indicated to give

$$\frac{2a}{bc^2}$$

EXAMPLE 1.5

Factorise this expression.

$$3a^2b + 6ab^2$$

SOLUTION

First you need to look for the highest common factor of the two terms, which is $3ab$ here.

$$3a^2b + 6ab^2 = 3ab(a + 2b)$$

Since $3a^2b = 3ab \times a$ and $6ab^2 = 3ab \times 2b$.

? Explain what the word *factorise* means.

EXAMPLE 1.6

Simplify this expression.

$$\frac{2x^2}{3yz} \div \frac{4xy^2}{5z^2}$$

SOLUTION

$$\text{Expression} = \frac{2x^2}{3yz} \times \frac{5z^2}{4xy^2}$$
$$= \frac{5xz}{6y^3}$$

EXAMPLE 1.7

Simplify this expression.

$$\frac{x}{4t} - \frac{2y}{5t} + \frac{z}{2t}$$

SOLUTION

$$\text{Expression} = \frac{5x}{20t} - \frac{8y}{20t} + \frac{10z}{20t}$$
$$= \frac{5x - 8y + 10z}{20t}$$

$20t$ is the common denominator for $4t$, $5t$ and $2t$.

EXERCISE 1A

1 Simplify the following expressions.

(i) $12a + 3b - 7c - 2a - 4b + 5c$

(ii) $4x - 5y + 3z + 2x + 2y - 7z$

(iii) $3(5x - y) + 4(x + 2y)$

(iv) $2(p + 5q) - (p - 4q)$

(v) $x(x + 3) - x(x - 2)$

(vi) $a(2a + 3) + 3(3a - 4)$

(vii) $3p(q - p) - 3q(p - q)$

(viii) $5f(g + 2h) - 5g(h - f)$

2 Factorise the following expressions by taking out the highest common factor.

(i) $8 - 10x^2$

(ii) $6ab + 8bc$

(iii) $2a^2 + 4ab$

(iv) $pq^3 - p^3q$

(v) $3x^2y + 6xy^4$

(vi) $6p^3q - 4p^2q^2 + 2pq^3$

(vii) $15lm^2 - 9l^3m^3 + 12l^2m^4$

(viii) $84a^5b^4 - 96a^4b^5$

3 Simplify the following expressions and factorise the answers.

(i) $4(3x + 2y) + 8(x - 3y)$

(ii) $x(x - 2) - x(x - 8) + 6$

(iii) $x(y + z) - y(x + z)$

(iv) $p(2q - r) + r(p - 2q)$

(v) $k(l + m + n) - km$

(vi) $a(a-2)-a(a+4)+2(a-4)$
(vii) $3x(x+y)-3y(x-2y)$
(viii) $a(a-2)-a(a-4)+8$

4 Simplify the following expressions as much as possible.
(i) $2a^2b\times 5ab^3$
(ii) $6p^3q\times 2q^3r$
(iii) $lm\times mn\times np$
(iv) $3r^3\times 6s^2\times 2rs$
(v) $ab\times 2bc\times 4cd\times 8de$
(vi) $3xy^2\times 4yz^2\times 5x^2z$
(vii) $2ab^3\times 6a^4\times 7b^6$
(viii) $6p^2q^3r\times 7pq^5r^4$

5 Simplify the following fractions as much as possible.
(i) $\dfrac{4a^2b}{2ab}$
(ii) $\dfrac{p^2}{q}\times\dfrac{q^2}{p}$
(iii) $\dfrac{8a}{3b^2}\times\dfrac{6b^3}{4a^2}$
(iv) $\dfrac{3ab}{2c^2}\times\dfrac{4cd}{6a^2}$
(v) $\dfrac{8xy^3z^2}{12yz}$
(vi) $\dfrac{3a^2}{9b^3}\div\dfrac{2a^4}{15b}$
(vii) $\dfrac{5p^3q}{8rs^2}\div\dfrac{15pq^5}{28r^4}$

6 Simplify the following expressions as single fractions.
(i) $\dfrac{2a}{3}+\dfrac{a}{4}$
(ii) $\dfrac{2x}{5}-\dfrac{x}{2}+\dfrac{3x}{4}$
(iii) $\dfrac{4p}{3}-\dfrac{3p}{4}$
(iv) $\dfrac{2s}{5}-\dfrac{s}{3}+\dfrac{4s}{15}$
(v) $\dfrac{3b}{8}-\dfrac{b}{6}+\dfrac{5b}{24}$
(vi) $\dfrac{3a}{b}-\dfrac{2a}{3b}$
(vii) $\dfrac{5}{2p}-\dfrac{3}{2q}$
(viii) $\dfrac{2x}{3y}-\dfrac{3x}{2y}$

Solving linear equations

What is an equation?
What does solving an equation mean?

Since both sides of an equation are equal, you may do what you wish to the equation, provided that you do exactly the same thing to both sides. The examples that follow illustrate this in great detail. In practice you would expect to omit some of the working.

EXAMPLE 1.8

Solve this equation.

$$3(3x-17) = 2(x-1)$$

SOLUTION

Open the brackets	$\Rightarrow\ 9x-51$	$= 2x-2$
Subtract $2x$ from both sides	$\Rightarrow\ 9x-51-2x$	$= 2x-2-2x$
Tidy up	$\Rightarrow\ 7x-51$	$= -2$
Add 51 to both sides	$\Rightarrow\ 7x-51+51$	$= -2+51$
Tidy up	$\Rightarrow\ 7x$	$= 49$
Divide both sides by 7	$\Rightarrow\ x$	$= 7$

EXAMPLE 1.9

Solve this equation.

$$\tfrac{1}{2}(x+8) = 2x + \tfrac{1}{3}(4x-5)$$

SOLUTION

Start by clearing the fractions by multiplying by 6 (the least common multiple of 2 and 3).

Multiply both sides by 6	$\Rightarrow\ 6\times\tfrac{1}{2}(x+8)$	$= 6\times 2x + 6\times\tfrac{1}{3}(4x-5)$
Tidy up	$\Rightarrow\ 3(x+8)$	$= 12x + 2(4x-5)$
Open the brackets	$\Rightarrow\ 3x+24$	$= 12x+8x-10$
Tidy up	$\Rightarrow\ 3x+24$	$= 20x-10$
Subtract $3x$ from both sides	$\Rightarrow\ 24$	$= 17x-10$
Add 10 to both sides	$\Rightarrow\ 34$	$= 17x$
Divide both sides by 17	$\Rightarrow\ x$	$= 2$

? Why have the variable and the number changed sides on the last line?

Sometimes you will need to set up the equation as well as solve it. When you are doing this, make sure that you define any variables that you introduce.

EXAMPLE 1.10

In a triangle, the largest angle is nine times as big as the smallest. The third angle is 60°.

(i) Write this information in the form of an equation for a, the size in degrees of the smallest angle.

(ii) Solve the equation to find the sizes of the three angles.

SOLUTION

Let the smallest angle = $a°$.
So the largest angle is $9a°$.
The sum of all three angles is 180°,

$$\begin{aligned} \Rightarrow\quad a + 9a + 60 &= 180 \\ \Rightarrow\quad 10a &= 120 \\ \Rightarrow\quad a &= 12 \end{aligned}$$

This gives $9a = 108$, so the angles are 12°, 60° and 108°.

EXERCISE 1B

1 Solve the following equations.

(i) $2x - 3 = x + 4$

(ii) $5a + 3 = 2a - 3$

(iii) $2(x + 5) = 14$

(iv) $7(2y - 5) = -7$

(v) $5(2c - 8) = 2(3c - 10)$

(vi) $3(p + 2) = 4(p - 1)$

(vii) $3(2x - 1) = 6(x + 2) + 3x$

(viii) $\frac{x}{3} + 7 = 5$

(ix) $\frac{5y - 2}{11} = 3$

(x) $\frac{k}{2} + \frac{k}{3} = 35$

(xi) $\frac{2t}{3} - \frac{3t}{5} = 4$

(xii) $\frac{5p - 4}{6} - \frac{2p + 3}{2} = 7$

(xiii) $p + \frac{1}{3}(p + 1) + \frac{1}{4}(p + 2) = \frac{5}{6}$

2 The length, l metres, of a field is 80 m greater than the width. The perimeter is 600 m.

(i) Write the information in the form of an equation for l.

(ii) Solve the equation and so find the area of the field.

3 Louise and Molly are twins and their brother Jonathan is four years younger. The total of their three ages is 17 years.

(i) Write this information in the form of an equation in j, Jonathan's age in years.

(ii) What are all their ages?

4 In a multiple-choice examination of 20 questions, four marks are given for each correct answer and one mark is deducted for each wrong answer. There is no penalty for not attempting a question. A candidate attempts a questions and gets c correct.

(i) Write down, and simplify, an expression for the candidate's total mark in terms of a and c.

(ii) A candidate attempts three-quarters of the questions and scores 40. Write down, and solve, an equation for the number of correct questions.

5 John is three times as old as his son, Michael, and in x years' time he will be twice as old as him.

(i) Write down expressions for John's and Michael's age in x years' time.

(ii) Write down, and solve, an equation in x.

6 A square has sides of length $2a$ metres, and a rectangle has length $3a$ metres and breadth 3 metres.

(i) Find, in terms of a, the perimeter of the square.

(ii) Find, in terms of a, the perimeter of the rectangle.

(iii) The perimeters of the square and the rectangle are equal. Find a.

7 The sum of five consecutive numbers is equal to 105. Let m represent the middle number.

(i) Write down the five numbers in terms of m.

(ii) Form an equation in m and solve it.

(iii) What are the five consecutive numbers?

8 One rectangle has a length of $(x + 2)$ cm and a breadth of 2 cm, and another, of equal area, has a length of 5 cm and a breadth of $(x - 3)$ cm.

(i) Write down an equation in x and solve it.

(ii) What is the area of each of the rectangles?

Changing the subject of an equation

The circumference of a circle is given by

$$C = 2\pi r$$

where r is the radius. An equation such as this is often called a formula.

C is called the *subject* of the formula. Explain what this means.

In some cases, you want to calculate r directly from C. You want r to be the subject of the formula.

EXAMPLE 1.11

Make r the subject of $C = 2\pi r$.

SOLUTION

Divide both sides by 2π $\Rightarrow$ $\dfrac{C}{2\pi} = r$

$$\Rightarrow \quad r = \frac{C}{2\pi}$$

Notice how the new subject should be on its own on the left-hand side of the new formula.

EXAMPLE 1.12

Make x the subject of this formula.

$$h = \sqrt{(x^2 + y^2)}$$

SOLUTION

Square both sides	$\Rightarrow$	$h^2 = x^2 + y^2$
Subject y^2 from both sides	$\Rightarrow$	$h^2 - y^2 = x^2$
Lead with the x^2 term	$\Rightarrow$	$x^2 = h^2 - y^2$
Take the square root of both sides	$\Rightarrow$	$x = \pm\sqrt{(h^2 - y^2)}$

? What would you do with the ± sign in the case where h is the hypotenuse of a right-angled triangle with x and y as the other two sides?

EXAMPLE 1.13

Make a the subject of this formula.

$$v = u + at$$

SOLUTION

Subtract u from both sides	$\Rightarrow$	$v - u = at$
Divide both sides by t	$\Rightarrow$	$\dfrac{v-u}{t} = a$
Write the answer with a on the left-hand side	$\Rightarrow$	$a = \dfrac{v-u}{t}$

EXERCISE 1C

In this exercise all the equations refer to real situations. How many of them can you recognise?

1 Make **(i)** u **(ii)** a the subject of $v = u + at$.

2 Make b the subject of $A = \frac{1}{2}bh$.

3 Make l the subject of $P = 2(l + b)$.

4 Make r the subject of $A = \pi r^2$.

5 Make c the subject of $A = \frac{1}{2}(b + c)h$.

6 Make h the subject of $A = \pi r^2 + 2\pi rh$.

7 Make l the subject of $T = \dfrac{\lambda e}{l}$.

8 Make **(i)** u **(ii)** a the subject of $s = ut + \frac{1}{2}at^2$.

9 Make x the subject of $v^2 = \omega^2(a^2 - x^2)$.

10 Make g the subject of $T = 2\pi\sqrt{\frac{l}{g}}$.

11 Make f the subject of $\frac{1}{f} = \frac{1}{u} + \frac{1}{v}$.

12 Make g the subject of $E = mgh + \frac{1}{2}mv^2$.

Quadratic expressions

Expansion

A quadratic expression is one in which the highest power of its terms is 2. For example,

$x^2 + 3$
a^2
$2y^2 - 3y + 5$

are all quadratic expressions.

Why is $(x + 5)(2x - 3)$ a quadratic expression?

EXAMPLE 1.14

Expand $(x + 5)(2x - 3)$.

SOLUTION

$$\begin{aligned}\text{Expression} &= x(2x - 3) + 5(2x - 3)\\ &= 2x^2 - 3x + 10x - 15\\ &= 2x^2 + 7x - 15\end{aligned}$$

This method has multiplied everything in the second bracket by each term in the first bracket. An alternative way of setting this out is used in the next example.

EXAMPLE 1.15

Expand $(3x - 5)^2$.

SOLUTION

$$\text{Expression} = (3x - 5)(3x - 5)$$

Write the square as the product of two brackets so you don't forget the middle term.

$$\begin{array}{r} 3x - 5\\ 3x - 5\\ \hline 9x^2 - 15x \qquad\\ -15x + 25\\ \hline 9x^2 - 30x + 25\\ \hline \end{array}$$

Multiply the top line by $3x$.

Multiply the top line by -5.

Add these two lines together.

Factorisation

The reverse process is called factorisation. Although you will factorise here according to a set of rules, these are sometimes time consuming to apply. If you have already learnt another method, and use it quickly and accurately, then you should stick with it. With practice, you may be able to factorise some of these expressions *by inspection.*

EXAMPLE 1.16

Factorise $xa + xb + ya + yb$.

SOLUTION

First take out a common factor of each pair of terms.

$$\Rightarrow \quad xa + xb + ya + yb = x(a + b) + y(a + b)$$

Next notice that $(a + b)$ is now a common factor.

$$\Rightarrow \quad x(a + b) + y(a + b) = (a + b)(x + y)$$

In practice this relates to areas of rectangles.

$\longleftarrow x + y \longrightarrow$

$a + b$	x	y
a	xa	ya
b	xb	yb

Figure 1.1

The idea illustrated in figure 1.1 can be used to factorise a quadratic expression containing three terms, but first you must decide how to split up the middle term.

EXAMPLE 1.17

Factorise $x^2 + 6x + 8$.

SOLUTION

Splitting the $6x$ as $4x + 2x$ gives

$$\begin{aligned} x^2 + 6x + 8 &= x^2 + 4x + 2x + 8 \\ &= x(x + 4) + 2(x + 4) \\ &= (x + 4)(x + 2). \end{aligned}$$

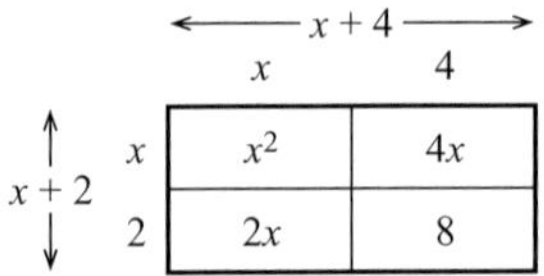

Figure 1.2

The crucial step is knowing how to split up the middle term.

Is the illustration in figure 1.2 the only possibility?

To answer this question, notice that:

- the numbers 4 and 2 have a sum of 6, which is the *coefficient* of x (i.e. the number multiplying x) in $x^2 + 6x + 8$.
- the numbers 4 and 2 have a product of 8 which is the *constant term* in $x^2 + 6x + 8$.

There is only one pair of numbers that satisfies both of these conditions.

EXAMPLE 1.18

Factorise $x^2 - 7x - 18$.

SOLUTION

Pairs of numbers with a product of (–18) are:

1 and (–18)
2 and (–9)
3 and (–6)
6 and (–3)
9 and (–2)
18 and (–1)

There is only one pair, 2 and (–9) with a sum of (–7), so use these.

$$\begin{aligned} x^2 - 7x - 18 &= x^2 + 2x - 9x - 18 \\ &= x(x + 2) - 9(x + 2) \\ &= (x + 2)(x - 9) \end{aligned}$$

Notice the sign change due to the – sign in front of the 9.

Do you get the same factors if the order in which you use the 2 and the (–9) is reversed so that you write it $x^2 - 9x + 2x - 18$?

NOTE

Since the pair of numbers that you are looking for is unique, you can stop listing products when you find one that has the correct sum.

EXAMPLE 1.19 Factorise $x^2 - 16$.

SOLUTION

First write

$$x^2 - 16 = x^2 + 0x - 16.$$

Pairs of numbers with a product of (–16) are:

1 and (–16)
2 and (–8)
4 and (–4),...(stop here)

The only pair with a sum of 0 is 4 and (–4).

$$\begin{aligned} x^2 - 16 &= x^2 + 4x - 4x - 16 \\ &= x(x + 4) - 4(x + 4) \\ &= (x + 4)(x - 4) \end{aligned}$$

This is an example of a special case called *the difference of two squares* since you have

$$x^2 - 4^2 = (x + 4)(x - 4)$$

In general

$$a^2 - b^2 = (a + b)(a - b)$$

Most people recognise this when it occurs and write down the answer straight away.

EXAMPLE 1.20 Factorise $4x^2 - 9y^2$.

SOLUTION

$$\begin{aligned} 4x^2 - 9y^2 &= (2x)^2 - (3y)^2 \\ &= (2x + 3y)(2x - 3y) \end{aligned}$$

The technique for finding how to split the middle term needs modifying for examples where the expression starts with a multiple of x^2. The difference is that you now multiply the two outside numbers together to give the product you want.

EXAMPLE 1.21 Factorise $2x^2 - 11x + 15$.

SOLUTION

A negative sum and a positive product means that both numbers are negative.

Here the sum is (-11) and the product is $2 \times 15 = 30$.
Options are:

(-1) and (-30)
(-2) and (-15)
(-3) and (-10)
(-5) and (-6)
(-6) and (-5) … (repeats)

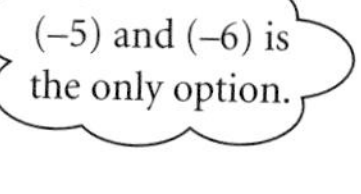

$$\begin{aligned} 2x^2 - 11x + 15 &= 2x^2 - 5x - 6x + 15 \\ &= x(2x - 5) - 3(2x - 5) \\ &= (2x - 5)(x - 3) \end{aligned}$$

EXERCISE 1D

1 Expand the following expressions.

(i) $(x + 5)(x + 4)$	**(ii)** $(x + 3)(x + 1)$
(iii) $(a + 5)(2a - 1)$	**(iv)** $(2p + 3)(3p - 2)$
(v) $(x + 3)^2$	**(vi)** $(2x + 3)(2x - 3)$
(vii) $(2 - 3m)(m - 4)$	**(viii)** $(6 + 5t)(2 - t)$
(ix) $(4 - 3x)^2$	**(x)** $(m - 3n)^2$

2 Factorise the following expressions.

(i) $x^2 + 5x + 6$	**(ii)** $y^2 - 5y + 4$
(iii) $m^2 - 8m + 16$	**(iv)** $m^2 - 8m + 15$
(v) $x^2 + 3x - 10$	**(vi)** $a^2 + 20a + 96$
(vii) $x^2 - x - 6$	**(viii)** $y^2 - 16y + 48$
(ix) $k^2 + 10k + 24$	**(x)** $k^2 - 10k - 24$

3 Each of these is a difference of two squares. Factorise them.

(i) $x^2 - 4$	**(ii)** $a^2 - 25$
(iii) $9 - p^2$	**(iv)** $x^2 - y^2$
(v) $t^2 - 64$	**(vi)** $4x^2 - 1$
(vii) $4x^2 - 9$	**(viii)** $4x^2 - y^2$
(ix) $16x^2 - 25$	**(x)** $9a^2 - 4b^2$

4 Factorise the following expressions.

(i) $2x^2 + 5x + 2$	**(ii)** $2a^2 + 11a - 21$
(iii) $15p^2 + 2p - 1$	**(iv)** $3x^2 + 8x - 3$
(v) $5a^2 - 9a - 2$	**(vi)** $2p^2 + 5p - 3$
(vii) $8x^2 + 10x - 3$	**(viii)** $2a^2 - 3a - 27$
(ix) $9x^2 - 30x + 25$	**(x)** $4x^2 + 4x - 15$

Solving a quadratic equation that factorises

EXAMPLE 1.22

Solve $x^2 + 3x - 18 = 0$.

SOLUTION

$$x^2 + 3x - 18 = 0$$
$$\Rightarrow \quad x^2 + 6x - 3x - 18 = 0$$
$$\Rightarrow \quad x(x + 6) - 3(x + 6) = 0$$
$$\Rightarrow \quad (x + 6)(x - 3) = 0$$

The only way this expression can ever equal zero is if one of the brackets equals zero.

$$\Rightarrow \quad \text{either } (x + 6) = 0 \text{ or } (x - 3) = 0$$
$$\Rightarrow \quad x = -6 \text{ or } x = 3$$

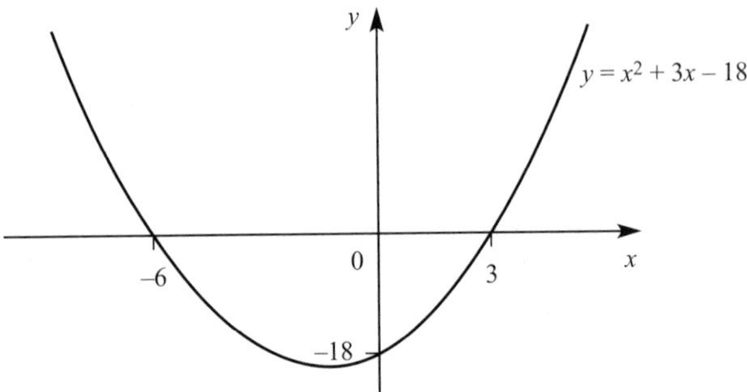

Figure 1.3

NOTE

The *solution* of the equation is the *pair* of values $x = -6$ or $x = 3$. The *roots* of the equation are the *individual* values $x = -6$ and $x = 3$.

Before starting to solve a quadratic equation, you must make sure that all terms of the quadratic expression are on the left-hand side of the equation.

EXAMPLE 1.23

Solve $8x^2 + 10x = 3$.

SOLUTION

First, rewrite the equation as

$$8x^2 + 10x - 3 = 0.$$
$$\Rightarrow \quad 8x^2 + 12x - 2x - 3 = 0$$
$$\Rightarrow \quad 4x(2x + 3) - 1(2x + 3) = 0$$

Sum of + 10 and product $8 \times (-3) = -24$ gives (12) and (–2).

Common factor of $-2x$ and -3 is -1.

$$\Rightarrow \quad (2x+3)(4x-1)=0$$
$$\Rightarrow \quad (2x+3)=0 \text{ or } (4x-1)=0$$
$$\Rightarrow \quad 2x=-3 \text{ or } 4x=1$$
$$\Rightarrow \quad x=-\tfrac{3}{2} \text{ or } x=\tfrac{1}{4}$$

Completing the square

Sometimes you need to solve a quadratic equation that does not factorise.

EXAMPLE 1.24

Solve $x^2 - 8x + 3 = 0$.

SOLUTION

Subtract the constant term from both sides of the equation.

$$\Rightarrow \quad x^2 - 8x = -3$$

Take the coefficient of x -8
Halve it -4
Square the answer $+16$

Add 16 to both sides of the equation $\Rightarrow \quad x^2 - 8x + 16 = -3 + 16$

Factorise the left-hand side $\Rightarrow \quad (x-4)^2 = 13$

This will always be a perfect square.

Take the square root of both sides $\Rightarrow \quad (x-4) = \pm\sqrt{13}$

Add 4 to both sides $\Rightarrow \quad x = 4 + \sqrt{13}$ or $x = 4 - \sqrt{13}$

$\Rightarrow \quad x = 7.6065\ldots$ or $x = 0.3944\ldots$

This technique is called completing the square.

Quadratic graphs

Completing the square is also useful when you are sketching the graph of a quadratic expression. It identifies the line of symmetry and the vertex.

The equation

$$y = x^2 - 8x + 3$$

can be written as

$$y = (x-4)^2 - 13.$$

The line of symmetry is $x = 4$, and the vertex is $(4, -13)$.

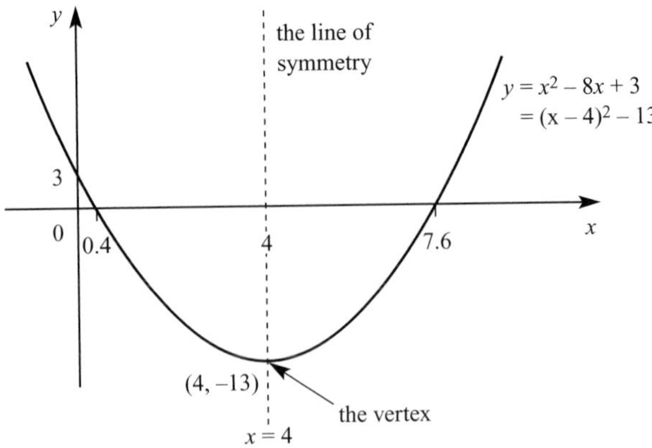

Figure 1.4

? How can this method be adapted when the coefficient of x^2 is not 1?

Example 1.25 shows how to deal with this case, and also how to generalise it to give a formula for solving quadratic equations.

EXAMPLE 1.25

Solve $2x^2 + 3x - 7 = 0$

SOLUTION

$$2x^2 + 3x - 7 = 0$$

$$\Rightarrow \quad x^2 + \tfrac{3}{2}x - \tfrac{7}{2} = 0$$

$$\Rightarrow \quad x^2 + \tfrac{3}{2}x = \tfrac{7}{2}$$

$$\Rightarrow \quad x^2 + \tfrac{3}{2}x + \left(\tfrac{3}{4}\right)^2 = \tfrac{7}{2} + \left(\tfrac{3}{4}\right)^2$$

$$\Rightarrow \quad \left(x + \tfrac{3}{4}\right)^2 = \tfrac{65}{16}$$

$$\Rightarrow \quad \left(x + \tfrac{3}{4}\right) = \pm\sqrt{\tfrac{65}{16}}$$

$$\Rightarrow \quad x = -\tfrac{3}{4} \pm \sqrt{\tfrac{65}{16}}$$

GENERALISATION

$$ax^2 + bx + c = 0$$

$$\Rightarrow \quad x^2 + \frac{b}{a}x + \frac{c}{a} = 0$$

$$\Rightarrow \quad x^2 + \frac{b}{a}x = -\frac{c}{a}$$

$$\Rightarrow \quad x^2 + \frac{b}{a}x + \left(\frac{b}{2a}\right)^2 = -\frac{c}{a} + \left(\frac{b}{2a}\right)^2$$

$$\Rightarrow \quad \left(x + \frac{b}{2a}\right)^2 = \frac{b^2}{4a^2} - \frac{c}{a} = \frac{b^2 - 4ac}{4a^2}$$

$$\Rightarrow \quad \left(x + \frac{b}{2a}\right) = \pm\sqrt{\frac{b^2 - 4ac}{4a^2}} = \frac{\sqrt{b^2 - 4ac}}{2a}$$

$$\Rightarrow \quad x = -\frac{b}{2a} \pm \frac{\sqrt{b^2 - 4ac}}{2a} = \frac{-b \pm \sqrt{b^2 - 4ac}}{2a}$$

The result

$$x = \frac{-b \pm \sqrt{b^2 - 4ac}}{2a}$$

is known as the *quadratic formula*. It allows you to solve any quadratic equation. One root is found by taking the + sign, and the other by taking the – sign. Figure 1.5 shows you the general curve and its line of symmetry.

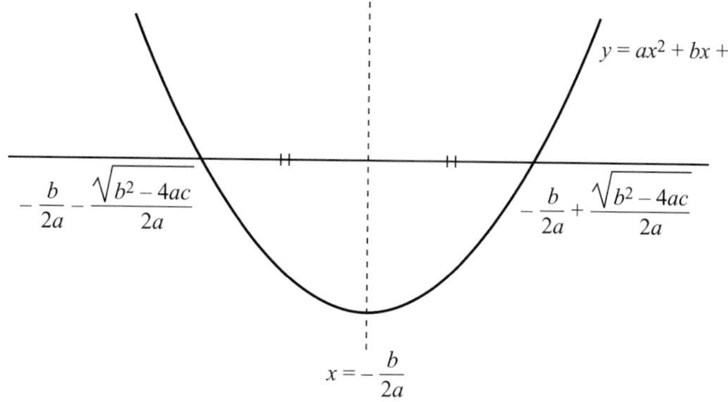

Figure 1.5

EXAMPLE 1.26

Use the quadratic formula to solve $2x^2 - 3x - 7 = 0$.

SOLUTION

Comparing

$$2x^2 - 3x - 7 = 0$$

with

$$ax^2 + bx + c = 0$$

gives

$$a = 2, b = -3, c = -7.$$

Using these values in the formula

$$x = \frac{-b \pm \sqrt{b^2 - 4ac}}{2a}$$

gives

$$x = \frac{-(-3) \pm \sqrt{(-3)^2 - 4 \times 2 \times (-7)}}{2 \times 2}$$

$$\Rightarrow \quad x = \frac{3 \pm \sqrt{65}}{4}$$

$$\Rightarrow \quad x = \frac{3 + 8.0622}{4} = 2.77 \text{ or } x = \frac{3 - 8.0622}{4} = -1.27.$$

When you use the quadratic formula and no level of accuracy is specified it is common practice to give the roots rounded to 2 decimal places.

EXAMPLE 1.27

The length of a carpet is 1 m more than its width. Its area is 9 m^2. Find the dimensions of the carpet to the nearest centimetre.

 The question asks for the dimensions *to the nearest centimetre.* This is a warning that you can expect to use the quadratic formula.

SOLUTION

Let the length be x metres, so the width is $(x - 1)$ metres.

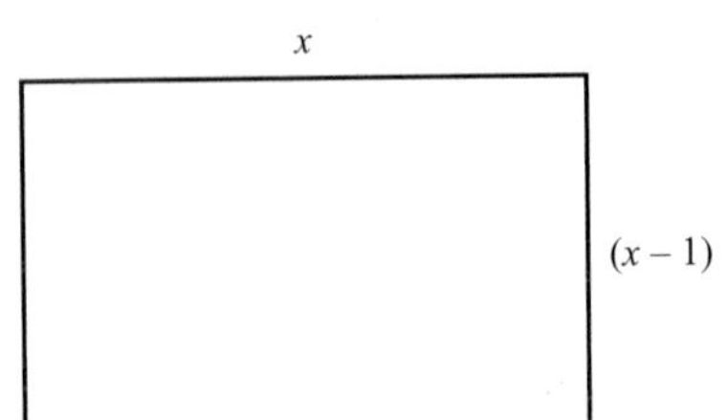

Figure 1.6

length × width = area

$$\Rightarrow \quad x(x - 1) = 9$$
$$\Rightarrow \quad x^2 - x = 9$$
$$\Rightarrow \quad x^2 - x - 9 = 0 \text{ collect everything on the left-hand side}$$

Putting $a = 1$, $b = -1$, $c = -9$ in

$$\Rightarrow \quad x = \frac{-b \pm \sqrt{b^2 - 4ac}}{2a}$$
$$\Rightarrow \quad x = \frac{-(-1) \pm \sqrt{(-1)^2 - 4 \times 1 \times (-9)}}{2 \times 1}$$
$$\Rightarrow \quad x = \frac{1 \pm \sqrt{37}}{2}$$
$$\Rightarrow \quad x = 3.541\ldots \text{ or } x = -2.541\ldots$$

Clearly a negative answer is not suitable here, so the dimensions are

length = 3.54 m
width = 2.54 m (to the nearest cm).

ACTIVITY 1.1

(i) Draw the graph of $y = x^2 - 4x + 13 = 0$.
(ii) What happens when you use the formula to solve $x^2 - 4x + 13 = 0$?

EXAMPLE 1.28

Show that the equation $x^2 - 6x + 20$ has no solution.

SOLUTION

Substituting the values $a = 1$, $b = -6$, $c = 20$ into the formula

$$x = \frac{-b \pm \sqrt{b^2 - 4ac}}{2a}$$
$$\Rightarrow \quad x = \frac{6 \pm \sqrt{36 - 4 \times 1 \times 20}}{2 \times 1}$$
$$\Rightarrow \quad x = \frac{6 \pm \sqrt{-44}}{2}.$$

Since it is not possible to find a value for $\sqrt{-44}$, there are no values of x and so the equation has no solution. Figure 1.7 shows the graph of

$$y = x^2 - 6x + 20$$

It does not cross the x axis.

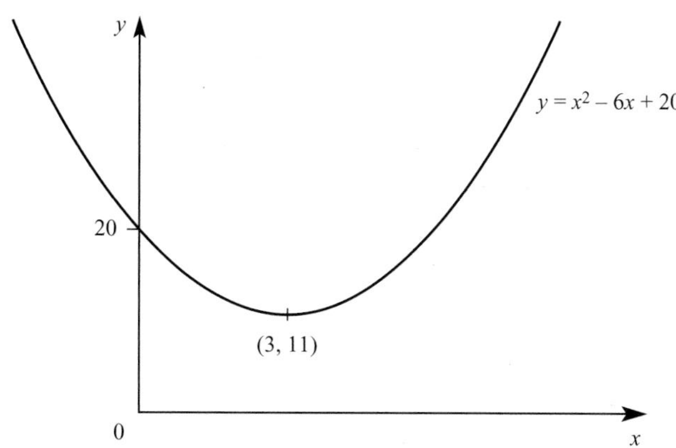

Figure 1.7

NOTE

1 You didn't need the whole of the formula to show that there was no solution, only the part under the square root sign. This part, $b^2 - 4ac$, is called the *discriminant*, because it discriminates between equations with roots and those with none.

2 By extending the definition of numbers to include $\sqrt{-1}$, you can write $\sqrt{-44} = \sqrt{44} \times \sqrt{-1}$. The letter i (or j) is usually used to denote $\sqrt{-1}$, and if you have a graphic calculator it will give $\sqrt{-44} = 6.633....\text{i}$. Square roots of negative numbers are called *imaginary* numbers, and those such as $\frac{6 \pm \sqrt{-44}}{2} = 3 + 3.166...\text{i}$ are called *complex* numbers.

EXERCISE 1E

1 Solve the following equations by factorising.

(i) $x^2 - 8x + 12 = 0$ **(ii)** $m^2 - 4m + 4 = 0$

(iii) $p^2 - 2p - 15 = 0$ **(iv)** $a^2 + 11a + 18 = 0$

(v) $2x^2 + 5x + 2 = 0$ **(vi)** $4x^2 + 3x - 7 = 0$

(vii) $15t^2 + 2t - 1 = 0$ **(viii)** $24r^2 + 19r + 2 = 0$

(ix) $3x^2 + 8x = 3$ **(x)** $3p^2 = 14p - 8$

2 Solve the following equations

(a) by completing the square

(b) by using the quadratic formula.

Give your answers correct to 2 decimal places.

(i) $x^2 - 2x - 10 = 0$ **(ii)** $x^2 + 3x - 6 = 0$

(iii) $x^2 + x - 8 = 0$ **(iv)** $2x^2 + x - 8 = 0$

(v) $2x^2 + 2x - 9 = 0$ **(vi)** $x^2 + x = 10$

(vii) $x^2 = 4x + 1$ **(viii)** $2x^2 - 8x + 5 = 0$

3 Solve the following equations by using the quadratic formula. Give your answers correct to 2 decimal places.

(i) $3x^2 + 5x + 1 = 0$ **(ii)** $4x^2 + 9x + 3 = 0$

(iii) $2x^2 + 11x - 4 = 0$ **(iv)** $4x^2 - 9x + 4 = 0$

(v) $5x^2 - 10x + 1 = 0$ **(vi)** $3x^2 + 11x + 9 = 0$

4 Try to solve the following equations. Where there is a solution, give your answers correct to 2 decimal places.

(i) $2x^2 - 7x + 2 = 0$ **(ii)** $2x^2 + 2x + 7 = 0$

(iii) $(x + 1)^2 = 15$ **(iv)** $3x^2 + x = 7$

(v) $x(2x - 1) + 8 = 0$ **(vi)** $(2x + 1)^2 = 5$

5 The sides of a right-angled triangle, in centimetres, are x, $2x - 2$, and $x + 2$, where $x + 2$ is the hypotenuse.

Use Pythagoras' theorem to find their lengths.

6 A rectangular lawn measures 8 m by 10 m and is surrounded by a path of uniform width x m.

The total area of the path is $63\,\text{m}^2$. Find x.

7 The difference between two positive numbers is 2 and the difference between their squares is 40.

Taking x to be the smaller of the two numbers, form an equation in x and solve it.

8 The formula $h = 15t - 5t^2$ gives the height h metres of a ball, t seconds after it is thrown up into the air.

(i) Find the times when the height is 10 m.

(ii) After how long does the ball hit the ground?

9 The area of this triangle is $68\,\text{cm}^2$.

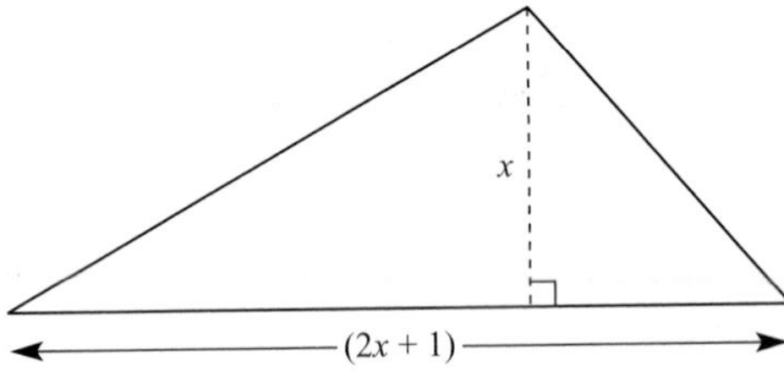

(i) Show that x satisfies the equation
$2x^2 + x - 136 = 0$.

(ii) Solve the equation to find the length of the base of the triangle.

10 Boxes are made by cutting 8 cm squares from the corners of sheets of cardboard and then folding. The sheets of cardboard are 6 cm longer than they are wide.

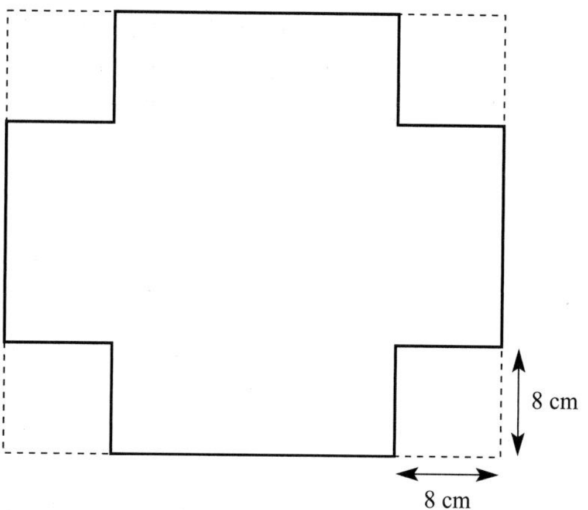

(i) For a sheet of cardboard whose width is x cm, find expressions, in terms of x, for

(a) the length of the sheet

(b) the length of the finished box

(c) the width of the finished box.

(ii) Show that the volume of the box is $8x^2 - 208x + 1280\,\text{cm}^3$.

(iii) Find the dimensions of the sheet of cardboard needed to make a box with a volume of $1728\,\text{cm}^3$.

Simultaneous equations

The equations you have met so far in this chapter have only involved one variable, for example

$$2x + 2 = x - 5 \quad \text{or} \quad a^2 - 3a + 2 = 0.$$

❓ When an equation involves two variables, for example $x + y = 4$, how many possible pairs of values are there for x and y?

Figure 1.8 shows the line $x + y = 4$.

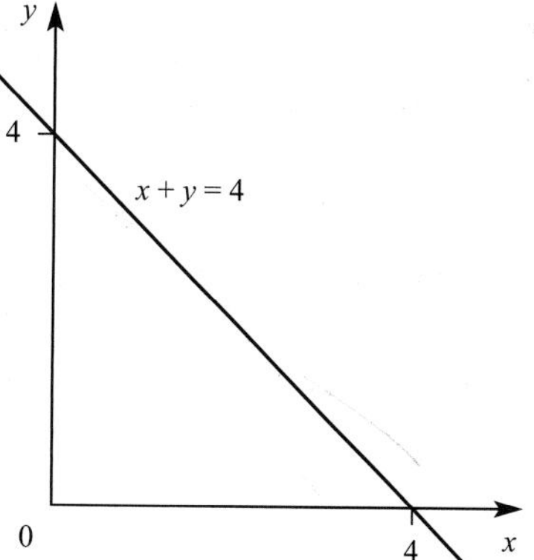

Figure 1.8

The co-ordinates of every point on that line give a pair of possible values for x and y. If you add the line $y = 2x + 1$, as in figure 1.9, you will see that the two lines intersect at a single point.

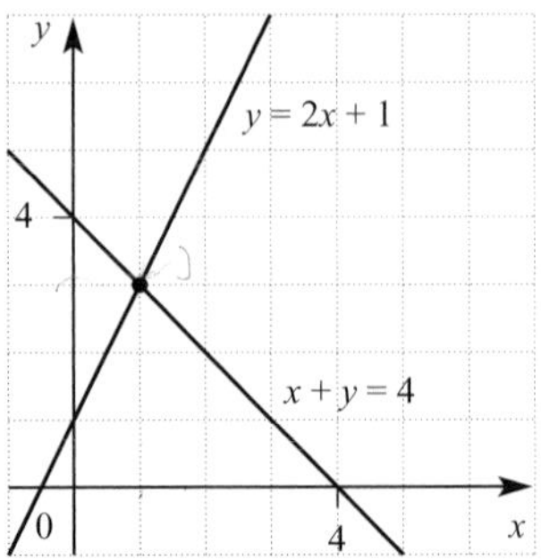

Figure 1.9

The co-ordinates of this point are the solution to the *simultaneous equations*

$$x + y = 4$$
and $$y = 2x + 1.$$

There are several ways of solving simultaneous equations. You have just seen one method, that of drawing graphs. This is valid, but it has two drawbacks

(i) it is tedious
(ii) it may not be very accurate, particularly if the solution does not have integer values.

Solving simultaneous equations by substitution

EXAMPLE 1.29

Solve the simultaneous equations

$$x + y = 4$$
$$y = 2x + 1$$

by substitution.

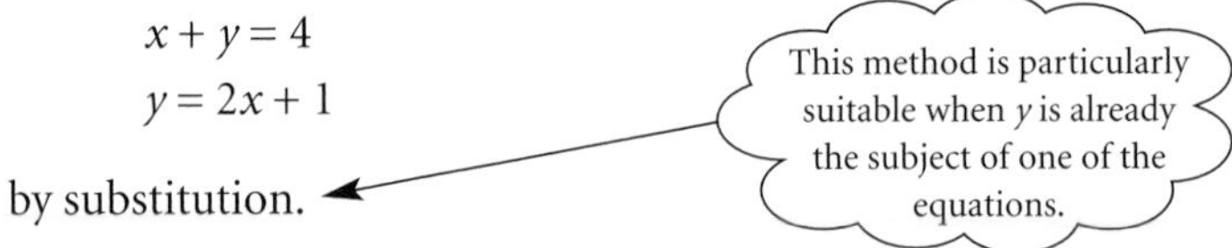

SOLUTION

Take the expression for y from the second equation and substitute it into the first. This gives

$$x + (2x + 1) = 4$$
$$\Rightarrow \quad 3x = 3$$
$$\Rightarrow \quad x = 1.$$

Since $y = 2x + 1$, when $x = 1$, $y = 3$, as before.

EXAMPLE 1.30

Figure 1.10 shows the graphs of $y = x^2 + x$ and $2x + y = 4$.

Solve the simultaneous equations

$$y = x^2 + x$$

and $\quad 2x + y = 4$

using the method of substitution.

This method is also particulary suitable when one of the equations represents a curve.

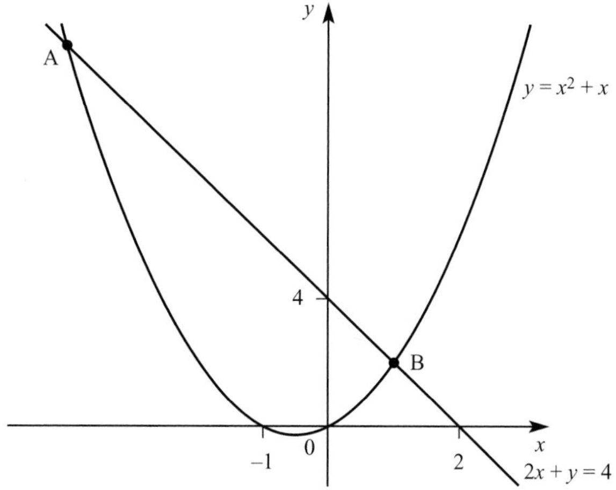

Figure 1.10

SOLUTION

⚠ Notice from figure 1.10 that there are two points of intersection, A and B, so expect the solution to be two pairs of values for x and y.

$$y = x^2 + x \qquad ①$$
$$2x + y = 4 \qquad ②$$

Substitute for y from equation ① into equation ②.

$$2x + (x^2 + x) = 4$$
$$\Rightarrow \quad x^2 + 3x - 4 = 0$$
$$\Rightarrow \quad (x + 4)(x - 1) = 0$$
$$\Rightarrow \quad x = -4 \text{ or } x = 1$$

Substituting in $2x + y = 4$ ②

Always substitute back into the linear equation.

$$x = -4 \quad \Rightarrow \quad -8 + y = 4 \quad \Rightarrow \quad y = 12$$
$$x = 1 \quad \Rightarrow \quad 2 + y = 4 \quad \Rightarrow \quad y = 2$$

The solution is $x = -4$, $y = 12$ (the point A) and $x = 1$, $y = 2$ (the point B).

Check your solution also fits equation ①.

The solution must always be given as pairs of values. It is wrong to write $x = -4$ or 1, $y = 12$ or 2, since not all pairs of values are possible.

Having found the values of x in the example above, the values of y were found by substituting into the equation of the line. What would happen if, instead, you were to substitute into the equation of the curve?

Solving linear simultaneous equations by elimination

An alternative method that you may prefer to use when neither equation has y as the subject is called *solution by elimination.*

EXAMPLE 1.31

Solve the simultaneous equations

$$2x + y = 8 \quad ①$$
$$5x + 2y = 21 \quad ②$$

SOLUTION

Notice that multiplying throughout equation ① by 2 gives you another equation containing $2y$.

$$5x + 2y = 21 \qquad \text{equation ②}$$
$$4x + 2y = 16 \qquad 2 \times \text{equation ①}$$

Subtracting $\Rightarrow \quad x \qquad = 5$

Substitute $x = 5$ into equation ①

$$10 + y = 8 \quad \Rightarrow \quad y = -2.$$

The solution is $x = 5$, $y = -2$.

Sometimes you need to manipulate both equations to eliminate one of the variables, as in the following example.

EXAMPLE 1.32

Solve the simultaneous equations

$$2x + 3y = -1 \quad ①$$
$$3x - 2y = 18. \quad ②$$

SOLUTION

It is equally easy to eliminate x or y. It is up to you to choose which. The following method eliminates y.

$$
\begin{array}{llll}
 & & 4x + 6y = -2 & 2 \times \text{equation ①} \\
 & & 9x - 6y = 54 & 3 \times \text{equation ②} \\
\text{Adding} & \Rightarrow & 13x \quad = 52 & \\
 & \Rightarrow & x = 4 &
\end{array}
$$

Substitute $x = 4$ into equation ①

$$8 + 3y = -1 \quad \Rightarrow \quad y = -3.$$

The solution is $x = 4$, $y = -3$.

? In Example 1.31 the equations were subtracted; in Example 1.32 they were added. How do you decide whether to add or subtract?

Simultaneous equations may arise in everyday problems.

EXAMPLE 1.33

Nisha has £2.20 to spend on fruit for a picnic and can buy either five apples and four pears or two apples and six pears.

(i) Write this information as a pair of simultaneous equations.
(ii) Solve your equations to find the cost of each type of fruit.

SOLUTION

Make sure you introduce your variables.

Let a pence be the cost of an apple and p pence be the cost of a pear.

The cost of each piece of fruit will be a number of pence, so writing £2.20 as 220 pence avoids working with decimals.

(i) $5a + 4p = 220$ ①
$2a + 6p = 220$ ②

$$
\begin{array}{lll}
\Rightarrow & 15a + 12p = 660 & 3 \times \text{equation ①} \\
 & 4a + 12p = 440 & 2 \times \text{equation ②} \\
\text{Subtracting} & 11a \quad = 220 & \\
\Rightarrow & a \quad = 20 &
\end{array}
$$

Substitute $a = 20$ into equation ①

$$100 + 4p = 220$$

$$\Rightarrow \quad p = 30.$$

An apple costs 20 pence and a pear costs 30 pence.

EXAMPLE 1.34

A flag consists of a purple cross on a white background. Each white rectangle measures $2x$ cm by x cm, and the cross is y cm wide.

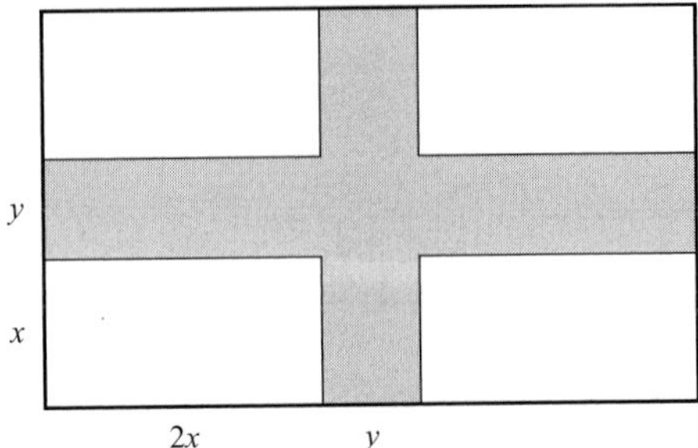

Figure 1.11

(i) Write down the total area of the flag in terms of x and y.

(ii) Show that the area of the cross is $6xy + y^2$.

(iii) The total area of the flag is $4500\,\text{cm}^2$ and the area of the cross is $1300\,\text{cm}^2$. Find x and y.

SOLUTION

(i) length $= 4x + y$ and width $= 2x + y$.

$$\Rightarrow \quad \text{area} = (4x + y)(2x + y)$$
$$= 8x^2 + 6xy + y^2 \qquad ①$$

(ii) Each white rectangle has an area of

$2x \times x = 2x^2$.

$$\Rightarrow \quad \text{area of cross} = 8x^2 + 6xy + y^2 - (4 \times 2x^2)$$
$$= 6xy + y^2 \qquad ②$$

(iii)

$$8x^2 + 6xy + y^2 = 4500 \qquad ①$$
$$6xy + y^2 = 1300 \qquad ②$$

Subtracting $\quad 8x^2 \qquad = 3200$

$\Rightarrow \quad x^2 = 400$

$\Rightarrow \quad x = 20 \quad$ (positive answer only)

Substitute $x = 20$ into equation ②

$120y + y^2 = 1300$

$\Rightarrow \quad y^2 + 120y - 1300 = 0$

$\Rightarrow \quad (y + 130)(y - 10) = 0$

$\Rightarrow \quad y = -130$ (reject since y is a length) or $y = 10$

$\Rightarrow \quad x = 20$ and $y = 10$

EXERCISE 1F

1 Solve the following pairs of simultaneous equations using the substitution method.

(i) $y = x - 3$
$3x + 2y = 19$

(ii) $y = 2x - 9$
$4x - y = 17$

(iii) $y = 11 - 2x$
$2x + 5y = 37$

(iv) $y = 3x + 3$
$x - 2y = 4$

(v) $y = 7 - 2x$
$2x + 3y = 15$

(vi) $y = 3x - 5$
$x + 3y = -20$

2 Solve the following pairs of simultaneous equations using the elimination method.

(i) $3x + 2y = 12$
$4x - y = 5$

(ii) $3x - 2y = 6$
$5x + 6y = 38$

(iii) $3x + 2y = 22$
$4x - 3y = 18$

(iv) $5x + 4y = 11$
$2x + 3y = 9$

(v) $4x + 5y = 33$
$3x + 2y = 16$

(vi) $4x - 3y = 2$
$5x - 7y = 9$

3 Solve the following pairs of simultaneous equations.

(i) $x + y = 5$
$x^2 + y^2 = 17$

(ii) $x - y + 1 = 0$
$3x^2 - 4y = 0$

(iii) $x^2 + xy = 8$
$x - y = 6$

(iv) $2x - y + 3 = 0$
$y^2 - 5x^2 = 20$

(v) $x = 2y$
$x^2 - y^2 + xy = 20$

(vi) $x + 2y = -3$
$x^2 - 2x + 3y^2 = 11$

4 For each of the following situations, form a pair of simultaneous equations and solve them to answer the question.

(i) Three chews and four lollipops cost 72p. Five chews and two lollipops cost 64p. Find the cost of a chew and the cost of a lollipop.

(ii) A taxi firm charges a fixed amount plus so much per mile. A journey of five miles costs £5.00 and a journey of seven miles costs £6.60. How much does a journey of three miles cost?

(iii) Three packets of crisps and two packets of nuts cost £1.45. Two packets of crisps and five packets of nuts cost £2.25. How much does one packet of crisps and four packets of nuts cost?

(iv) Two adults and one child paid £37.50 to go to the theatre. The cost for one adult and three children was also £37.50. How much does it cost for two adults and five children?

5 The diagram shows the circle $x^2 + y^2 = 25$ and the line $x + y = 7$. Find the co-ordinates of A and B.

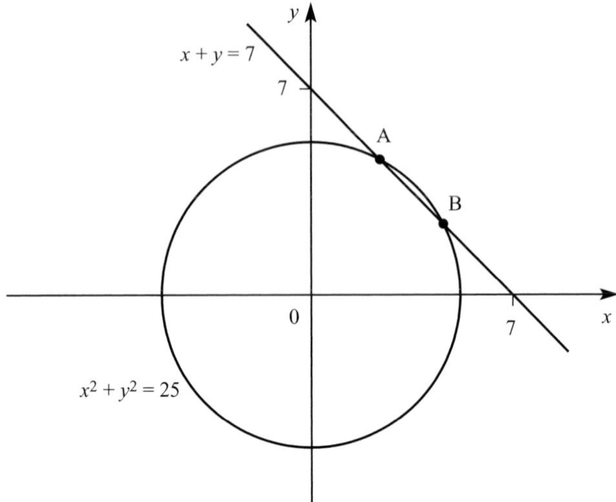

Figure 1.12

KEY POINTS

Each chapter in this book ends with KEY POINTS, a summary of the essential ideas that you should have understood in the chapter. Chapter 1, however, is fundamental to mathematics and you will need to be confident on all of the techniques covered in this chapter if you are to understand the rest of your course. These are:

1 Simplifying algebraic expressions by
- collecting like terms
- removing brackets
- cancelling by common factors
- factorising
- expressing them as a single fraction.

2 Solving linear equations.

3 Changing the subject of an equation.

4 Factorising quadratic expressions.

5 Solving a quadratic equation by
- factorising
- completing the square
- using the quadratic formula.

6 Sketching graphs of quadratic expressions.

7 Solving simultaneous equations by
- drawing graphs
- substitution
- elimination.

2 Algebra II – techniques

Others have done it before me. I can, too.

Corporal John Faunce (American soldier)

❓ What is the difference between an equation and an inequality?

❓ The radius of the Earth's orbit round the Sun is approximately 1.5×10^8 km; that of Mars is about 2.3×10^8 km. The Earth takes 365 days for one orbit and Mars takes 687 days.

At some time the distance from Earth to Mars is x km. What can you say about x?

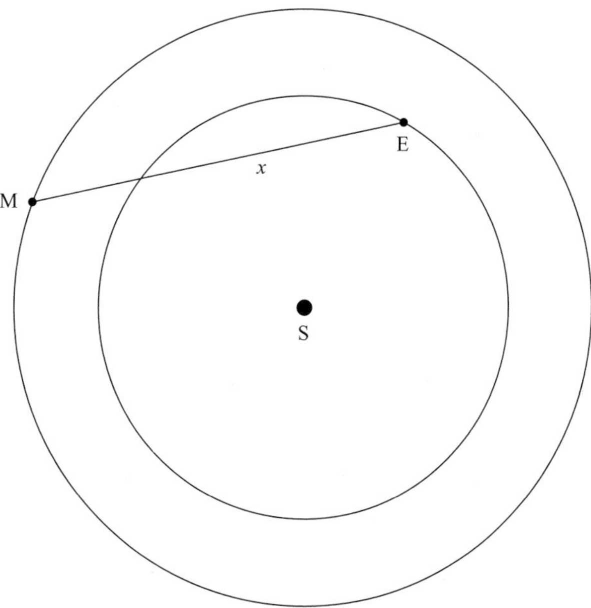

Figure 2.1

Linear inequalities

EXAMPLE 2.1

Solve $3x - 12 \leqslant x - 4$ and illustrate the solution on a number line.

SOLUTION

		$3x - 12 \leqslant x - 4$
Add 12	$\Rightarrow$	$3x \leqslant x + 8$
Subtract x	$\Rightarrow$	$2x \leqslant 8$
Divide by 2	$\Rightarrow$	$x \leqslant 4$

This can be represented by a section of the number line.

A solid circle at the end of the line segment means that $x = 4$ *is included* in the solution.

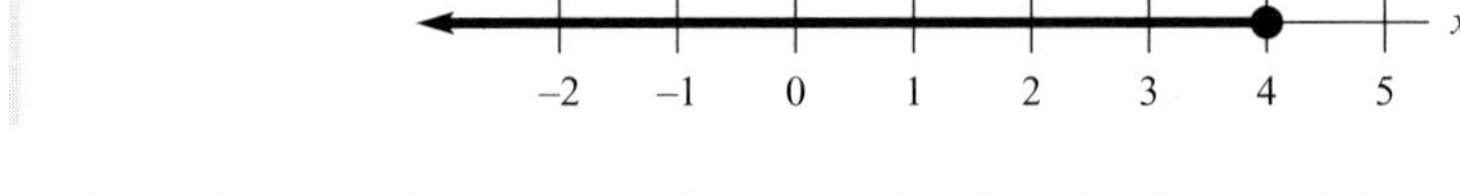

In what ways is solving an inequality like solving an equation? In what ways is it different?

EXAMPLE 2.2

Solve $2y + 6 < 5y + 12$ and illustrate the solution on a number line.

SOLUTION

Method 1

		$2y + 6 < 5y + 12$
Subtract $2y$	$\Rightarrow$	$6 < 3y + 12$
Subtract 12	$\Rightarrow$	$-6 < 3y$
Divide by 3	$\Rightarrow$	$-2 < y$
Make y the subject	$\Rightarrow$	$y > -2$

Here the open circle shows that $y = -2$ *is not* included in the solution.

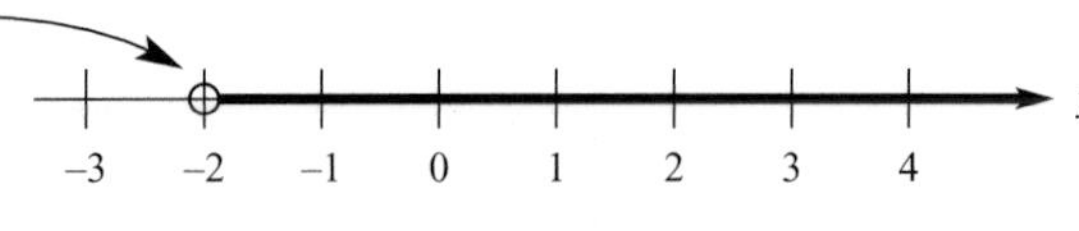

Method 2

		$2y + 6 < 5y + 12$
Subtract $5y$	$\Rightarrow$	$-3y + 6 < 12$
Subtract 6	$\Rightarrow$	$-3y < 6$
Divide by (-3)	$\Rightarrow$	$y > -2$

Explain, with examples, why you need to reverse the inequality sign when you multiply or divide an inequality by a negative number.

EXAMPLE 2.3

Solve the inequality $5 < 3x - 1 \leqslant 17$ and illustrate the solution on a number line.

SOLUTION

Add 1 throughout $\Rightarrow$ $6 < 3x \leqslant 18$

Divided by 3 $\Rightarrow$ $2 < x \leqslant 6$

Notice how the different circles show that 6 *is* included in the solution, but 2 *is not.*

EXERCISE 2A

1 Solve the following inequalities and represent their solutions on a number line.

(i) $2x - 3 < 7$

(ii) $5 + 3x \geqslant 11$

(iii) $6y + 1 \leqslant 4y + 9$

(iv) $y - 4 > 3y - 12$

(v) $4x + 1 \geqslant 3x - 2$

(vi) $b - 3 \leqslant 5b + 9$

(vii) $\frac{x + 5}{2} > 1$

(viii) $\frac{2x - 3}{3} < 7$

(ix) $\frac{5 - 3x}{4} \leqslant 5$

(x) $\frac{2 - 4x}{3} \geqslant 6$

(xi) $4 \leqslant 5x - 6 \leqslant 14$

(xii) $11 \leqslant 3x + 5 \leqslant 20$

(xiii) $5 < 7 - 2x < 13$

(xiv) $5 > 9 - 4x > 1$

Solving quadratic inequalities

The quadratic inequalities in this section all involve quadratic expressions that factorise. This means that you can either find a solution by sketching the appropriate graph, or you can use line segments to reduce the quadratic inequality to two simultaneous linear inequalities.

EXAMPLE 2.4

Solve

(i) $x^2 - 2x - 3 < 0$

(ii) $x^2 - 2x - 3 \geqslant 0$.

SOLUTION

Method 1

$$x^2 - 2x - 3 = (x + 1)(x - 3)$$

So the graph of $y = x^2 - 2x - 3$ crosses the x axis when $x = -1$ and $x = 3$.

Look at the two graphs in figure 2.2

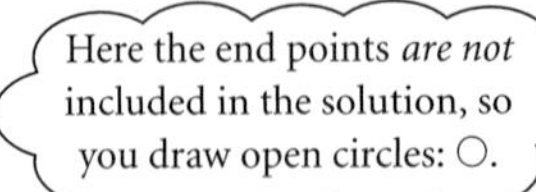

Here the end points *are* included in the solution, so you draw solid circles: ●.

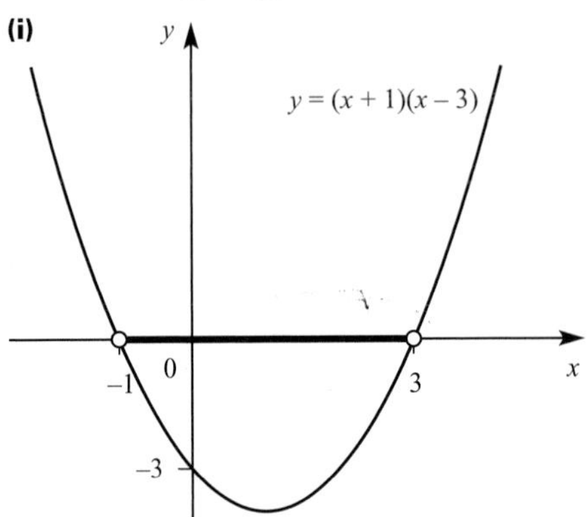

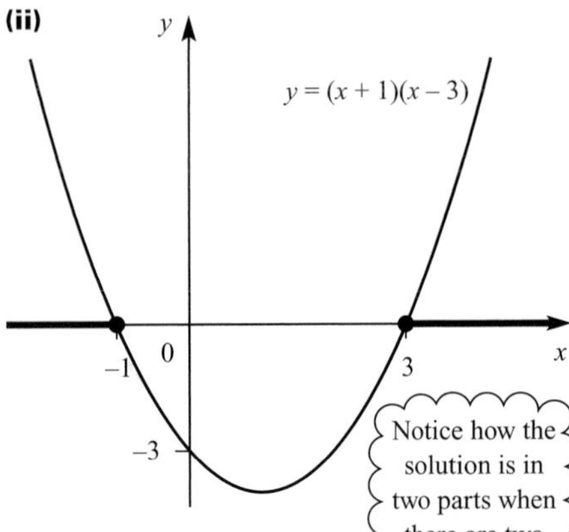

Figure 2.2

The solution is $-1 < x < 3$

The solution is $x \leqslant -1$ or $x \geqslant 3$

(i) You want the values of x for which $y < 0$, that is where the curve is below the x axis.

(ii) You want the values of x for which $y \geqslant 0$, that is where the curve crosses or is above the x axis.

An alternative method identifies the values of x for which each of the factors is zero and considers the sign of each factor in the intervals between these critical values.

Method 2

	$x < -1$	$x = -1$	$-1 < x < 3$	$x = 3$	$x > 3$
sign of $(x + 1)$	–	0	+	+	+
sign of $(x - 3)$	–	–	–	0	+
sign of $(x + 1)(x - 3)$	(–)×(–) = +	(0)×(–) = 0	(+)×(–) = –	(+)×(0) = 0	(+)×(+) = +

From the table the solution to $(x + 1)(x - 3) < 0$ is $-1 < x < 3$.

The solution to $(x + 1)(x - 3) \geqslant 0$ is $x \leqslant -1$ or $x \geqslant 3$.

NOTE

If the inequality to be solved contains > or <, then the solution is described using > and <, but if the original inequality contains $\geqslant$ or $\leqslant$, then the solution is described using $\geqslant$ and $\leqslant$.

Both of these are valid methods, and you should decide which you prefer. This may depend on how easily you sketch graphs, or if you have a graphic calculator which will do this for you.

If the quadratic inequality has terms on both sides, you must first collect everything on one side, as you would do when solving a quadratic equation.

EXAMPLE 2.5

Solve $2x + x^2 > 3$.

SOLUTION

$$2x + x^2 > 3 \quad \Rightarrow \quad x^2 + 2x - 3 > 0$$
$$\Rightarrow \quad (x-1)(x+3) > 0$$

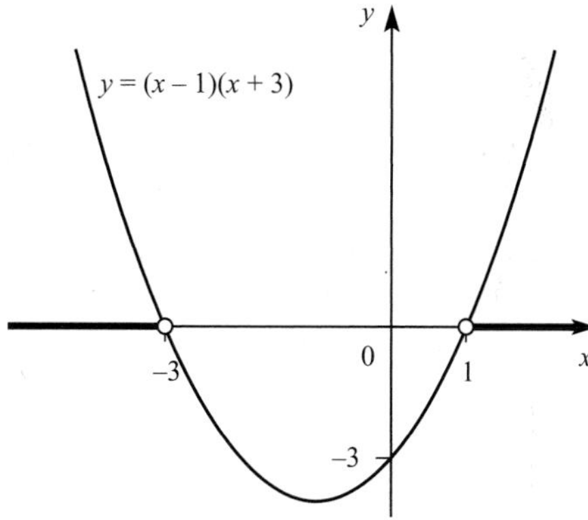

Figure 2.3

From figure 2.3 the solution is $x < -3$ or $x > 1$.

EXERCISE 2B

1 Solve the following inequalities.

(i) $x^2 - 6x + 5 > 0$ **(ii)** $a^2 + 3a - 4 \leqslant 0$

(iii) $2y^2 + y - 3 < 0$ **(iv)** $4 - y^2 \geqslant 0$

(v) $x^2 - 4x + 4 > 0$ **(vi)** $p^2 - 3p \leqslant -2$

(vii) $(a+2)(a-1) > 4$ **(viii)** $8 - 2a \geqslant a^2$

(ix) $3y^2 + 2y - 1 > 0$ **(x)** $y^2 \geqslant 4y + 5$

Simplifying algebraic fractions

What is a fraction in arithmetic?
What about in algebra?

Fractions in algebra obey the same rules as fractions in arithmetic. These cover two pairs of operations: $\times$ and $\div$, and $+$ and $-$.

? When can you cancel fractions in arithmetic?
What about in algebra?
What is a factor in arithmetic?
What about in algebra?

EXAMPLE 2.6

Simplify the following.

(i) $\frac{18}{24}$ **(ii)** $\frac{2x+2}{3x+3}$ **(iii)** $\frac{a^2-a-6}{a^2-8a+15}$

SOLUTION

(i) $\frac{18}{24} = \frac{\overset{1}{\cancel{6}} \times 3}{\underset{1}{\cancel{6}} \times 4} = \frac{3}{4}$

(ii) $\frac{2x+2}{3x+3} = \frac{2\overset{1}{\cancel{(x+1)}}}{3\underset{1}{\cancel{(x+1)}}} = \frac{2}{3}$

Look at the calculation for **(ii)**.
Why is it wrong?

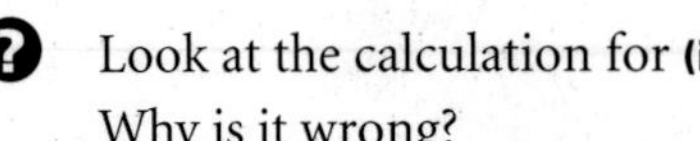

(iii) $\frac{a^2-a-6}{a^2-8a+15} = \frac{\overset{1}{\cancel{(a-3)}}(a+2)}{\underset{1}{\cancel{(a-3)}}(a-5)} = \frac{(a+2)}{(a-5)}$

Look at this answer to **(iii)**.
Why is it wrong?

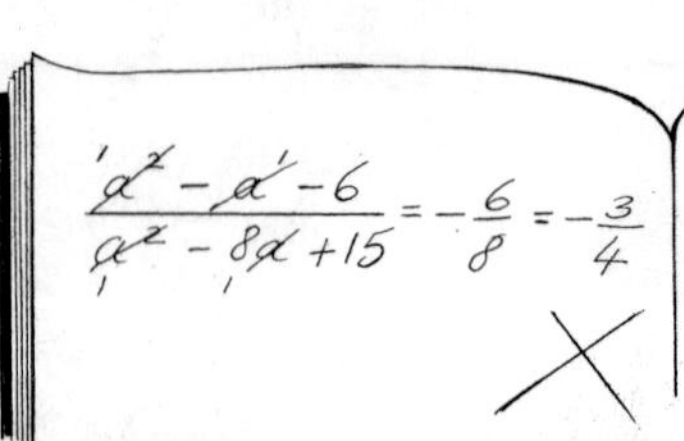

EXAMPLE 2.7 Simplify the following.

(i) $\frac{2}{3}\times\frac{9}{14}$ **(ii)** $\frac{3}{4}\div\frac{9}{16}$ **(iii)** $\frac{3a^2b}{2c}\times\frac{4c^3}{9ab}$ **(iv)** $\frac{4n^2-9}{n+1}\div\frac{2n+3}{n^2-1}$

SOLUTION

(i) $\frac{2}{3}\times\frac{9}{14}=\frac{1\times 3}{1\times 7}=\frac{3}{7}$

(ii) $\frac{3}{4}\div\frac{9}{16}=\frac{3}{4}\times\frac{16}{9}=\frac{4}{3}$

(iii) $\frac{3a^2b}{2c}\times\frac{4c^3}{9ab}=\frac{2ac^2}{3}$

(iv) $\frac{4n^2-9}{n+1}\div\frac{2n+3}{n^2-1}=\frac{\overset{1}{\cancel{(2n+3)}}(2n-3)}{\underset{1}{\cancel{(n+1)}}}\times\frac{\overset{1}{\cancel{(n+1)}}(n-1)}{\underset{1}{\cancel{(2n+3)}}}$

$=(2n-3)(n-1)$

Look at this answer to **(iv)**. Why is it wrong?

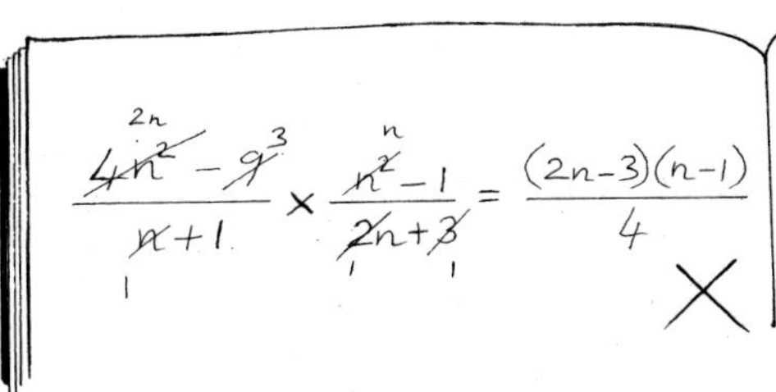

What is a common denominator?

To add or subtract fractions it is first necessary to find a *common denominator*.

EXAMPLE 2.8 Simplify the following.

(i) $\frac{2}{3}+\frac{3}{4}$ **(ii)** $\frac{5x}{6}+\frac{x}{4}$

(iii) $\frac{2}{(x+1)}+\frac{5}{(x-1)}$ **(iv)** $\frac{a}{a^2-1}-\frac{2}{a+1}$

SOLUTION

(i) $\frac{2}{3}+\frac{3}{4}=\frac{8}{12}+\frac{9}{12}=\frac{17}{12}$

What is the lowest common multiple of the following?

(a) 6 and 4

(b) (x^2-1) and (x^2-4x+3)

(ii) $\frac{5x}{6}+\frac{x}{4}=\frac{10x}{12}+\frac{3x}{12}=\frac{13x}{12}$

Notice that the common denominator is the lowest common multiple of the original denominators.

(iii) $$\frac{2}{(x+1)} + \frac{5}{(x-1)} = \frac{2(x-1)}{(x+1)(x-1)} + \frac{5(x+1)}{(x+1)(x-1)}$$
$$= \frac{2x-2+5x+5}{(x+1)(x-1)}$$
$$= \frac{7x+3}{(x+1)(x-1)}$$

(iv) $$\frac{a}{a^2-1} - \frac{2}{a+1} = \frac{a}{(a-1)(a+1)} - \frac{2}{a+1}$$
$$= \frac{a}{(a-1)(a+1)} - \frac{2(a-1)}{(a-1)(a+1)}$$
$$= \frac{a-2a+2}{(a-1)(a+1)}$$
$$= \frac{2-a}{(a-1)(a+1)}$$

EXERCISE 2C

1 Simplify the following.

(i) $\dfrac{2(x+3)}{4x+12}$ **(ii)** $\dfrac{4x-8}{(x-2)(x+8)}$

(iii) $\dfrac{3(x+y)}{x^2-y^2}$ **(iv)** $\dfrac{6x^2y^3}{9xy^4}$

(v) $\dfrac{2p}{6p-2p^2}$ **(vi)** $\dfrac{4ab^3}{10a^3b}$

(vii) $\dfrac{x^2-4x+3}{2x-6}$ **(viii)** $\dfrac{x^2+xy}{x^2-y^2}$

(ix) $\dfrac{a+2}{a^2-a-6}$ **(x)** $\dfrac{3x^2+15x}{10x+2x^2}$

(xi) $\dfrac{9x^2-1}{9x+3}$ **(xii)** $\dfrac{3x^2+3xy}{6xy+6y^2}$

2 Simplify the following.

(i) $\dfrac{3a}{b^2} \times \dfrac{b^3}{6a}$ **(ii)** $\dfrac{xy-y^2}{y} \times \dfrac{x}{x-y}$

(iii) $\dfrac{x+1}{2x} \div \dfrac{4x^2-4}{x^2}$ **(iv)** $\dfrac{3a^2+a-2}{2} \div \dfrac{6a^2-a-2}{8a+4}$

(v) $\dfrac{x^2-4x+4}{x^2-2x} \times \dfrac{x-2}{x^2-4}$ **(vi)** $\dfrac{2x-1}{x+1} \div \dfrac{2x^2-x-1}{x^2+3x+2}$

(vii) $\dfrac{4p^2+12}{p-3} \times \dfrac{p^2-9}{p^2+3}$ **(viii)** $\dfrac{3x^2-9}{x+2} \div \dfrac{x^2-6x+9}{x^2+x-2}$

3 Simplify the following.

(i) $\frac{3a}{5} - \frac{a}{4}$

(ii) $\frac{5}{3a} - \frac{4}{a}$

(iii) $\frac{2}{(m+n)} - \frac{1}{(m-n)}$

(iv) $\frac{4}{p-2} - \frac{3}{2p+1}$

(v) $\frac{2}{a^2+a} + \frac{3}{a^2-a}$

(vi) $\frac{2x}{x-y} + \frac{2y}{y-x}$

(vii) $\frac{p}{p^2-1} - \frac{1}{p+1}$

(viii) $\frac{a-b}{a+b} + \frac{a+b}{a-b}$

? What is the difference between *simplifying* fractions and *solving* an equation involving fractions?

Solving equations involving fractions

When you solved the equations earlier in this chapter you used mathematical operations such as +, –, × and ÷ to find the value of the variable. The same principle applies for solving equations involving fractions.

EXAMPLE 2.9

Solve the following.

$$\frac{x+2}{6} = \frac{x-6}{2}$$

SOLUTION

The LCM of 6 and 2 is 6, so multiply by 6.

$$^{1}\not{6} \times \frac{(x+2)}{\not{6}_1} = {}^{3}\not{6} \times \frac{(x-6)}{\not{2}_1}$$

⚠ ? When you multiply a fraction, you only multiply its numerator (top line). Why?

? How does this help?

$$\Rightarrow \quad x + 2 = 3x - 18$$
$$\Rightarrow \quad 20 = 2x$$
$$\Rightarrow \quad x = 10$$

EXAMPLE 2.10

Solve the following.

$$\frac{x+2}{6} + 3 = \frac{x}{5}$$

SOLUTION

The LCM of 6 and 5 is 30, so multiply by 30.

$$\overset{5}{\cancel{30}} \times \frac{(x+2)}{\cancel{6}_1} + 30 \times 3 = \overset{6}{\cancel{30}} \times \frac{x}{\cancel{5}_1}$$

Look at this version of the first stage of the solution. Why is it wrong?

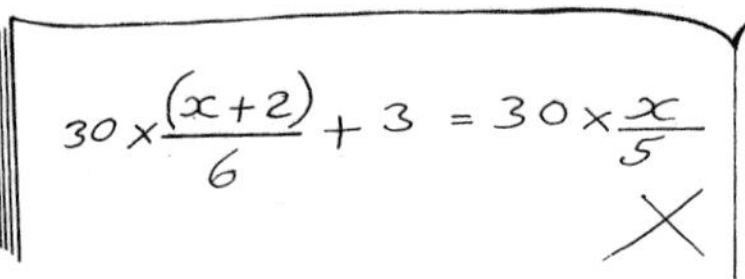

$$\Rightarrow \quad 5x + 10 + 90 = 6x$$
$$\Rightarrow \quad x = 100$$

EXAMPLE 2.11

Solve the following.

$$\frac{5}{a+1} - \frac{2a}{a^2-1} = \frac{1}{2}$$

SOLUTION

First factorise $(a^2 - 1)$ as $(a+1)(a-1)$.

How does this help?

$$\frac{5}{a+1} - \frac{2a}{(a+1)(a-1)} = \frac{1}{2}$$

Multiply by $2\,(a+1)(a-1)$

$$\Rightarrow \quad 2\overset{1}{\cancel{(a+1)}}(a-1) \times \frac{5}{\cancel{(a+1)}_1} - 2\overset{1}{\cancel{(a+1)}}\overset{1}{\cancel{(a-1)}} \times \frac{2a}{\cancel{(a+1)}_1\cancel{(a-1)}_1} = \overset{1}{\cancel{2}}(a+1)(a-1) \times \frac{1}{\cancel{2}_1}$$

$$\Rightarrow \quad 10(a-1) - 4a = (a+1)(a-1)$$
$$\Rightarrow \quad 10a - 10 - 4a = a^2 - 1$$
$$\Rightarrow \quad 0 = a^2 - 6a + 9$$
$$\Rightarrow \quad 0 = (a-3)(a-3)$$
$$\Rightarrow \quad a = 3 \text{ (repeated root)}$$

EXAMPLE 2.12

Leena and Simon take the same time to travel to school in the mornings. Leena travels 5 km while Simon travels 4 km. Leena's average speed is $3\,\text{km}\,\text{h}^{-1}$ faster than Simon's. Find the average speed of each student.

SOLUTION

Let $x\,\text{km}\,\text{h}^{-1}$ be Leena's average speed.
Then Simon's average speed is $(x-3)\,\text{km}\,\text{h}^{-1}$.
Leena's time is $\frac{5}{x}$ and Simon's is $\frac{4}{(x-3)}$.

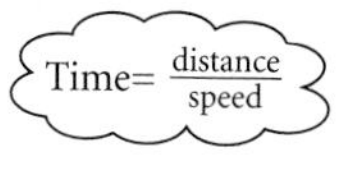

Both times are the same, so

$$\frac{5}{x}=\frac{4}{(x-3)}.$$

Multiply by $x(x-3)$

$$\Rightarrow \quad \cancel{x}(x-3)\times\frac{5}{\cancel{x}}=x\cancel{(x-3)}\times\frac{4}{\cancel{(x-3)}}.$$

$$\Rightarrow \quad 5x-15=4x$$

$$\Rightarrow \quad x=15$$

So Leena's average speed is $15\,\text{km}\,\text{h}^{-1}$ and Simon's is $12\,\text{km}\,\text{h}^{-1}$.

EXERCISE 2D

1 Solve the following equations.

(i) $x-\frac{x}{5}=\frac{2}{3}$

(ii) $\frac{2}{a}-\frac{3}{4a}=2$

(iii) $\frac{1}{x}=3-\frac{2}{x+1}$

(iv) $\frac{3x+2}{2}-\frac{x-1}{5}=3$

(v) $\frac{2}{3x-1}+\frac{1}{x+8}=\frac{1}{2}$

(vi) $\frac{2}{a}-\frac{5}{2a-1}=0$

(vii) $\frac{1}{p}+p+1=\frac{13}{3}$

(viii) $1+\frac{1}{x-1}=\frac{2x}{x+1}$

2 A formula used in physics is

$$\frac{1}{f}=\frac{1}{u}+\frac{1}{v}$$

where f is the focal length of a mirror, u is the distance of the object from the mirror, and v is the distance of the image from the mirror.
For a mirror with focal length 20 cm, find the distance of the object from the mirror when the image is twice as far away from the mirror as is the object.

3 Each day I travel 20 km from home to work. One day, because of road works, my average speed was $5\,\text{km}\,\text{h}^{-1}$ slower than usual and my journey took an extra 20 minutes.

Take $x\,\text{km}\,\text{h}^{-1}$ as my usual speed.

(i) Write down an expression in x that represents my usual time in hours.

(ii) Write down an expression in x that represents my time when I travelled $5\,\text{km}\,\text{h}^{-1}$ slower than usual.

(iii) Use these expressions to form an equation in x and solve it.

(iv) How long does my journey usually take?

4 Lamp-posts are to be placed, at equal spacings, along a 2.4 km length of straight road.

(i) If the lamp-posts are x metres apart, find an expression in terms of x for the number needed.

(ii) Find an expression in terms of x for the number needed if they are $(x + 2)$ metres apart.

(iii) Find the value of x if two more lamp-posts are needed when they are closer together.

(iv) How many lamp-posts would be needed in this case?

5 Oliver plays cricket for his school and so far this season he has made 184 runs. He has played n matches so far.

(i) Write down an expression in terms of n for his average number of runs per match.

In his next match he scores 41 runs and his batting average increases by 2.

(ii) Write down an expression in terms of n for his new average.

(iii) Form an equation in n and solve it.

(iv) What is his new batting average?

Simplifying expressions containing square roots

In mathematics there are times when it is helpful to be able to manipulate square roots, rather than just find their values from your calculator. This ensures that you are working with the exact value, not just a rounded version.

EXAMPLE 2.13

Simplify the following.

(i) $\sqrt{8}$ **(ii)** $\sqrt{6} \times \sqrt{3}$

(iii) $\sqrt{32} - \sqrt{18}$ **(iv)** $(4 + \sqrt{3})(4 - \sqrt{3})$

SOLUTION

(i)
$$\begin{aligned}\sqrt{8} &= \sqrt{2 \times 2 \times 2} \\ &= \sqrt{2} \times \sqrt{2} \times \sqrt{2} \\ &= (\sqrt{2})^2 \times \sqrt{2} \\ &= 2\sqrt{2}\end{aligned}$$

(ii)
$$\begin{aligned}\sqrt{6} \times \sqrt{3} &= \sqrt{6 \times 3} \\ &= \sqrt{2 \times 3 \times 3} \\ &= (\sqrt{3})^2 \times \sqrt{2} \\ &= 3\sqrt{2}\end{aligned}$$

(iii)
$$\begin{aligned}\sqrt{32} - \sqrt{18} &= \sqrt{16 \times 2} - \sqrt{9 \times 2} \\ &= 4\sqrt{2} - 3\sqrt{2} \\ &= \sqrt{2}\end{aligned}$$

Start by looking for the largest square number factors of 32 and 18.

(iv) $(4+\sqrt{3})(4-\sqrt{3}) = 16 - 4\sqrt{3} + 4\sqrt{3} - (\sqrt{3})^2$
$= 16 - 3$
$= 13$

Notice that in this last example there is no square root in the answer.

In the next example, all the numbers involve fractions with a square root on the bottom line. It is easier to work with numbers if any square roots are only on the top line. Manipulating a number to that form is called *rationalising the denominator.*

❓ What is a rational number?

EXAMPLE 2.14

Simplify the following by rationalising their denominators.

(i) $\dfrac{2}{\sqrt{3}}$ **(ii)** $\sqrt{\dfrac{3}{5}}$ **(iii)** $\sqrt{\dfrac{3}{8}}$

SOLUTION

(i)
$$\frac{2}{\sqrt{3}} = \frac{2}{\sqrt{3}} \times \frac{\sqrt{3}}{\sqrt{3}}$$
$$= \frac{2\sqrt{3}}{(\sqrt{3})^2}$$
$$= \frac{2\sqrt{3}}{3}$$

(ii)
$$\sqrt{\frac{3}{5}} = \frac{\sqrt{3}}{\sqrt{5}}$$
$$= \frac{\sqrt{3}}{\sqrt{5}} \times \frac{\sqrt{5}}{\sqrt{5}}$$
$$= \frac{\sqrt{3}\times\sqrt{5}}{(\sqrt{5})^2}$$
$$= \frac{\sqrt{15}}{5}$$

(iii)
$$\sqrt{\frac{3}{8}} = \frac{\sqrt{3}}{\sqrt{8}}$$
$$= \frac{\sqrt{3}}{2\sqrt{2}}$$
$$= \frac{\sqrt{3}}{2\sqrt{2}} \times \frac{\sqrt{2}}{\sqrt{2}}$$
$$= \frac{\sqrt{3}\times\sqrt{2}}{2(\sqrt{2})^2}$$
$$= \frac{\sqrt{6}}{4}$$

EXERCISE 2E

Do not use a calculator for this exercise.

1 Simplify the following.

(i) $\sqrt{32}$ **(ii)** $\sqrt{125}$

(iii) $\sqrt{5} \times \sqrt{15}$ **(iv)** $\sqrt{8} - \sqrt{2}$

(v) $3\sqrt{27} - 6\sqrt{3}$ **(vi)** $4(3 + \sqrt{2}) - 3(5 - \sqrt{2})$

(vii) $4\sqrt{32} - 3\sqrt{8}$ **(viii)** $5(6 - \sqrt{3}) + 2(3 + 4\sqrt{3})$

(ix) $2\sqrt{125} + 6\sqrt{5}$ **(x)** $3(2\sqrt{2} - 3\sqrt{3}) - 2(3\sqrt{2} - 5\sqrt{3})$

2 Simplify the following.

(i) $(\sqrt{2} - 1)^2$ **(ii)** $(4 - \sqrt{5})(2 + \sqrt{5})$

(iii) $(2 - \sqrt{7})(\sqrt{7} - 1)$ **(iv)** $(\sqrt{5} - \sqrt{3})(\sqrt{5} + \sqrt{3})$

(v) $(3 + \sqrt{2})(5 - 2\sqrt{2})$ **(vi)** $(\sqrt{7} - 3)(2\sqrt{7} + 3)$

(vii) $(3\sqrt{3} - 2)(2\sqrt{3} - 3)$ **(viii)** $(\sqrt{5} - \sqrt{3})^2$

(ix) $(5 - 3\sqrt{2})(2\sqrt{2} - 1)$ **(x)** $(2\sqrt{2} + 3)^2$

3 Simplify the following by rationalising the denominator.

(i) $\dfrac{1}{\sqrt{3}}$ **(ii)** $\dfrac{5}{\sqrt{5}}$

(iii) $\dfrac{8}{\sqrt{6}}$ **(iv)** $\sqrt{\dfrac{2}{3}}$

(v) $\dfrac{2\sqrt{2}}{\sqrt{8}}$ **(vi)** $\sqrt{\dfrac{3}{7}}$

(vii) $\dfrac{21}{\sqrt{7}}$ **(viii)** $\dfrac{5}{3\sqrt{5}}$

(ix) $\dfrac{\sqrt{75}}{\sqrt{125}}$ **(x)** $\dfrac{8}{\sqrt{128}}$

KEY POINTS

1 Linear inequalities are dealt with like equations *but* if you multiply or divide by a negative number you must reverse the inequality sign.

2 When solving a quadratic inequality it is advisable to sketch the graph.

3 When simplifying an algebraic fraction involving multiplication or division you can cancel by a common factor.

4 When simplifying an algebraic fraction involving addition or subtraction you need to find a common denominator.

5 When solving an equation involving fractions you start by multiplying through by the LCM of all the denominators to eliminate the fractions.

6 When simplifying expressions involving square roots you should
- make the number under the square root sign as small as possible
- rationalise the denominator.

3 Algebra III – polynomials

If *A* equals success, then the formula is *A* equals *X* plus *Y* plus *Z*, with *X* being work, *Y* play, and *Z* keeping your mouth shut.

Albert Einstein

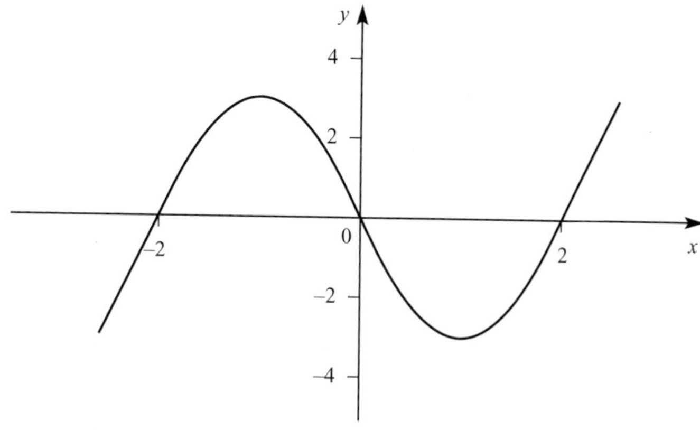

Figure 3.1

❓ Look at the graph in figure 3.1. Which of these is the equation of the curve?

- $y = x^2 - 4$
- $y = 4 - x^2$
- $y = x^3 - 4x$
- $y = -x^3 + 4x$
- $y = x^4 - 4x^2$
- $y = 4x^2 - x^4$

Operations with polynomials

A quadratic expression, such as

$$5x^2 - 3x - 4,$$

is an example of a *polynomial.*

The *order* of a polynomial is the highest power of x (or whatever letter is being used), which is 2 for a quadratic.

Similarly, a *cubic* polynomial, such as

$$x^3 - 6x + 2$$

is of order 3.

Polynomials can have any positive whole number as their order and, apart from the constant term, they only include positive whole number powers of the variable.

? Here are four expressions. Only two of them are polynomials. Which two are they? Why are the other two not polynomials?

- $x^3 + x^2 + \sqrt{x} + 4$
- $x^5 + 3x^3 + 2x - 1$
- $x^2 + x + 2 + \dfrac{3}{x}$
- $z^3 + z^2 + z + 1$

You will already have met the graphs of quadratic and some cubic functions. Graphs of higher order polynomials are also continuous and may have twists and turns.

A turning point can be a *maximum* or a *minimum*.

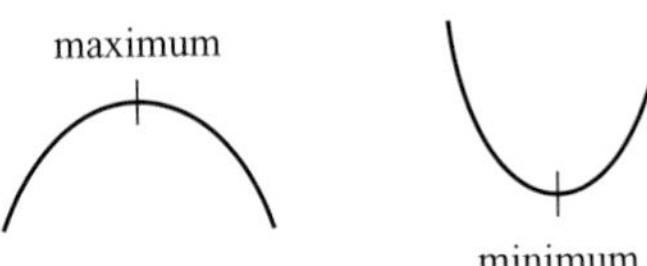

A twist is called a *point of inflection*.

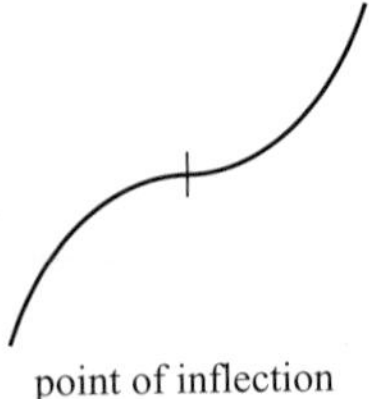

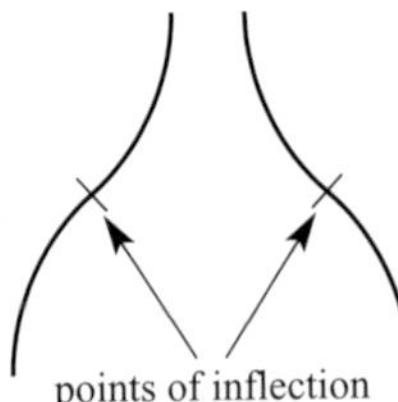

Some polynomial graphs are drawn in figure 3.2 below. Use a graphic calculator or a computer graph drawing package to check them for yourself.

(a)

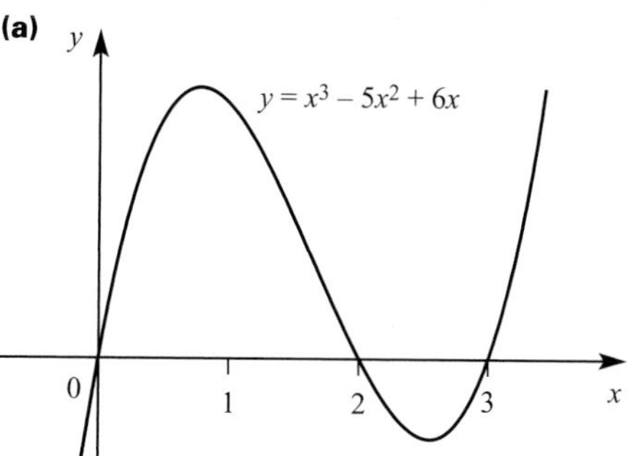

(b)

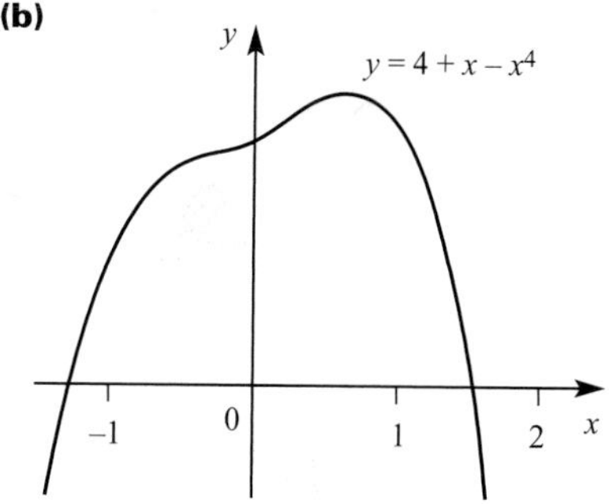

(c)

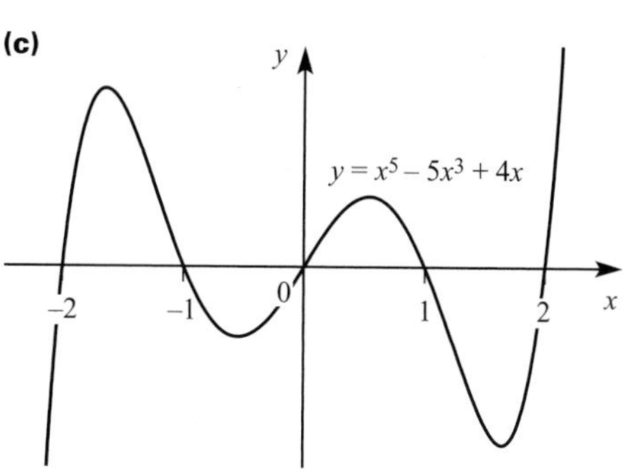

(d)

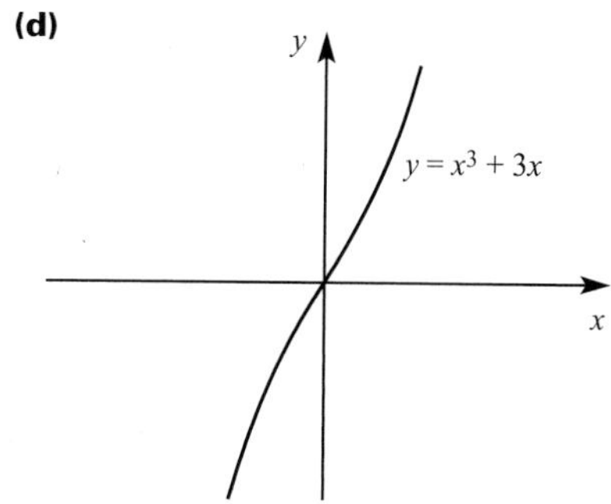

Figure 3.2

For the curves in figure 3.2, say which points are maxima, which minima and which points of inflection.

You can see the curve $y = x^3 - 5x^2 + 6x$ in figure 3.2. What does the curve $y = -(x^3 - 5x^2 + 6x)$ look like?

Addition and subtraction of polynomials

To add or subtract polynomials, collect like terms.

EXAMPLE 3.1

Add $(2x^3 + 3x + 3)$ to $(7x^2 - x + 4)$.

SOLUTION

$$\begin{array}{rrrrr} & 2x^3 & & +3x & +3 \\ + & & 7x^2 & -x & +4 \\ \hline & 2x^3 & +7x^2 & +2x & +7 \end{array}$$

Leave a space on the top row since there is no x^2 term.

EXAMPLE 3.2 Subtract $(x^3 - 5x^2 + 7x - 1)$ from $(3x^3 + x^2 - 2x - 4)$.

SOLUTION

$$\begin{array}{rr} & 3x^3 + x^2 - 2x - 4 \\ - & (x^3 - 5x^2 + 7x - 1) \\ \hline & 2x^3 + 6x^2 - 9x - 3 \\ \hline \end{array}$$

Notice that subtracting a negative quantity is the same as adding the positive value $-(-5x^2)$ is $+5x^2$.

Multiplication of polynomials

You met multiplying two linear polynomials in Chapter 1.

EXAMPLE 3.3 Multiply $(2x + 3)$ by $(x - 4)$.

SOLUTION

$$\begin{array}{lrr} & & 2x + 3 \\ & \times & x - 4 \\ \hline \text{Multiply top line by } x & & 2x^2 + 3x \\ \text{Multiply top line by } (-4) & & - 8x - 12 \\ \hline & & 2x^2 - 5x - 12 \\ \hline \end{array}$$

Multiplying higher order polynomials is just an extension of this method.

EXAMPLE 3.4 Multiply $(x^3 + 2x - 4)$ by $(x^2 - x + 3)$.

Leave a gap here, since there is no x^2 term.

SOLUTION

$$\begin{array}{lrrrrrrr} & & & & x^3 & & + 2x & - 4 \\ & \times & & & & x^2 & - x & + 3 \\ \hline \text{Multiply top line by } x^2 & & x^5 & & + 2x^3 & - 4x^2 & & \\ \text{Multiply top line by } (-x) & & & - x^4 & & - 2x^2 & + 4x & \\ \text{Multiply top line by } 3 & & & & 3x^3 & & + 6x & - 12 \\ \hline & & x^5 & - x^4 & + 5x^3 & - 6x^2 & + 10x & - 12 \\ \hline \end{array}$$

Keep a separate column for each power of x. Sometimes it is necessary to leave gaps. In arithmetic zeros are placed in the gaps. For example four thousand and five is written 4005.

Division of polynomials

This is usually set out rather like arithmetical long division.

EXAMPLE 3.5

Divide $(x^3 + 5x^2 + 7x + 2)$ by $(x + 2)$.

SOLUTION

$$\begin{array}{r} x^2 \\ x+2\,\overline{)\,x^3 + 5x^2 + 7x + 2} \\ x^3 + 2x^2 \end{array}$$

Found by dividing x^3 (the first term in $x^3 + 5x^2 + 7x + 2$), by x (the first term in $x + 2$).

$x^2(x+2)$

Now subtract $(x^3 + 2x^2)$ from $(x^3 + 5x^2)$ bringing down the next term (i.e. $7x$), and repeat the method above.

$$\begin{array}{r} x^2 + 3x \\ x+2\,\overline{)\,x^3 + 5x^2 + 7x + 2} \\ \underline{x^3 + 2x^2} \\ 3x^2 + 7x \\ 3x^2 + 6x \end{array}$$

$3x^2 \div x$

$3x(x+2)$

Continuing gives

$$\begin{array}{r} x^2 + 3x + 1 \\ x+2\,\overline{)\,x^3 + 5x^2 + 7x + 2} \\ \underline{x^3 + 2x^2} \\ 3x^2 + 7x \\ \underline{3x^2 + 6x} \\ x + 2 \\ \underline{x + 2} \\ 0 \end{array}$$

This is the answer.

The final remainder of zero means that $(x + 2)$ divides exactly into $x^3 + 5x^2 + 7x + 2$.

You now have the result that $(x^3 + 5x^2 + 7x + 2) \div (x + 2) = x^2 + 3x + 1$.

NOTE

Alternatively, if you know that there is no remainder, this may be set out as follows.

Let $(x^3 + 5x^2 + 7x + 2) \div (x + 2) = ax^2 + bx + c$.

This polynomial must be of order 2 because $x^3 \div x$ will give an x^2 term.

Multiplying both sides by $(x + 2)$

$$(x^3 + 5x^2 + 7x + 2) = (ax^2 + bx + c)(x + 2)$$

Multiplying out the expression on the right

$$x^3 + 5x^2 + 7x + 2 = ax^3 + (2a + b)x^2 + (2b + c)x + 2c$$

Comparing coefficients of x^3: $1 = a \quad \Rightarrow \quad a = 1$

Checking the constant term: $2 = 2c \quad \Rightarrow \quad c = 1$

Comparing coefficients of x^2: $5 = 2a + b$

$= 2 + b \quad \Rightarrow \quad b = 3$

Comparing coefficients of x allows you to check the values of b and c.
With practice you may be able to do this method *by inspection.*

3

Algebra

EXAMPLE 3.6 Divide $(x^4 - x^3 - 2)$ by $(x + 1)$.

SOLUTION

⚠ Keep a separate column for each power of x. Leave gaps if necessary.

$$
\begin{array}{r}
x^3 - 2x^2 + 2x - 2 \\
x+1 \overline{) x^4 - x^3 \qquad\qquad - 2} \\
\underline{x^4 + x^3} \qquad\qquad\qquad \\
-2x^3 \qquad\qquad\qquad \\
\underline{-2x^3 - 2x^2} \qquad\qquad \\
2x^2 \qquad\qquad \\
\underline{2x^2 + 2x} \qquad \\
-2x - 2 \\
\underline{-2x - 2} \\
0
\end{array}
$$

EXERCISE 3A

1 State the orders of the following polynomials.

(i) $2x^3 + 3x - 5$

(ii) $x^5 - 2$

(iii) $4x^6 + 2x^3 - x^2 + 1$

(iv) $7x^4 - 2x^3 + 3$

2 (i) Add $(2x^3 + 3x^2 - x + 4)$ to $(3x^3 - x^2 + 5x - 1)$.

(ii) Add $(5x^4 - 2x^2 + 3x + 1)$ to $(x^3 - x^2 + 5x - 5)$.

3 (i) Subtract $(2x^3 - 3x^2 + 4x + 1)$ from $(5x^3 - 2x^2 + 4x - 1)$.

(ii) Subtract $(3x^4 - 2x^2 - 3x + 2)$ from $(x^3 + 4x^2 + 3x - 2)$.

4 (i) Multiply $(x^3 - x^2 + x - 2)$ by $(x^2 + 1)$.

(ii) Multiply $(x^4 - 2x^2 + 3)$ by $(x^2 + 2x - 1)$.

(iii) Multiply $(2x^3 - 3x + 5)$ by $(x^2 - 2x + 1)$.

(iv) Multiply $(x^5 + x^4 + x^3 + x^2 + x + 1)$ by $(x - 1)$.

(v) Expand $(x + 2)(x - 1)(x + 3)$. Hint: expand two brackets first

(vi) Expand $(2x + 1)(x - 2)(x + 4)$.

(vii) Simplify $(2x^2 - 1)(x + 2) - 4(x + 2)^2$.

(viii) Simplify $(x^2 - 1)(x + 1) - (x^2 + 1)(x - 1)$.

(ix) Divide $(x^3 + 2x^2 - x - 2)$ by $(x - 1)$.

(x) Divide $(x^3 + x^2 - 5x + 3)$ by $(x + 3)$.

(xi) Divide $(2x^3 - 5x^2 - 11x - 4)$ by $(2x + 1)$.

(xii) Divide $(6x^3 - 7x^2 - 7x + 6)$ by $(2x - 3)$.

The factor theorem

The factor theorem is an important result which allows you to find factors of polynomials of any order (if they factorise), and so to solve polynomial equations.

Look at this quadratic equation.

$$x^2 - 5x - 6 = 0$$

Factorising $\Rightarrow (x-6)(x+1) = 0$

$\Rightarrow (x-6) = 0$ or $(x+1) = 0$

$\Rightarrow x = 6$ or $x = -1$

? What happens if you substitute $x = 6$ in $x^2 - 5x - 6$?
What about $x = -1$?

NOTE

If $f(x) = x^2 - 5x - 6$, you use the notation f(6) to mean the value of $f(x)$ when $x = 6$.

The factor theorem states this result in a general form.

> If $(x - a)$ is a factor of the polynomial $f(x)$, then
> - $f(a) = 0$
> - $x = a$ is a root of the equation $f(x) = 0$.
>
> Conversely, if $f(a) = 0$, then $(x - a)$ is a factor of $f(x)$.

EXAMPLE 3.7

Given that

$$f(x) = x^3 + 2x^2 - x - 2$$

(i) find $f(-2)$, $f(-1)$, $f(0)$, $f(1)$, $f(2)$
(ii) factorise $x^3 + 2x^2 - x - 2$
(iii) use the values from **(i)** to plot the curve.

SOLUTION

(i) $f(-2) = (-2)^3 + 2(-2)^2 - (-2) - 2$
$= -8 + 8 + 2 - 2$
$= 0 \quad\Rightarrow\quad (x+2)$ is a factor

$f(-1) = (-1)^3 + 2(-1)^2 - (-1) - 2$
$= -1 + 2 + 1 - 2$
$= 0 \quad\Rightarrow\quad (x+1)$ is a factor

$f(0) = 0 + 0 - 0 - 2$
$= -2$

$f(1) = 1 + 2 - 1 - 2$
$= 0 \quad\Rightarrow\quad (x-1)$ is a factor

$f(2) = 8 + 8 - 2 - 2$
$= 12$

(ii) Since $(x-1)(x+1)(x+2)$ would expand to give a polynomial of order 3, you can say that

$$x^3 + 2x^2 - x - 2 = k(x-1)(x+1)(x+2)$$

where k is a constant.

? The value of k is 1. How can you show this?

---

so

$$x^3 + 2x^2 - x - 2 = (x-1)(x+1)(x+2)$$

(iii) Using the values from **(i)** gives the curve shown in figure 3.3.

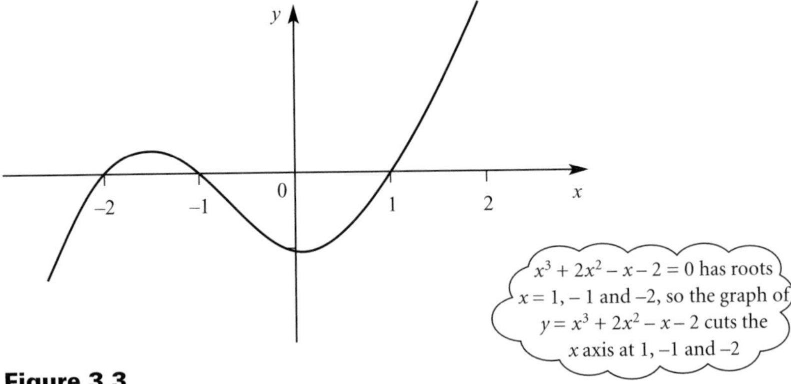

Figure 3.3

EXAMPLE 3.8

Given that

$$f(x) = x^3 + 3x^2 - x - 3$$

(i) show that $(x+1)$ is a factor of $f(x)$
(ii) what other values of x should you try when you are looking for another factor?
(iii) solve the equation $f(x) = 0$.

SOLUTION

(i) $(x+1)$ is a factor if $f(-1) = 0$.

$$f(-1) = -1 + 3 + 1 - 3$$
$$= 0$$
$$\Rightarrow (x+1) \text{ is a factor of } x^3 + 3x^2 - x - 3.$$

(ii) Any other linear factor will be of the form $(x - a)$ where a is a factor of the constant term (−3).

This means that the only other values of x which are worth trying are 1, 3 and −3.

(iii) $f(1) = 1 + 3 - 1 - 3$
$= 0 \quad \Rightarrow \quad (x - 1)$ is a factor
$f(3) = 27 + 27 - 3 - 3$
$= 48$
$f(-3) = -27 + 27 + 3 - 3$
$= 0 \quad \Rightarrow \quad (x + 3)$ is a factor

$f(-1) = 0$, $f(1) = 0$ and $f(-3) = 0$ and the equation is a cubic, so has at most three roots. The solution is $x = -1$, 1 or -3.

Sometimes you may only be able to find one linear factor for the cubic and, in this case, you then need to use long division.

EXAMPLE 3.9

Given that

$$f(x) = x^3 - x^2 - 3x - 1$$

(i) show that $(x + 1)$ is a factor
(ii) factorise $f(x)$
(iii) solve $f(x) = 0$.

SOLUTION

(i) $f(-1) = (-1)^3 - (-1)^2 - 3(-1) - 1$
$= -1 - 1 + 3 - 1$
$= 0$
$\Rightarrow (x + 1)$ is a factor of $x^3 - x^2 - 3x - 1$

? What happens when you try $x = 1$?
Is there any other value you should try?

(ii) Since $(x + 1)$ is a factor, divide $f(x)$ by $(x + 1)$.

$$\begin{array}{r} x^2 - 2x - 1 \\ x + 1 \overline{)\, x^3 - x^2 - 3x - 1} \\ \underline{x^3 + x^2 } \\ -2x^2 - 3x \\ \underline{-2x^2 - 2x } \\ -x - 1 \\ \underline{-x - 1} \\ 0 \end{array}$$

So

$$f(x) = (x + 1)(x^2 - 2x - 1)$$

(iii) $f(x) = 0 \Rightarrow (x + 1)(x^2 - 2x - 1) = 0$
$\Rightarrow$ either $x = -1$ or $x^2 - 2x - 1 = 0$

Using the quadratic formula on

$$x^2 - 2x - 1 = 0$$

gives

$$\begin{aligned} x &= \frac{2 \pm \sqrt{4 - (4 \times 1 \times (-1))}}{2} \\ &= \frac{2 \pm \sqrt{8}}{2} \\ &= 2.414 \text{ or } -0.414 \end{aligned}$$

The complete solution is $x = -1, -0.414$ or 2.414 (to 3 d.p.).

The remainder theorem

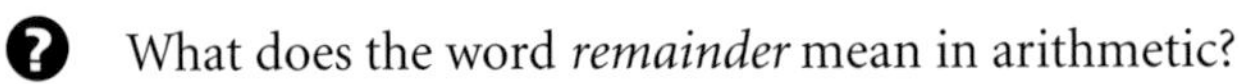

? What does the word *remainder* mean in arithmetic?

Look at these divisions.

$49 \div 7 = 7$ remainder 0
$49 \div 8 = 6$ remainder 1
$49 \div 5 = 9$ remainder 4

? In these examples the remainder is always smaller than the number that you are dividing by. Is this always the case in arithmetic? Explain your answer.

ACTIVITY 3.1

(i) Let $f(x) = (x^3 + 3x^2 - 2x + 1)$ and divide $f(x)$ by $(x + 1)$. You will see that you have a remainder.

(ii) Calculate $f(-1)$.

(iii) What do you notice?

(iv) Now try dividing $f(x)$ by $(x - 2)$. Is the remainder the same as $f(2)$?

The answers to Activity 3.1 suggest a possible idea. This can be stated generally as the *remainder theorem.*

> The remainder theorem states that, for a polynomial $f(x)$, $f(a)$ is the remainder when $f(x)$ is divided by $(x - a)$.

This means that f(x) can be written as

$$f(x) = (x - 1)Q(x) + f(a)$$

Compare this with the corresponding numerical results $49 \div 8 = 6$ rem 1 $\Rightarrow 49 = 8 \times 6 + 1$.

where Q(x) is the quotient and f(a) is the remainder.

You can show it is true by substituting $x = a$.

? You divide a polynomial, f(x), by a linear function like $(x - 2)$. Can the remainder have any terms in x?

EXAMPLE 3.10

Find the remainder when $x^3 - 2x + 3$ is divided by $(x + 2)$.

SOLUTION

The remainder is found by substituting $x = -2$.

$$(-2)^3 - 2(-2) + 3$$
$$= -8 + 4 + 3$$
$$= -1, \text{ so the remainder is } -1$$

EXAMPLE 3.11

A polynomial is given by $f(x) = 2x^3 + ax^2 + bx + 4$.

When f(x) is divided by $(x - 1)$ the remainder is 1, and when it is divided by $(x + 1)$ the remainder is 3.

Find a and b.

SOLUTION

Division by $(x - 1)$ gives a remainder of 1.

$\Rightarrow \quad f(1) = 2 \times 1^3 + a \times 1^2 + b \times 1 + 4 = 1$

$\Rightarrow \quad a + b + 6 = 1$

$\Rightarrow \quad a + b = -5$ ①

Division by $(x + 1)$ gives a remainder of 3.

$\Rightarrow \quad f(-1) = 2 \times (-1)^3 + a \times (-1)^2 + b \times (-1) + 4 = 3$

$\Rightarrow \quad a - b + 2 = 3$

$\Rightarrow \quad a - b = 1$ ②

Solving equations ① and ② simultaneously

$$\begin{aligned} a + b &= -5 \\ + \quad a - b &= \ \ 1 \\ \hline 2a \quad &= -4 \quad \Rightarrow \quad a = -2 \end{aligned}$$

Substitute in ①

$$-2 + b = -5$$
$$\Rightarrow \quad b = -5 + 2$$
$$\Rightarrow \quad b = -3$$

so $a = -2$, $b = -3$

EXERCISE 3B

1 Determine whether the following linear functions are factors of the given polynomials.

(i) $x^3 - 8x + 7 \quad (x - 1)$ **(ii)** $2x^3 + 3x^2 - 4x + 2 \quad (x + 1)$
(iii) $2x^3 + x^2 - 5x - 4 \quad (x + 2)$ **(iv)** $2x^4 - x^3 - 4 \quad (x - 2)$
(v) $x^3 + 27 \quad (x + 3)$ **(vi)** $x^3 - ax^2 + a^2x - a^3 \quad (x - a)$

2 Factorise the following functions as a product of three linear factors.

(i) $x^3 - 3x^2 - x + 3$ **(ii)** $x^3 - 7x - 6$
(iii) $x^3 - x^2 - 2x$ **(iv)** $2x^3 + 7x^2 + 7x + 2$
(v) $2x^3 - 3x^2 - 3x + 2$ **(vi)** $x^3 + 5x^2 - x - 5$
(vii) $x^3 + 2x^2 - 4x - 8$ **(viii)** $x^3 + 5x^2 + 3x - 9$
(ix) $2x^3 + 5x^2 - x - 6$ **(x)** $3x^3 - 8x^2 - 20x + 16$

3 Factorise the following functions as far as possible.

(i) $x^3 - x^2 - x - 2$ **(ii)** $x^3 - x^2 + x - 1$
(iii) $x^3 - 5x^2 + 10x - 8$ **(iv)** $x^3 - 8$
(v) $x^3 + 3x^2 + 3x + 2$ **(vi)** $2x^3 - 5x^2 - 2x - 3$
(vii) $x^3 + 6x^2 + 13x + 10$ **(viii)** $3x^3 + 5x^2 + 3x + 1$
(ix) $2x^3 + 9x^2 + 25$ **(x)** $x^3 + 10x^2 + 10x + 9$

4 Find the remainder when the following functions are divided by the linear factors indicated.

(i) $x^3 + 3x^2 - 6x + 2 \quad (x + 2)$ **(ii)** $2x^3 - 3x^2 + 2 \quad (x - 3)$
(iii) $x^3 + x^2 + 12x + 10 \quad (x + 1)$ **(iv)** $x^3 + x^2 + x + 2 \quad (x + 2)$
(v) $4x^4 - 2x^3 + x^2 - 2 \quad (2x - 1)$ **(vi)** $2x^4 + 3x^3 + 5 \quad (x + 3)$
(vii) $x^3 + a^3 \ (x - a)$ **(viii)** $x^3 - a^3 \quad (x + a)$

5 $(x + 2)$ is a factor of $ax^2 + 3x - 2$.
Find a.

6 For what value of k is $(x + 1)$ a factor of $x^4 - 2x^2 + k$?

7 Find the remainder when $x^3 - 3x^2 + 5x + 10$ is divided by $x - 4$.

8 When $x^3 + 4x^2 + ax + 6$ is divided by $(x + 5)$ the remainder is -4.
What is a?

9 The expression $x^4 + px^3 + qx + 36$ is exactly divisible by $(x - 2)$ and $(x + 3)$.

Find, and simplify, two simultaneous equations for p and q and hence find p and q.

10 **(i)** Find the value of k for which $x = 2$ is a root of $x^3 + kx + 6 = 0$.
(ii) Find the other roots when k has this value.

11 When $f(x) = 2x^3 + ax^2 + bx + 6$ is divided by $(x - 1)$ there is no remainder, and when $f(x)$ is divided by $(x + 1)$ the remainder is 10.

(i) Find the values of a and b.

(ii) Solve the equation $f(x) = 0$ when a and b have these values.

12 Given that $f(x)$, where $f(x) = x^2 + ax + 5$ and a is a constant, is such that the remainder on dividing $f(x)$ by $(x - 1)$ is three times the remainder on dividing $f(x)$ by $(x + 1)$, find the value of a.

13 A hilly road can be modelled by the part of the cubic curve

$$y = 0.0002x^3 - 0.04x^2 + 2x \quad \text{for } 0 \leqslant x \leqslant a.$$

The curve is illustrated in the graph.

Find the value of a.

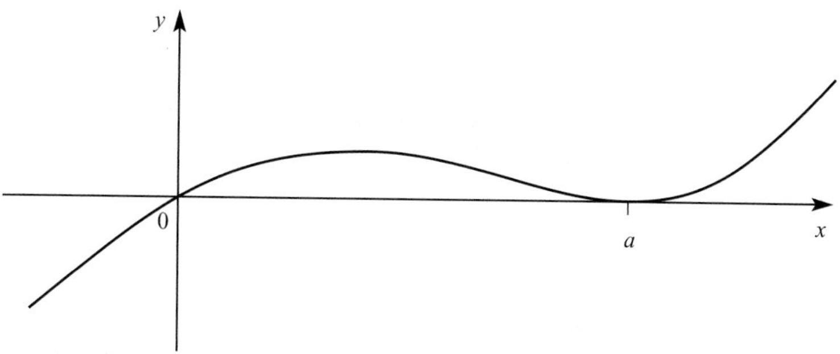

14 The diagram shows an open rectangular tank whose base is a square of side x metres and whose volume is $8\,\text{m}^3$.

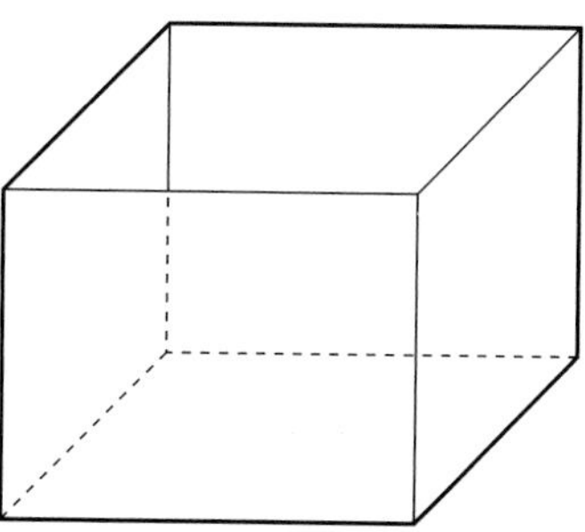

(i) Write down an expression in terms of x for the height of the tank.

(ii) Show that the surface area of the tank is $\left(x^2 + \frac{32}{x}\right)\text{m}^2$.

(iii) Given that the surface area is $24\,\text{m}^2$ show that

$$x^3 - 24x + 32 = 0.$$

(iv) Solve $x^3 - 24x + 32 = 0$ to find the possible values of x.

[MEI]

KEY POINTS

1 A polynomial in x has terms in positive integer powers of x, and may also have a constant term.

2 The order of a polynomial in x is the highest power of x that appears in the polynomial.

3 The factor theorem states that if $(x - a)$ is a factor of a polynomial $f(x)$, then $f(a) = 0$ and $x = a$ is a root of the equation $f(x) = 0$. Conversely if $f(a) = 0$, then $(x - a)$ is a factor of $f(x)$.

4 The remainder theorem states that $f(a)$ is the remainder when $f(x)$ is divided by $(x - a)$.

4 Algebra IV – applications

Is probability probable? ... It is annoying to dwell upon such trifles, but there is a time for trifling.

Blaise Pascal (1623–62)

The map in figure 4.1 shows part of an American town. Peter lives at P, Queenie lives at Q and Reena lives at R. When Peter visits Queenie or Reena he travels along the roads and always goes East or North.

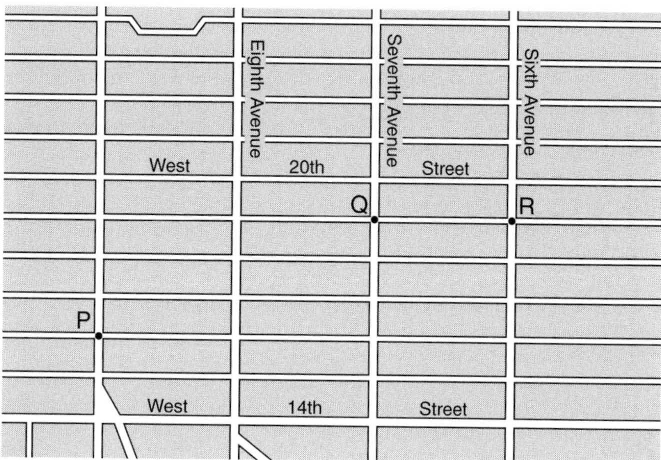

Figure 4.1

How many different routes can Peter take when he visits Queenie?
How can you describe each one of them?
How many different routes can he take when he visits Reena?
Can you generalise this?

The binomial expansion

A *binomial expansion* is obtained when a *binomial* (i.e. two-term) expression, like $(a + b)$, is raised to a power. To see what is going on, start by looking at the simpler expression $(1 + x)$.

ACTIVITY 4.1 Work out the following powers of $(1 + x)$ and see how many patterns you can find in the coefficients (these are the numbers multiplying the powers of x).

Notice that each line is $(1 + x)$ times the previous one.

(i) $(1 + x)^0$
(ii) $(1 + x)^1$
(iii) $(1 + x)^2$
(iv) $(1 + x)^3$
(v) $(1 + x)^4$

? The pattern in figure 4.2 is called *Pascal's triangle* and the numbers in each line are called *binomial coefficients.*

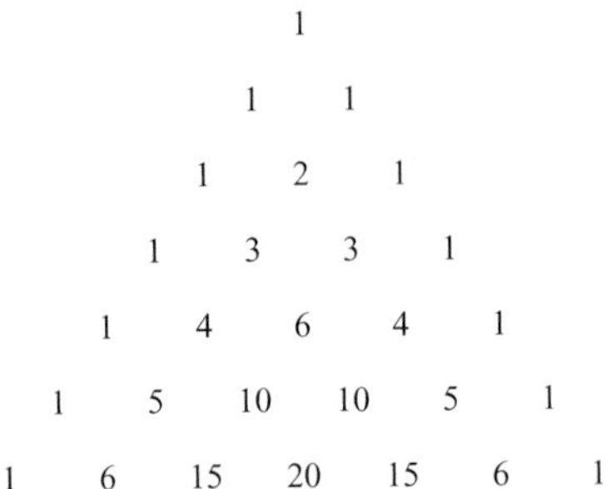

Figure 4.2

How is it connected to the discussion point at the beginning of the chapter?

Historical note

Blaise Pascall was born in France in 1623. A gifted mathematician, he was making discoveries in geometry by the age of 16 and at the age of 18 he invented and made the first calculating machine in history. In 1654 he worked with Fermat, by correspondence, on the theory of probability in response to a problem proposed by the Chevalier de Mere, a professional gambler.

Pascal's triangle (and the binomial theorem) had actually been discovered by Chinese mathematicians several centuries earlier, and can be found in the works of Yang Hui (around AD 1270) and Chu Shi-Kie (in AD 1303). Pascal is remembered for his application of the triangle to elementary probability, and for his study of the relationships between the coefficients. After a lifetime of ill health, he died when he was only 39.

What is the next row in Pascal's triangle?
What symmetry does it have?
What sequences can you see in the numbers?

A more general binomial expansion is for powers of $(a + b)$. Writing down the first few of these gives:

$$(a+b)^0 = 1$$
$$(a+b)^1 = 1a^1 + 1b^1$$
$$(a+b)^2 = 1a^2 + 2a^1b^1 + 1b^2$$
$$(a+b)^3 = 1a^3 + 3a^2b^1 + 3a^1b^2 + 1b^3$$
$$(a+b)^4 = 1a^4 + 4a^3b^1 + 6a^2b^2 + 4a^1b^3 + 1b^4$$

Notice how in each row the powers of a decrease and those of b increase so that their sum is always the same as the power of $(a + b)$.

EXAMPLE 4.1

Write out the binomial expansion of $(1 - 2x)^4$.

SOLUTION

The binomial coefficients for power 4 are

1 4 6 4 1.

Taking $a = 1$ and $b = (-2x)$ in the standard expansion:

$$\begin{aligned}(1-2x)^4 &= 1 + 4(-2x) + 6(-2x)^2 + 4(-2x)^3 + 1(-2x)^4\\ &= 1 - 8x + 6(4x^2) + 4(-8x^3) + 16x^4\\ &= 1 - 8x + 24x^2 - 32x^3 + 16x^4.\end{aligned}$$

EXAMPLE 4.2

Write out the binomial expansion of $(2p + 3q)^3$.

SOLUTION

The binomial coefficients for power 3 are

1 3 3 1.

Replacing a by $2p$ and b by $3q$ gives:

$$\begin{aligned}(2p+3q)^3 &= 1(2p)^3 + 3(2p)^2(3q) + 3(2p)(3q)^2 + 1(3q)^3\\ &= 8p^3 + 3(4p^2)(3q) + 3(2p)(9q^2) + 27q^3\\ &= 8p^3 + 36p^2q + 54pq^2 + 27q^3.\end{aligned}$$

How do the x terms arise when you multiply out $(1 + x)(1 + x)$?
How do the x^2 terms arise when you multiply out $(1 + x)(1 + x)(1 + x)$?
Can you generalise this?
Can you find a connection between this and the example at the beginning of the chapter (the number of different routes when Peter visits Queenie and Reena)?

Similarly in the expansion of $(1 + x)^7$ the coefficient of x^3 is the number of ways you can choose x from three of the brackets, and 1 from the remaining four.

It is not always practical to write out Pascal's triangle when you need more than a few lines, and modern calculators provide an alternative. If you are using a scientific calculator, look for the key labelled *nCr*. On a graphic calculator this function is available through the option menu.

Notice that pressing

gives you the coefficient of x^3 in the expansion of $(1 + x)^7$.

NOTE

In print you will see this written as ${}^7C_3$ and this is how you should write it.

Describe

(i) ${}^{10}C_4$
(ii) ${}^{10}C_0$
(iii) ${}^{10}C_{10}$

Using this notation, the binomial expansion can be written as

$$(1 + x)^n = {}^nC_0 + {}^nC_1\,x + {}^nC_2\,x^2 + {}^nC_3x^3 + \ldots + {}^nC_nx^n$$

An alternative notation for ${}^nC_r$ is $\binom{n}{r}$.

ACTIVITY 4.2

(i) Show that the numbers in line 4 of Pascal's triangle are the same as ${}^4C_0$, ${}^4C_1$, ${}^4C_2$, ${}^4C_3$, and ${}^4C_4$.

(ii) 4 *factorial*, written as 4!, is $4 \times 3 \times 2 \times 1$, or 24.

(a) Using this definition work out $\dfrac{4!}{1! \times 3!}$ $\dfrac{4!}{2! \times 2!}$ and $\dfrac{4!}{3! \times 1!}$.

(b) What is the connection between these values and the answers in **(i)**?

(c) Can you extend **(a)** to give similar expressions for ${}^4C_0$ and ${}^4C_4$?

(d) What does this tell you about 0!?

Activity 4.2 shows that

$${}^nC_r = \frac{n!}{r!\,(n-r)!}.$$

EXAMPLE 4.3

Find the coefficient of x^4 in the expansion of $(1 - 2x)^{10}$.

SOLUTION

$$\begin{aligned}
(1 - 2x)^{10} &= (1 + (-2x))^{10} \\
&= 1 + {}^{10}C_1(-2x) + \ldots + {}^{10}C_4(-2x)^4 + \ldots + (-2x)^{12} \\
\Rightarrow \quad & x^4 \text{ term is } {}^{10}C_4 \times 16x^4 \\
&= 3360x^4 \\
\Rightarrow \quad & \text{the coefficient of } x^4 \text{ is } 3360.
\end{aligned}$$

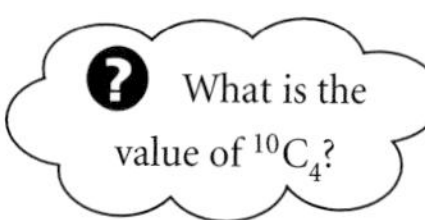

The more general expansion can also be written using the ${}^nC_r$ notation.

$$(a + b)^n = {}^nC_0a^n + {}^nC_1a^{n-1}b + {}^nC_2a^{n-2}b^2 + \ldots + {}^nC_nb^n$$

EXAMPLE 4.4

Find the coefficient of p^5q^3 in the expansion of $(p + q)^8$.

SOLUTION

$$(p + q)^8 = {}^8C_0p^8 + {}^8C_1p^7q + {}^8C_2p^6q^2 + {}^8C_3p^5q^3 + \ldots$$

The coefficient of p^5q^3 is ${}^8C_3 = 56$.

EXERCISE 4A

1 Write out these binomial expansions.

(i) $(1 - x)^4$ **(ii)** $(1 + 3x)^3$ **(iii)** $(1 - 2x)^5$
(iv) $(1 + 2x)^4$ **(v)** $(1 - 3x)^5$ **(vi)** $(1 + x^2)^6$

2 Write out these binomial expansions.

(i) $(2x - 1)^3$ **(ii)** $(x + 3)^5$ **(iii)** $(2x - 3)^4$
(iv) $(3x + y)^3$ **(v)** $(x - 2y)^3$ **(vi)** $(3x + 4y)^3$

3 Find

(i) the coefficient of x^5 in the expansion of $(1 + x)^9$
(ii) the coefficient of x^4 in the expansion of $(1 + 2x)^7$
(iii) the coefficient of x^7 in the expansion of $(1 - 3x)^{11}$
(iv) the coefficient of a^5b^3 in the expansion of $(a + 2b)^8$
(v) the coefficient of p^7q^2 in the expansion of $(p - q)^9$.

4 Simplify $(1 + x)^4 - (1 - x)^4$.

5 **(i)** Expand $(1 + 2x)^3$.
(ii) Hence expand $(1 - x)(1 + 2x)^3$.

6 **(i)** Expand $(a + bx)^3$.

The expansion of $(1 - x)(a + bx)^3$ in ascending powers of x begins $8 + 28x + \ldots$.
(ii) Find the values of a and b.

7 Calculate 1.06^5 by putting $x = 0.06$ in the expansion of $(1 + x)^5$.

8 **(i)** Expand $(1 + x)^3$.
(ii) Hence write down and simplify the expansion of $(1 + y + y^2)^3$ by replacing x by $(y + y^2)$.

The binomial distribution

Alistair, Brian, Collette, Dharmi and Ellen want to play a board game.

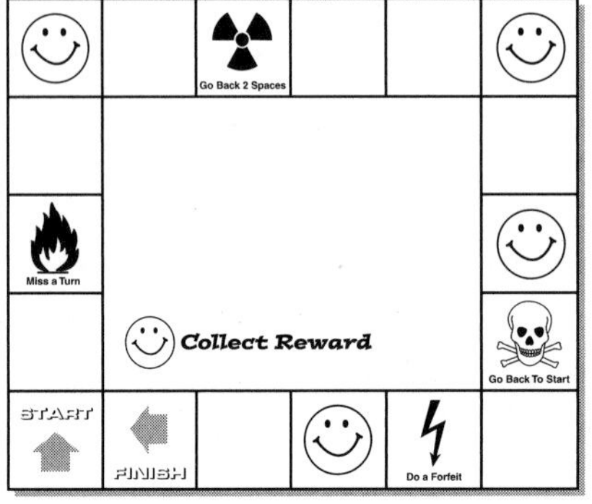

Rules
- You need to throw a 6 to start.
- Each player throws the die in turn.
- …

What is the probability that exactly two out of the five of them start playing with their first throw of the die?

You can use the binomial expansion of $(p + q)^5$ to find the probabilities of different numbers of 6s when five people throw a die, where:

$p = \frac{1}{6}$ is the probability that the die shows 6

$q = \frac{5}{6}$ is the probability that the die does not show 6.

$$(p + q)^5 = p^5 + 5p^4q^1 + 10p^3q^2 + 10p^2q^3 + 5p^1q^4 + q^5$$

This term represents two 6s and three not-6s.

The probability is $10 \times (\frac{1}{6})^2 \times (\frac{5}{6})^3 = 0.161$.

 What is the probability that four people get 6s and just one does not?

The number of people getting a 6 is an example of a *binomial distribution*, and the answer has been found using an appropriate binomial expansion.

Look at the features that characterise such a distribution:

- there is a fixed number, n, of independent trials.

In the example above, a trial is one person throwing the die ⇒ five trials and $n = 5$.

- each trial must only have two possible outcomes, success or failure.

Here, success is *throwing a 6* and failure is *throwing any other number*, i.e. not throwing a 6.

- the probability of success must be the same for each trial.

The probability of throwing a 6 is $\frac{1}{6}$ each time the die is tossed.

There are two possible outcomes, success or failure. By convention

p = the probability of success
$q = 1 - p$ = the probability of failure.

Revision of basic probability results

- If a number of events are *independent* (i.e. the outcome of one does not affect the outcome of any other) then the probability that they all occur is found by multiplying the probabilities for each one.
- If two events are *mutually exclusive* (i.e. there is no overlap) then the probability of one or the other occurring is the sum of the individual probabilities.

In probability questions, the words *and* and *or* are used quite often. For example, 'What is the probability that Alan *and* Beth both play tennis?' or 'What is the probability that either it rains *or* it snows tomorrow?'

 What mathematical operations do these key words relate to?

The dice problem (on page 64) is one example of a binomial experiment. The method can be generalised as follows.

In an experiment where

- the probability of a successful outcome is p
- the probability that the outcome is a failure is q, where $q = 1 - p$
- there are n trials
- the number of successes is denoted by X,

then

$$P(X = r) = {}^nC_r p^r q^{n-r}$$

the probability of r successes

where $r = 0, 1, 2, \ldots, n$

You can describe the situation more briefly using the notation

$$X \sim B(n, p)$$

~ means 'has the distribution'.

Read this as 'X has the binomial distribution for n trials where p is the probability of success in each trial'.

? Describe a real-life situation for which $X \sim B\left(20, \frac{1}{2}\right)$.

EXAMPLE 4.5

The probability that a pen drawn at random from a large box of pens is defective is 0.1.

A sample of eight pens is taken. Find the probability that it contains

(i) no defective pens
(ii) one defective pen
(iii) at least two defective pens.

SOLUTION

Let X = the number of defective pens in a sample of eight
p = the probability of a defective pen
$= 0.1$
so $q = 0.9$

(i) $P(X = 0) = {}^8C_0\, p^0\, q^8$
$= 1 \times 0.1^0 \times 0.9^8$
$= 0.430$

(ii) $P(X = 1) = {}^8C_1\, p^1\, q^7$
$= 8 \times 0.1^1 \times 0.9^7$
$= 0.383$

(iii) $P(X \geqslant 2) = P(X = 2) + P(X = 3) + \ldots + P(X = 8)$
$= 1 - [P(X = 0) + P(X = 1)]$
$= 1 - 0.4305 - 0.3826$
$= 0.187$

? Explain this line.

EXAMPLE 4.6

A multiple-choice test consists of ten questions with a choice of four answers for each, only one of which is correct.

A student guesses at each of the answers.

Find the probability that he gets

(i) none correct

(ii) one correct.

He needs at least eight correct answers to pass.

(iii) Find the probability that he passes.

SOLUTION

Let X = the number of correct answers from ten questions

p = the probability of a correct answer

$= \frac{1}{4}$ since he is guessing

so $q = \frac{3}{4}$

(i) $$\begin{aligned} P(X=0) &= {}^{10}C_0\, p^0\, q^{10} \\ &= 1 \times \left(\tfrac{1}{4}\right)^0 \times \left(\tfrac{3}{4}\right)^{10} \\ &= 0.056 \end{aligned}$$

(ii) $$\begin{aligned} P(X=1) &= {}^{10}C_1\, p^1\, q^9 \\ &= 10 \times \left(\tfrac{1}{4}\right)^1 \times \left(\tfrac{3}{4}\right)^9 \\ &= 0.751 \end{aligned}$$

(iii) $$\begin{aligned} P(\text{passes}) &= P(X=8) + P(X=9) + P(X=10) \\ &= {}^{10}C_8 \left(\tfrac{1}{4}\right)^8 \left(\tfrac{3}{4}\right)^2 + {}^{10}C_9 \left(\tfrac{1}{4}\right)^9 \left(\tfrac{3}{4}\right)^1 + {}^{10}C_{10} \left(\tfrac{1}{4}\right)^{10} \left(\tfrac{3}{4}\right)^0 \\ &= 0.000\,386\,24 + 0.000\,028\,61 + 0.000\,000\,95 \\ &= 0.000\,42 \quad \text{(2 s.f.)} \end{aligned}$$

EXAMPLE 4.7

Simon has been practising his archery and he is now able to hit the target with a probability of 0.3.

Find the least number of arrows needed if the probability that the target is hit at least once is greater than 90%.

SOLUTION

Let X = the number of hits in n trials

p = the probability of a hit = 0.3

so $q = 0.7$

$$\begin{aligned} \text{P (the target is hit at least once)} &= 1 - \text{P (the target is not hit at all)} \\ &= 1 - P(X=0) \\ &= 1 - {}^nC_0\,(0.3)^0\,(0.7)^n \\ &= 1 - (0.7)^n \end{aligned}$$

since $(0.3)^0 = 1$ and ${}^nC_0 = 1$ for all values of n

This must be at least 90%. So

$$1-(0.7)^n \geqslant 0.9$$
$$\Rightarrow \quad (0.7)^n \leqslant 0.1$$

Using your calculator to check different powers of 0.7 gives

$$0.7^4 = 0.2401$$
$$0.7^5 = 0.1681$$
$$0.7^6 = 0.1176$$
$$0.7^7 = 0.0824 \quad \Rightarrow \quad n \text{ must be at least } 7$$

Seven arrows is the least number needed.

EXERCISE 4B

1 Find the probability of throwing at least five 6s in eight throws of an unbiased die.

2 A typist has a probability of 0.99 of typing a letter correctly. What is the probability of exactly two mistakes in a sentence containing 180 letters, if mistakes are made at random.

3 $X \sim \text{B}(6, 0.2)$. Find

(i) $\text{P}(X = 4)$

(ii) $\text{P}(X \leqslant 1)$.

4 In a box of smarties there are eight different colours which normally occur in equal proportions.

Rachael is given 24 smarties, and orange ones are her favourite. Assume these come from a very large box.

(i) How many orange ones would she expect to get?

(ii) What is the probability that she gets this number?

(iii) What is the probability that she gets fewer orange ones?

(iv) What is the probability that she gets more orange ones than she expects?

5 A company manufactures cheap drinking glasses and normally expects 10% of them to be defective.

Regular samples of eight glasses are taken at random and if more than one glass is defective the machine is stopped and checked.

(i) What is the probability that a sample of size eight contains no defective glasses?

(ii) What is the probability that the machine is stopped and checked after a sample is taken?

6 35% of the students in a particular school travel to school by bus. From a sample of 12 students chosen at random, find the probability that

(i) only two travel by bus

(ii) half of them travel by bus.

7 A bag contains tulip bulbs and daffodil bulbs in the ratio 1 : 4. 20 bulbs are selected at random.

Find the probability that

(i) they are all daffodil bulbs

(ii) the sample contains the two types of bulbs in the same ratio as the bag

(iii) there are equal numbers of the two types of bulbs in the sample.

8 In a large town one person in five is left-handed.

(i) Find the probability that in a class of 30 children

(a) exactly three will be left-handed

(b) one-third will be left-handed.

(ii) How large must a sample be if the probability of at least one left-handed person is to be greater than 90%

9 A bag contains 12 red and 8 blue balls that are identical except for their colour.

A ball is chosen from the bag, its colour noted, and then it is replaced. Eight balls are chosen.

(i) Find

(a) the probability of obtaining five red balls

(b) the probability of obtaining at least two blue balls

(c) the probability of obtaining equal numbers of red and blue balls.

(ii) How many balls need to be chosen if the probability of obtaining at least one red ball is to be at least 99%?

10 In a certain mathematics examination, the probability of being awarded a grade A is believed to be 0.3.

For a group of ten candidates, calculate

(i) the probability that three achieve grade A

(ii) the probability that at least half achieve grade A.

The probability of being awarded a grade B is also 0.3.

(iii) Find the probability that all ten candidates achieve either grade A or grade B.

11 Eggs are sold in boxes of six and it is likely that 1% of the eggs will be broken when they are unpacked. Find

(i) the probability that a box contains no broken eggs

(ii) the probability that a box contains no more than one broken egg.

I buy four boxes of eggs. Find

(iii) the probability that I get no broken eggs

(iv) the probability that I have at least two broken eggs.

12 It is thought that, on average, 3% of light bulbs produced by a certain company last for less than 250 hours. This will be referred to as being defective.

In an inspection scheme, a sample of 25 light bulbs is selected at random from a large batch, they are tested for 250 hours and the number of defective bulbs is noted.

If this number is more than two, the whole batch is rejected; if it is less than two, the whole batch is accepted.

If there are exactly two defective bulbs in this batch, a further sample of size ten is taken.

The whole batch is rejected if there are any defective bulbs in this sample; otherwise the batch is accepted.

Find

(i) the probability that the batch is accepted after taking the first sample

(ii) the probability that the batch is accepted after taking the second sample

(iii) the probability that the batch is rejected.

KEY POINTS

1 Binomial coefficients, denoted by

$$^nC_r \text{ or } \binom{n}{r}$$

can be found

- using Pascal's triangle
- using the formula

$$^nC_r \frac{n!}{r!(n-r)!}.$$

2 The binomial expansion can be written as

$$(1+x)^n = {}^nC_0 + {}^nC_1x + {}^nC_2x^2 + {}^nC_3x^3 + \ldots + {}^nC_nx^n.$$

This generalises to

$$(a+b)^n = {}^nC_0\,a^n + {}^nC_1a^{n-1}\,b + {}^nC_2a^{n-2}\,b^2 + \ldots + {}^nC_nb^n.$$

3 The binomial distribution may be used to model a situation in which:

- the probability of a successful outcome is p
- the probability that the outcome is a failure is q, where $q = 1 - p$
- there are n trials
- the number of successes is denoted by X.

Then

$$P(X = r) = {}^nC_rp^rq^{n-r}$$

where $r = 0, 1, 2, \ldots, n$.

SECTION 2
Co-ordinate geometry

Co-ordinate geometry I

Cogito ergo sum. (I think, therefore I am.)

René Descartes (1596–1650)

❓ How many pieces of information do you need to define a point
(i) in two dimensions **(ii)** in three dimensions?

Figure 5.1

Historical note

René Descartes was born near Tours in France in 1596. This period was one of great intellectual advancement: he was a contemporary of Shakespeare, Pascal, Newton, Milton, Galileo and many other famous people. A sickly child, he was allowed to lie in bed as late as he pleased and throughout the rest of his life he spent his mornings in bed when he wished to think. He is probably best known for his application of algebra to geometry and the familiar system of Cartesian co-ordinates.

Co-ordinates

Co-ordinates are a means of describing the position of a point in relation to a fixed origin. In the *Cartesian system*, this is done using a set of perpendicular axes, *x*, *y* in two dimensions and *x*, *y*, *z* in three dimensions.

This is not the only system of co-ordinates. *Polar co-ordinates* describe position by stating the distance of a point from the origin and its direction.

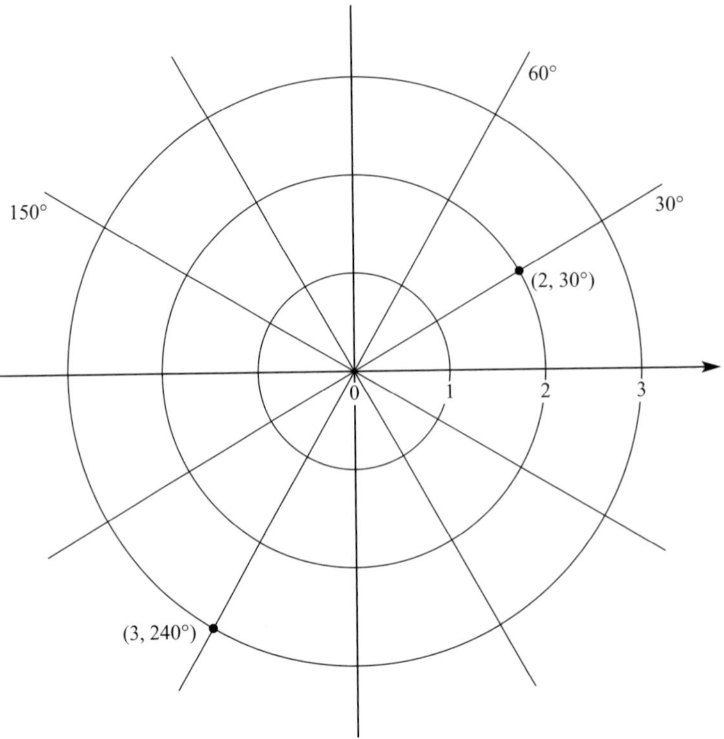

Figure 5.2

This chapter concentrates exclusively on two-dimensional Cartesian co-ordinates.

The gradient of a line

In mathematics the word *line* refers to a straight line. The slope of a line is measured by its *gradient* and the letter *m* is often used to represent this.

❓ What information do you need to have in order to fix the position of a line?

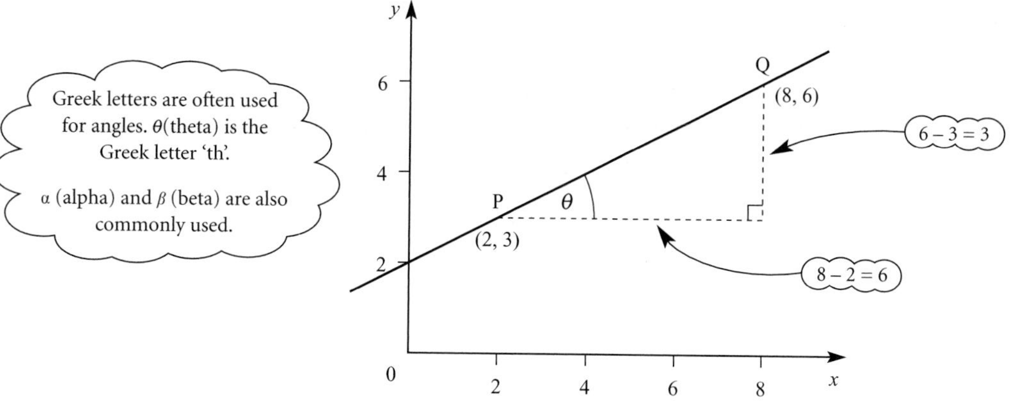

Figure 5.3

$$\text{gradient} = \frac{\text{increase in } y \text{ co-ordinate from P to Q}}{\text{increase in } x \text{ co-ordinate from P to Q}}$$

In figure 5.3, gradient $= \frac{6-3}{8-2} = \frac{3}{6} = \frac{1}{2}$.

ACTIVITY 5.1 On each line in figure 5.4, take any two points and use them to calculate the gradient of the line.

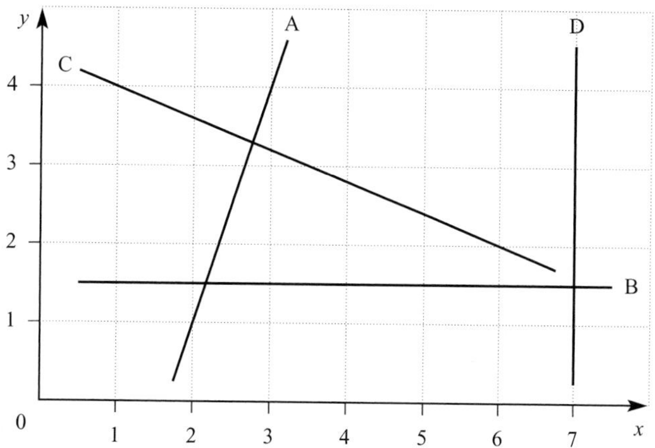

Figure 5.4

You can generalise the previous activity to find the gradient m of the line joining (x_1, y_1) to (x_2, y_2).

$$m = \frac{y_2 - y_1}{x_2 - x_1}$$

? Does it matter which point you call (x_1, y_1) and which (x_2, y_2)?

If the same scale is used on both axes, then

$m = \tan\theta$ (see figure 5.3).

You can easily tell by looking at a line if its gradient is positive, negative, zero or infinite.

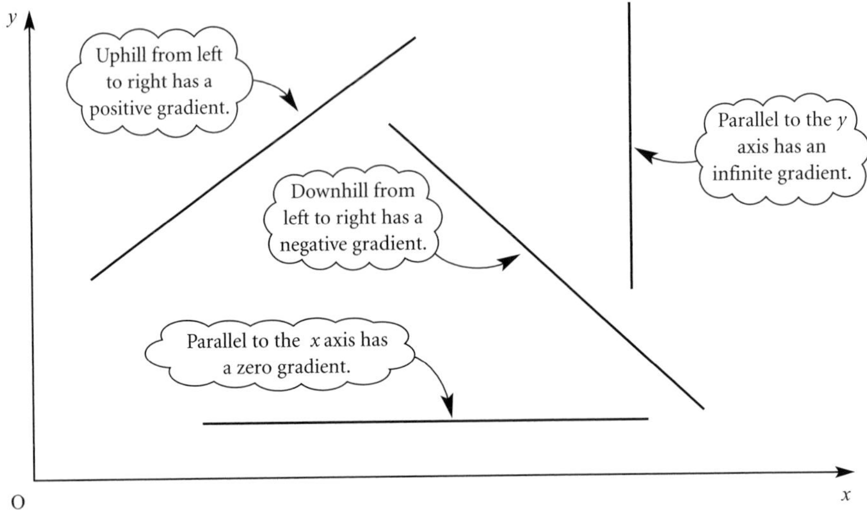

Figure 5.5

Parallel and perpendicular lines

If you know the gradients m_1 and m_2 of two lines, you can tell at once if they are parallel or perpendicular.

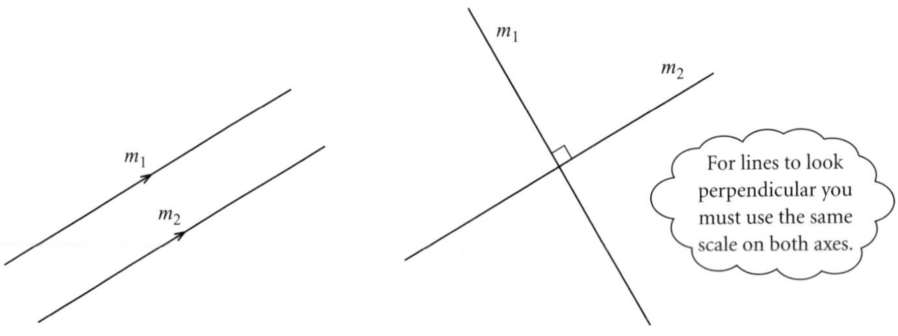

Figure 5.6

Parallel lines: $m_1 = m_2$

Perpendicular lines: $m_1 m_2 = -1$

❓ How would you explain the result for parallel lines?

To illustrate the result for perpendicular lines try activity 5.2 on squared paper.

ACTIVITY 5.2

(i) Draw two congruent right-angled triangles in the positions shown in figure 5.7. p and q can take any value.

(ii) Explain why $\angle ABC = 90°$.

(iii) Calculate the gradient of AB (m_1) and the gradient of BC (m_2).

(iv) Show that $m_1 m_2 = -1$.

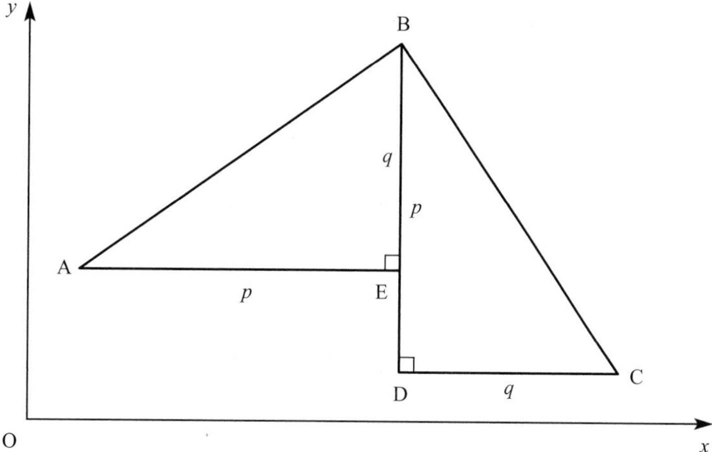

Figure 5.7

The distance between two points

You can use Pythagoras' theorem to find the distance between two points if you know their co-ordinates. Look at figure 5.8. P is (3, 1) and Q = (6, 5).

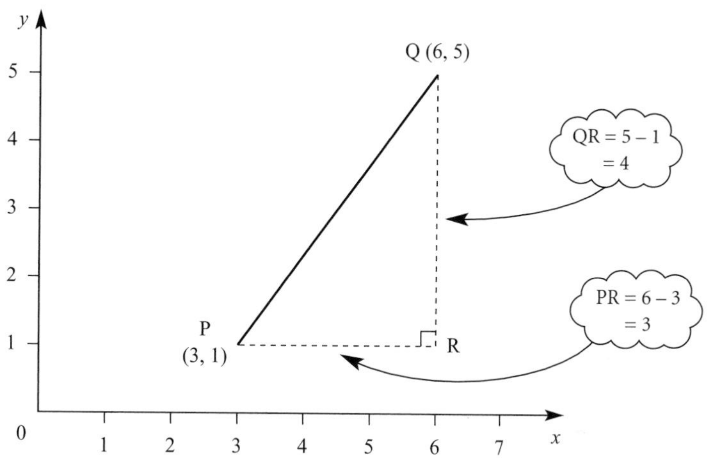

Figure 5.8

This can be generalised. If P has co-ordinates (x_1, y_1) and Q has co-ordinates (x_2, y_2), then

$$\text{length PQ} = \sqrt{(x_2 - x_1)^2 + (y_2 - y_1)^2}.$$

The midpoint of a line joining two points

Look at the line joining the points P(1, 2) and Q(7, 4) in figure 5.9. The point M is the midpoint of PQ.

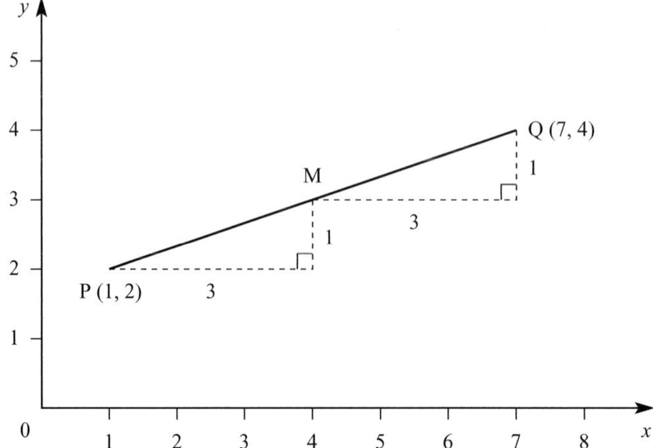

Figure 5.9

The co-ordinates of M are the means of the co-ordinates of P and Q.

$\frac{1}{2}(1 + 7) = 4$ and $\frac{1}{2}(2 + 4) = 3$

M is (4, 3).

Again, if P has co-ordinates (x_1, y_1) and Q has co-ordinates (x_2, y_2), then the co-ordinates of the midpoint of PQ are given by

$$\text{midpoint} = \left(\frac{x_1 + x_2}{2}, \frac{y_1 + y_2}{2}\right).$$

EXAMPLE 5.1

A and B are the points (–4, 2) and (2, 5). Find

(i) the gradient of AB

(ii) the gradient of the line perpendicular to AB

(iii) the length of AB

(iv) the co-ordinates of the midpoint of AB.

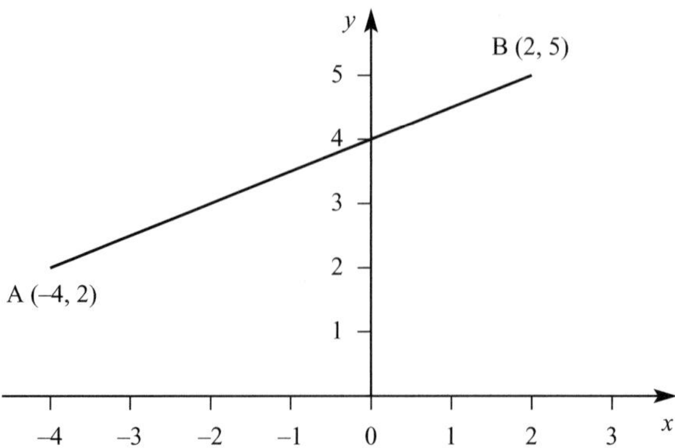

Figure 5.10

SOLUTION

(i) Taking (–4, 2) as (x_1, y_1) and (2, 5) as (x_2, y_2)

$$\text{gradient} = \frac{5-2}{2-(-4)} = \frac{3}{6} = \frac{1}{2}.$$

(ii) $m_1 = \frac{1}{2}$ and $m_1 m_2 = -1$

$\Rightarrow \quad \frac{1}{2}m_2 = -1$

$\Rightarrow \quad m_2 = -2$

The line perpendicular to AB has gradient –2.

(iii) $\text{length} = \sqrt{(2-(-4))^2 + (5-2)^2}$

$= \sqrt{36+9}$

$= \sqrt{45}$

$= 6.71$ (3 s.f.)

(iv) $\text{midpoint} = \left(\frac{-4+2}{2}, \frac{2+5}{2}\right)$

$= (-1, 3.5)$

EXAMPLE 5.2

P is the point (a, b) and Q is the point $(3a, 5b)$.

Find, in terms of a and b,

(i) the gradient of PQ

(ii) the length of PQ

(iii) the midpoint of PQ.

SOLUTION

Taking (a, b) as (x_1, y_1) and $(3a, 5b)$ as (x_2, y_2)

(i) $\text{gradient} = \frac{5b-b}{3a-a}$

$= \frac{4b}{2a} = \frac{2b}{a}$

(ii) $\text{length} = \sqrt{(3a-a)^2 + (5b-b)^2}$

$= \sqrt{4a^2 + 16b^2}$

? How can this result be simplified further?

(iii) $\text{midpoint} = \left(\frac{a+3a}{2}, \frac{b+5b}{2}\right)$

$= (2a, 3b)$

EXAMPLE 5.3

A, B and C are the points (1, 2), (5, b) and (6, 2). $\angle ABC = 90°$.

(i) Find two possible values of b.

(ii) Show all four points on a sketch and describe the shape of the figure you have drawn.

SOLUTION

(i) Gradient of AB $= \dfrac{b-2}{5-1} = \dfrac{b-2}{4}$

Gradient of BC $= \dfrac{2-b}{6-5} = 2 - b$

$$\angle ABC = 90° \Rightarrow \text{AB and BC are perpendicular.}$$
$$\Rightarrow \frac{(b-2)}{4} \times (2-b) = -1$$
$$\Rightarrow (b-2)(2-b) = -4$$
$$\Rightarrow 2b - b^2 - 4 + 2b = -4$$
$$\Rightarrow 4b - b^2 = 0$$
$$\Rightarrow b(4-b) = 0$$

So $b = 0$ or $b = 4$.

(ii) See figure 5.11.

AB_1CB_2 is a quadrilateral with diagonals that are perpendicular, since AC is parallel to the x axis and B_1B_2 is parallel to the y axis.
This makes AB_1CB_2 a kite.

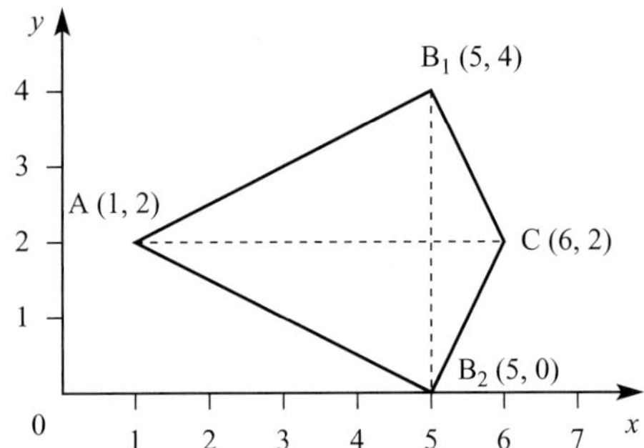

Figure 5.11

EXERCISE 5A

1 For each of the following pairs of points A and B, calculate

(a) the gradient of the line AB

(b) the gradient of the line perpendicular to AB

(c) the length of AB

(d) the co-ordinates of the midpoint of AB.

(i)	A(4, 3)	B(8, 11)	**(ii)**	A(3, 4)	B(0, 13)
(iii)	A(5, 3)	B(10, –8)	**(iv)**	A(–6, –14)	B(1, 7)
(v)	A(6, 0)	B(8, 15)	**(vi)**	A(–2, –4)	B(3, 9)
(vii)	A(–3, –6)	B(2, –7)	**(viii)**	A(4, 7)	B(7, –4)

2 A(0, 5), B(4, 1) and C(2, 7) are the vertices of a triangle. Show that the triangle is right-angled

(i) by finding the gradients of the sides

(ii) by finding the lengths of the sides.

3 A(3, 6), B(7, 4) and C(1, 2) are the vertices of a triangle. Show that ABC is a right-angled isosceles triangle.

4 A(3, 5), B(3, 11) and C(6, 2) are vertices of a triangle.

(i) Find the perimeter of the triangle.

(ii) Using AB as the base, find the area of the triangle.

5 A quadrilateral PQRS has vertices at P(–2, –5), Q(11, – 7), R(9, 6) and S(–4, 8).

(i) Find the lengths of the four sides of PQRS.

(ii) Find the midpoints of the diagonals PR and QS.

(iii) Without drawing a diagram, show why PQRS cannot be a square. What is it?

6 The points A, B and C have co-ordinates (2, 3), (6, 12) and (11, 7).

(i) Draw the triangle ABC.

(ii) Show by calculation that the triangle is isosceles and name the two equal sides.

(iii) Find the midpoint of the third side.

(iv) By calculating appropriate lengths find the area of triangle ABC.

7 A parallelogram WXYZ has three of its vertices at W(2, 1), X(–1, 5) and Y(–3, 3).

(i) Find the midpoint of WY.

(ii) Use this information to find the co-ordinates of Z.

8 A triangle ABC has vertices at A(3, 2), B(4, 0) and C(8, 2).

(i) Show that the triangle is right-angled.

(ii) Find the co-ordinates of the point D such that ABCD is a rectangle.

9 The three points P(–2, 3), Q(1, q) and R(7, 0) are collinear (i.e. they lie on the same straight line).

(i) Find the value of q.

(ii) Find the ratio of the lengths PQ : QR.

10 A quadrilateral has vertices A(–2, 8), B(–5, 5), C(5, 3) and D(3, 7).

(i) Draw the quadrilateral.

(ii) Show by calculation that it is a trapezium.

(iii) Find the co-ordinates of E when ABCE is a parallelogram.

The equation of a straight line

When you talk about a straight line, what does the word *straight* mean?

EXAMPLE 5.4

Find the equation of the straight line with gradient 2 through the point with co-ordinates (0, 1).

SOLUTION

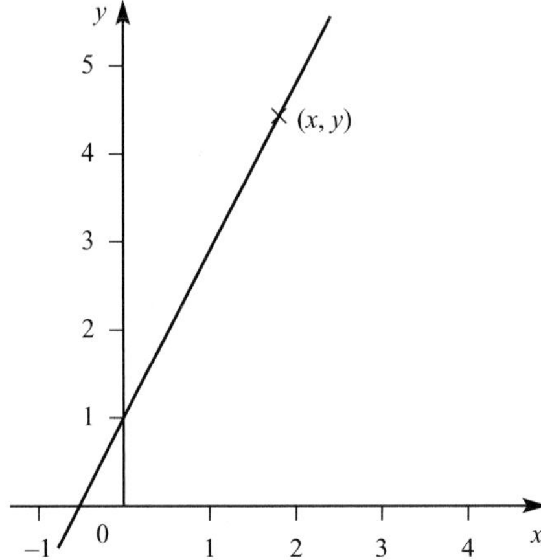

Figure 5.12

Take a general point (x, y) on the line, as shown in figure 5.12. The gradient of the line joining (0, 1) to (x, y) is given by

$$\text{gradient} = \frac{y-1}{x-0} = \frac{y-1}{x}.$$

Since you are given that the gradient of the line is 2, you have

$$\frac{y-1}{x} = 2 \quad \Rightarrow \quad y = 2x + 1.$$

Since (x, y) is a general point on the line, this holds for any point on the line and is therefore the equation of the line.

This example can be generalised to give the result that the equation of the line with gradient m cutting the y axis at the point $(0, c)$ is

$$\frac{y-c}{x-0} = m$$

$$\Rightarrow \quad y = mx + c.$$

This is a well known standard form for the equation of a straight line.

Drawing a line given its equation

There are several standard forms for the equation of a straight line. When you need to draw a line, look at its equation and see if it fits one of these.

Equations of the form $x = a$

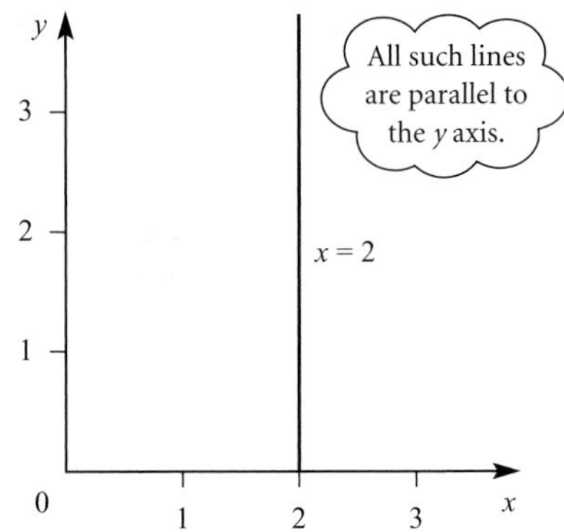

Equations of the form $y = b$

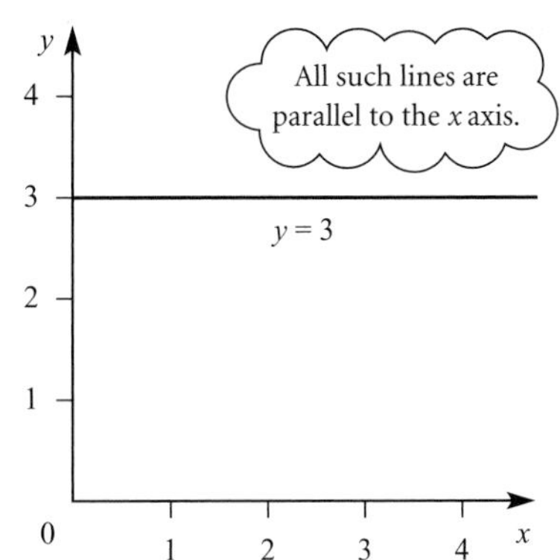

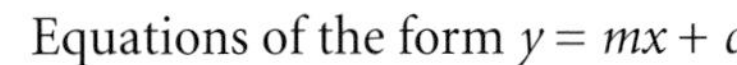

Equations of the form $y = mx$

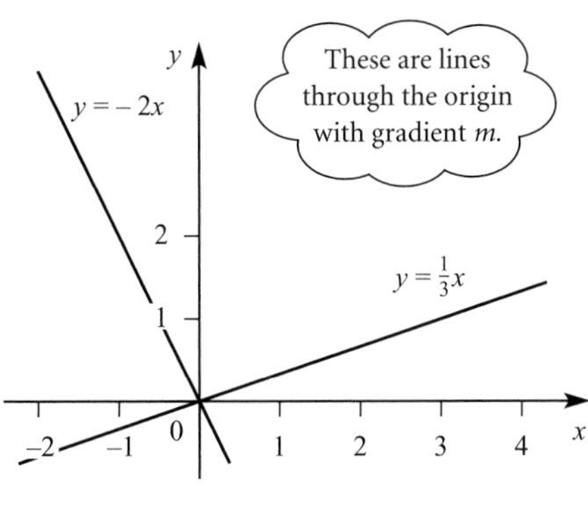

Equations of the form $y = mx + c$

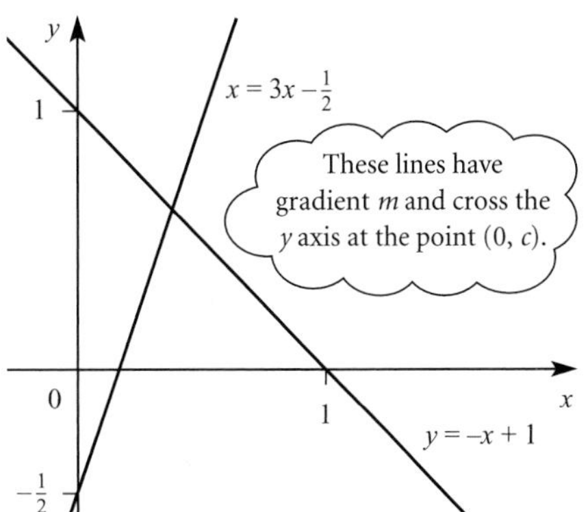

Equations of the form $px + qy + r = 0$

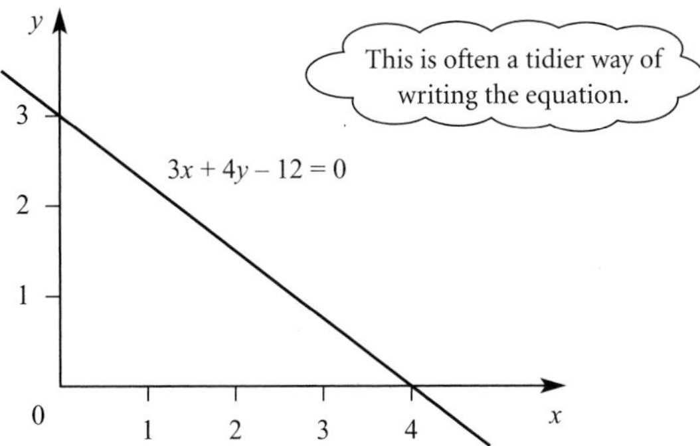

Figure 5.13

Graphs of equations in this form will usually be sketched by finding the co-ordinates of the points where the line crosses the x and y axes.

? **(i)** Rearrange the equation $3x + 4y - 12 = 0$ into the form $\frac{x}{a} + \frac{y}{b} = 1$.
(ii) What are the values of a and b?
(iii) What do these numbers represent?

EXAMPLE 5.5

(i) Sketch the lines $y = -2$, $y = 3x - 2$ and $x + 3y - 9 = 0$ on the same axes.
(ii) Show that these are the sides of a right-angled triangle.

SOLUTION

(i) The line $y = -2$ is parallel to the x axis and passes through $(0, -2)$.
The line $y = 3x - 2$ has gradient 3 and passes through $(0, -2)$.
To sketch the line $x + 3y - 9 = 0$ find two points on it.

$$x = 0 \Rightarrow 3y - 9 = 0 \Rightarrow y = 3 \quad (0, 3) \text{ is on the line}$$
$$y = 0 \Rightarrow x - 9 = 0 \Rightarrow x = 9 \quad (9, 0) \text{ is on the line}$$

Figure 5.14 shows the triangle ABC formed by the three lines.

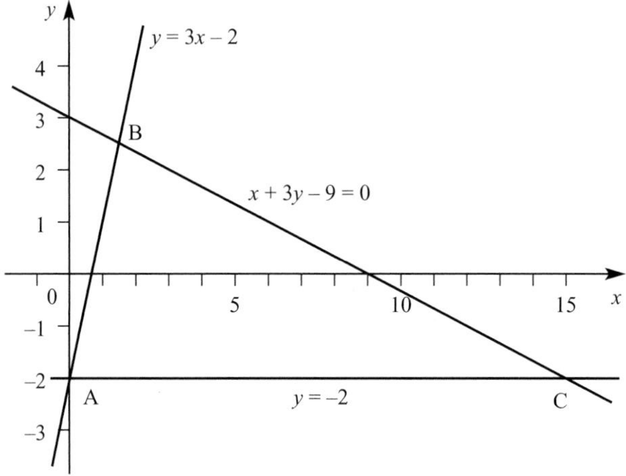

Figure 5.14

(ii) Look at the gradients of lines AB and BC.
Line AB has gradient 3.
Using the points $(0, 3)$ and $(9, 0)$

$$\text{the gradient of BC} = \frac{0-3}{9-0} = -\frac{1}{3}.$$

Since $3 \times \left(-\frac{1}{3}\right) = -1$ it follows that the lines AB and BC are perpendicular ($m_1 m_2 = -1$). The triangle has a right angle at B.

EXERCISE 5B

In this exercise you should sketch the lines by hand. If you have access to a graphic calculator, you can use it to check your results.

1 Sketch these lines.

(i)	$x = 5$	**(ii)**	$y = -3$	**(iii)**	$y = 4x$
(iv)	$y = -3x$	**(v)**	$y = 2x + 3$	**(vi)**	$y = -x + 2$
(vii)	$y = \frac{1}{2}x - 1$	**(viii)**	$y = -2x - 1$	**(ix)**	$y = \frac{1}{3}x + \frac{2}{3}$
(x)	$y = 3x - 2$	**(xi)**	$y = 2x - 3$	**(xii)**	$x + y - 1 = 0$
(xiii)	$2x + y - 4 = 0$	**(xiv)**	$x - 3y + 6 = 0$	**(xv)**	$2x - 3y = 12$
(xvi)	$y - 3x + 9 = 0$	**(xvii)**	$3x = 2y - 6$	**(xviii)**	$5x - 4y + 3 = 0$

2 By calculating the gradients of the following pairs of lines, state whether they are parallel, perpendicular or neither.

(i)	$x = 2$ $y = -2$	**(ii)**	$y = 2x$ $y = -2x$	**(iii)**	$x + 2y = 1$ $2x - y = 1$
(iv)	$y = x - 3$ $x - y + 4 = 0$	**(v)**	$y = 3 - 4x$ $y = 4 - 3x$	**(vi)**	$x + y = 5$ $x - y = 5$
(vii)	$x - 2y = 3$ $y = \frac{1}{2}x - 1$	**(viii)**	$x + 3y - 4 = 0$ $y = 3x + 4$	**(ix)**	$2y = x$ $2x + y = 4$
(x)	$2x + 3y - 4 = 0$ $2x + 3y - 6 = 0$	**(xi)**	$x + 3y = 1$ $y + 3x = 1$	**(xii)**	$2x = 5y$ $5x + 2y = 0$

Finding the equation of a line

The simplest way of finding the equation of a straight line depends on what information you have been given.

Given the gradient, m, and the co-ordinates (x_1, y_1) of one point on the line

Take the general point (x, y) on the line, as shown in figure 5.15.

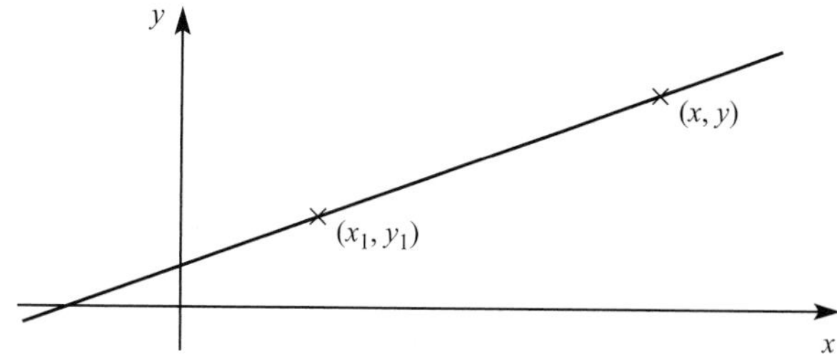

Figure 5.15

The gradient m of the line joining (x_1, y_1) to (x, y) is given by

$$m = \frac{y - y_1}{x - x_1}$$

$$\Rightarrow \quad y - y_1 = m(x - x_1).$$

This is a standard result, and one you will find very useful.

EXAMPLE 5.6

Find the equation of the line with gradient 2 which passes through the point (–1, 3).

SOLUTION

Using $y - y_1 = m(x - x_1)$

$$\Rightarrow \quad y - 3 = 2(x - (-1))$$
$$\Rightarrow \quad y - 3 = 2x + 2$$
$$\Rightarrow \quad y = 2x + 5.$$

In the formula

$$y - y_1 = m(x - x_1)$$

two positions of the point (x_1, y_1) lead to results you have met already.

(x_1, y_1) is at (0, 0) (x_1, y_1) is at $(0, c)$

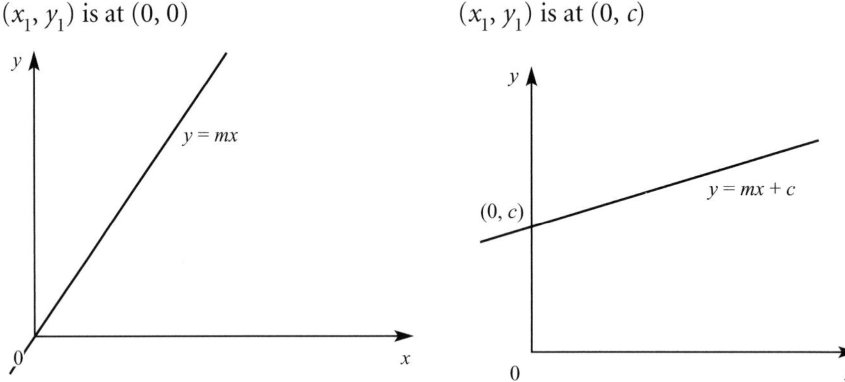

Figure 5.16

Given two points (x_1, y_1) and (x_2, y_2)

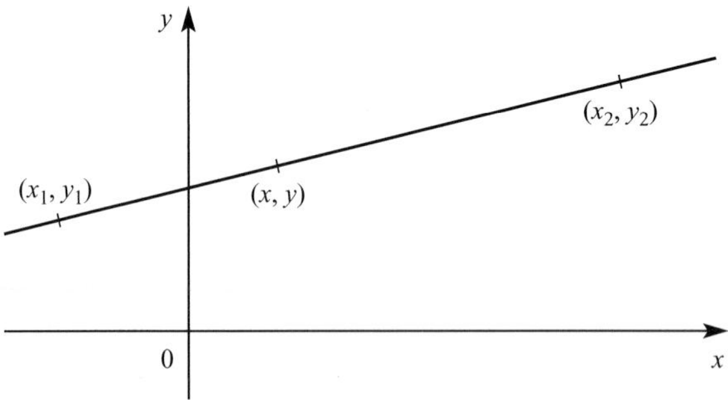

Figure 5.17

The two points are used to find the gradient

$$m = \frac{y_2 - y_1}{x_2 - x_1}.$$

This value is then substituted in the equation

$$y - y_1 = m(x - x_1).$$

This gives

$$y - y_1 = \frac{y_2 - y_1}{x_2 - x_1}(x - x_1).$$

Rearranging this gives

$$\frac{y - y_1}{y_2 - y_1} = \frac{x - x_1}{x_2 - x_1} \quad \text{or}$$

$$\frac{y - y_1}{x - x_1} = \frac{y_2 - y_1}{x_2 - x_1}.$$

EXAMPLE 5.7

Find the equation of the line joining (–1, 4) to (2, –3).

SOLUTION

Let (x_1, y_1) be (–1, 4) and (x_2, y_2) be (2, –3).

Substituting these values in $= \dfrac{y - y_1}{y_2 - y_1} = \dfrac{x - x_1}{x_2 - x_1}$

gives $\quad \dfrac{y-4}{(-3)-4} = \dfrac{x-(-1)}{2-(-1)}$

$\Rightarrow \quad \dfrac{y-4}{(-7)} = \dfrac{x+1}{3}$

$\Rightarrow \quad 3(y-4) = (-7)(x+1)$

$\Rightarrow \quad 7x + 3y - 5 = 0.$

Applying the different techniques

The following examples illustrate the different techniques, and show how these can be used to solve a practical problem.

EXAMPLE 5.8

Find the equations of the lines **(a)** to **(e)** in figure 5.18.

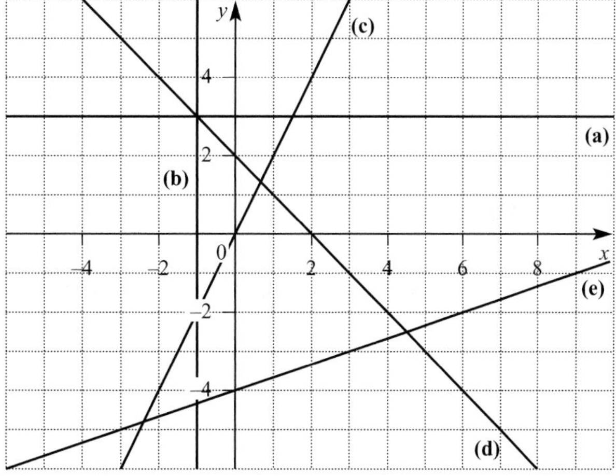

Figure 5.18

SOLUTION

Line **(a)** is parallel to the x axis and passes though $(0, 3)$

$\Rightarrow$ equation of **(a)** is $y = 3$.

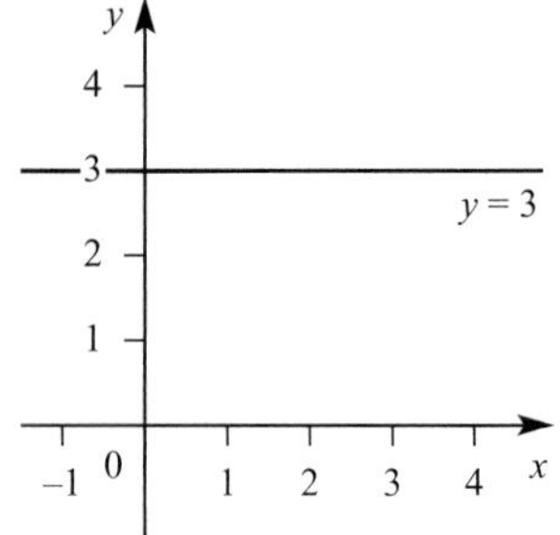

Line **(b)** is parallel to the y axis and passes through $(-1, 0)$

$\Rightarrow$ equation of **(b)** is $x = -1$.

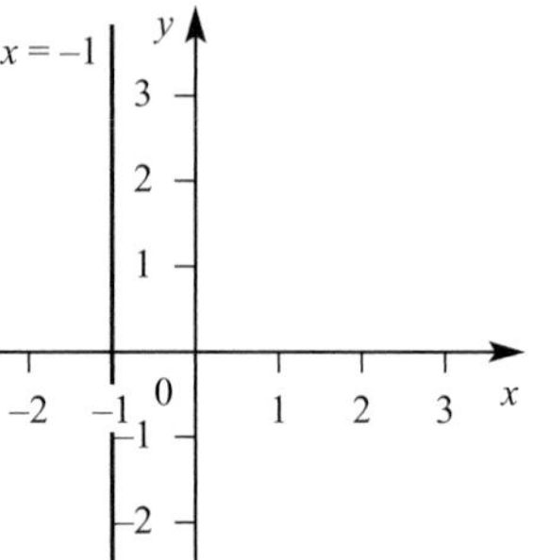

Line **(c)** passes through $(0, 0)$ and has gradient 2

$\Rightarrow$ equation of **(c)** is $y = 2x$.

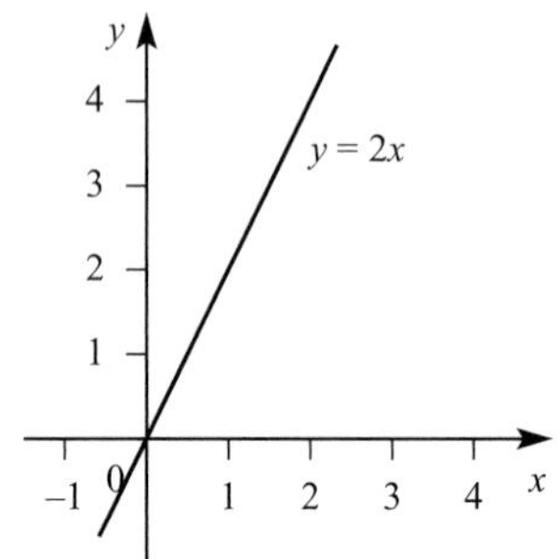

Line **(d)** passes through $(0, 2)$ and has gradient (-1)

$\Rightarrow$ equation of **(d)** is $y = -x + 2$.

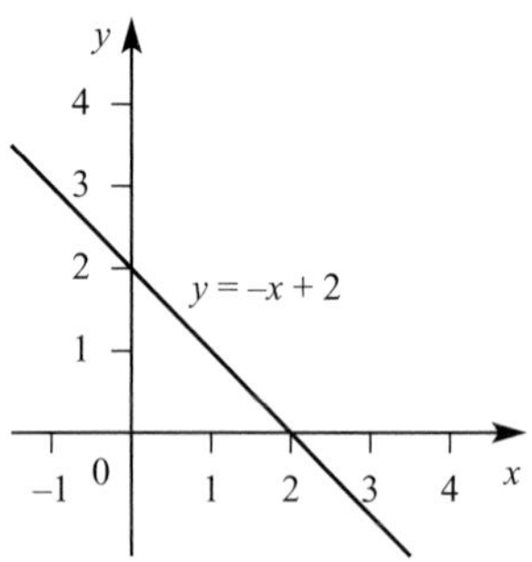

Line **(e)** passes through $(0, -4)$ and has gradient $\frac{1}{3}$

$\Rightarrow$ equation of **(e)** is $y = \frac{1}{3}x - 4$.

This can be rearranged to give

$x - 3y - 12 = 0$.

EXAMPLE 5.9

An isosceles triangle with AB = AC has vertices at A(2, 3), B(8, 5) and C(4, 9). Find the equation of the line of symmetry.

SOLUTION

Figure 5.19 shows the triangle ABC with the line of symmetry joining A to the midpoint of BC.

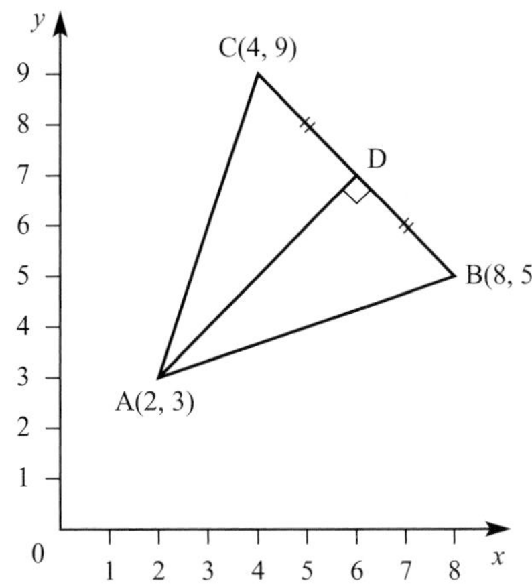

Figure 5.19

The co-ordinates of D are $\left(\frac{8+4}{2}, \frac{5+9}{2}\right) = (6, 7)$.

Let (x_1, y_1) be (2, 3) and (x_2, y_2) be (6, 7).

$$\frac{y - y_1}{y_2 - y_1} = \frac{x - x_1}{x_2 - x_1}$$

$$\Rightarrow \quad \frac{y-3}{7-3} = \frac{x-2}{6-2}$$

$$\Rightarrow \quad \frac{y-3}{4} = \frac{x-2}{4}$$

$$\Rightarrow \quad y = x + 1$$

EXAMPLE 5.10

Celsius and Fahrenheit are two different scales for measuring temperature. The freezing point of water can be written as 0 °C or 32 °F and the boiling point as 100 °C or 212 °F.

(i) Use this information to find a formula that will convert degrees Celsius (c) to degrees Fahrenheit (f).

(ii) What temperature is the same in both scales?

SOLUTION

(i) The formula is the equation of the line joining (0, 32) to (100, 212).

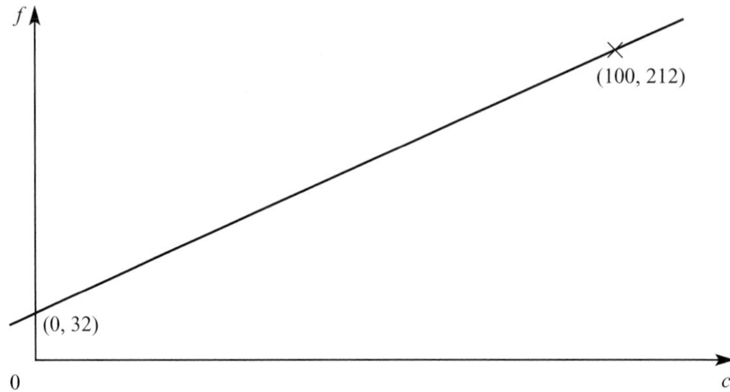

Figure 5.20

Let (0, 32) be (c_1, f_1) and (100, 212) be (c_2, f_2).

Substituting these values in

$$\frac{f-f_1}{f_2-f_1} = \frac{c-c_1}{c_2-c_1}$$

gives $$\frac{f-32}{212-32} = \frac{c-0}{100-0}$$

$$\Rightarrow \quad \frac{f-32}{180} = \frac{c-0}{100}$$

$$\Rightarrow \quad f-32 = \tfrac{9}{5}c$$

$$\Rightarrow \quad f = \tfrac{9}{5}c + 32.$$

(ii) Let t be the temperature which is the same in both scales.

Then $t = \tfrac{9}{5}t + 32$

$$\Rightarrow \quad 5t = 9t + 160$$

$$\Rightarrow \quad t = -40,$$

giving −40 °C = −40 °F.

EXERCISE 5C

1 Find the equations of the lines **(a)** – **(j)** in these diagrams.

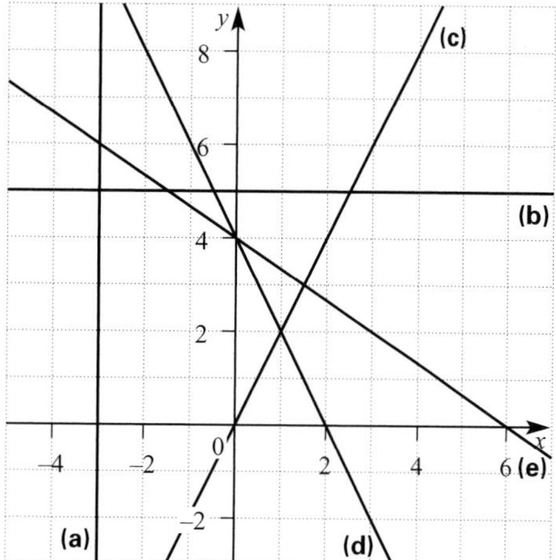

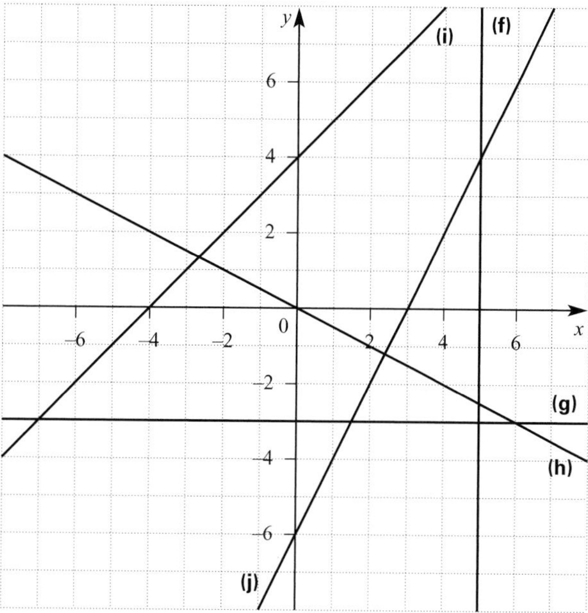

2 Find the equations of these lines.

(i) Parallel to $y = 3x$ and passing through $(2, -1)$.

(ii) Parallel to $y = 2x + 3$ and passing through $(0, 0)$.

(iii) Parallel to $y = 3x - 4$ and passing through $(2, -7)$.

(iv) Parallel to $4x - y + 2 = 0$ and passing through $(4, 0)$.

(v) Parallel to $3x + 2y - 1 = 0$ and passing through $(1, -2)$.

(vi) Parallel to $2x + 4y - 5 = 0$ and passing through $(0, 6)$.

3 Find the equations of these lines.

(i) Perpendicular to $y = 2x$ and passing through (0, 0).

(ii) Perpendicular to $y = 3x - 1$ and passing through (0, 4).

(iii) Perpendicular to $y + x = 2$ and passing through (3, –1).

(iv) Perpendicular to $2x - y + 4 = 0$ and passing through (1, –1).

(v) Perpendicular to $3x + 2y + 4 = 0$ and passing through (3, 0).

(vi) Perpendicular to $2x + y - 1 = 0$ and passing through (4, 1).

4 Find the equation of the line AB in each of these cases.

(i) A(2, 0) B(3, 1)

(ii) A(3, –1) B(0, 4)

(iii) A(2, –3) B(3, –2)

(iv) A(–1, 3) B(4, 0)

(v) A(3, –5) B(10, –6)

(vi) A(–1, –2) B(–4, –8)

5 Points P and Q have co-ordinates P(3, – 1) and Q(5, 7).

(i) Find the gradient of PQ.

(ii) Find the co-ordinates of the midpoint of PQ.

(iii) Find the equation of the perpendicular bisector of PQ.

6 A triangle has vertices P(2, 5), Q(–2, –2) and R(6, 0).

(i) Sketch the triangle.

(ii) Find the co-ordinates of L, M and N, which are the midpoints of PQ, QR and RP respectively.

(iii) Find the equations of the lines LR, MP and NQ (these are the medians of the triangle).

(iv) Show that the point (2, 1) lies on all three of these lines. (This shows that the medians of a triangle are concurrent.)

7 The straight line with equation $2x + 3y - 12 = 0$ cuts the x axis at A and the y axis at B.

(i) Sketch the line.

(ii) Find the co-ordinates of A and B.

(iii) Find the area of triangle OAB where O is the origin.

(iv) Find the equation of the line which passes through O and is perpendicular to AB.

(v) Find the length of AB and, using the result in **(iii)**, calculate the shortest distance from O to AB.

8 A quadrilateral has vertices at the points A(–7, 0), B(2, 3), C(5, 0) and D(–1, –6).

(i) Sketch the quadrilateral.

(ii) Find the gradient of each side.

(iii) Find the equation of each side.

(iv) Find the length of each side.

(v) Find the area of the quadrilateral.

9 When the mortgage interest rate was 7% a small building society lent £75 million, but when the rate dropped to 4.6% it lent £105 million. Assume that the graph of amount lent against interest rate is linear for interest rates between 2% and 10%.

(i) Sketch the graph of amount lent (in £million) against interest rates (%) in this interval with interest rates on the horizontal axis.

(ii) Find the equation of the line.

(iii) Find the amount lent if the interest rate is 6%.

The building society has only a total of £130 million available for loans.

(iv) Find the least interest rate for which there is no need to refuse potential borrowers.

10 A student is carrying out an experiment to measure the elasticity of an elastic band. His instructions were to hang different masses on one end, to measure the stretched length, and to repeat this six times with different masses.

However, because of shortage of time, he only does the experiment twice and uses those pairs of data instead of finding a line of best fit for more points. Here are his results.

mass in grams (x)	50	100
length in mm (y)	180	270

Assume that the graph of length against mass is a straight line.

(i) Sketch the graph of length against mass.

(ii) Find the equation of the line.

(iii) Find the unstretched length of the elastic band.

(iv) Find the load which gives an *extension* of 150 mm.

(v) What do you think will happen when a load of 1 kg is attached to the elastic band?

The intersection of two lines

You can find the point of intersection of any two lines (or curves) by solving their equations simultaneously.

EXAMPLE 5.11

(i) Sketch the lines $x + 3y - 6 = 0$ and $y = 2x - 5$ on the same axes.

(ii) Find the co-ordinates of the point where they intersect.

SOLUTION

(i) The line $x + 3y - 6 = 0$ passes through $(0, 2)$ and $(6, 0)$.
The line $y = 2x - 5$ passes through $(0, -5)$ and has a gradient of 2.

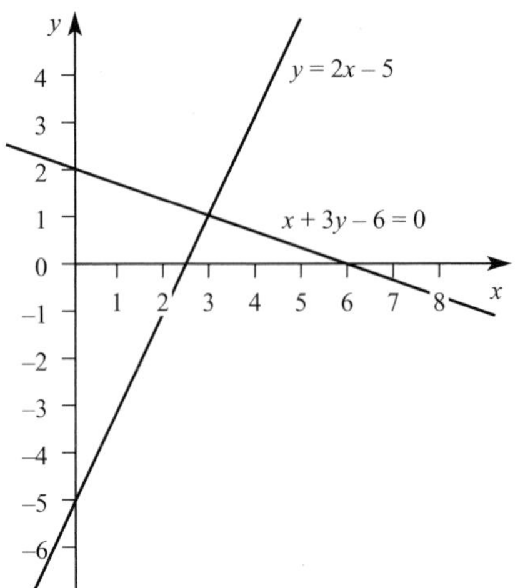

Figure 5.21

(ii)

$$\begin{aligned} x + 3y - 6 = 0 &\Rightarrow 2x + 6y - 12 = 0 \quad ① \quad \text{(multiplying by 2)} \\ y = 2x - 5 &\Rightarrow 2x - y - 5 = 0 \quad ② \\ ① - ② &\Rightarrow 7y - 7 = 0 \\ &\Rightarrow y = 1 \end{aligned}$$

Substituting $y = 1$ in ① gives $2x + 6 - 12 = 0$

$$\Rightarrow \quad x = 3$$

The co-ordinates of the point of intersection are therefore (3, 1).

An alternative method for solving these equations simultaneously would be to plot both lines on graph paper and read off the co-ordinates of the point of intersection.

? Graphical methods such as this will have limited accuracy. What factors would affect the accuracy of your solution in this case?

EXAMPLE 5.12

(i) Plot the lines $x + y - 2 = 0$ and $4y - x = 4$ on the same set of axes, for $-4 \leqslant x \leqslant 4$, using 1 cm to represent 1 unit on both axes.

(ii) Read off the solution to the simultaneous equations

$$x + y - 2 = 0$$
$$4y - x = 4$$

SOLUTION

(i) For each line choose three values of x and calculate the corresponding values of y. Then plot the lines and read off the co-ordinates of the point of intersection.

$x + y - 2 = 0$

x	–2	0	2
y	4	2	0

$4y - x = 4$

x	–4	0	4
y	0	1	2

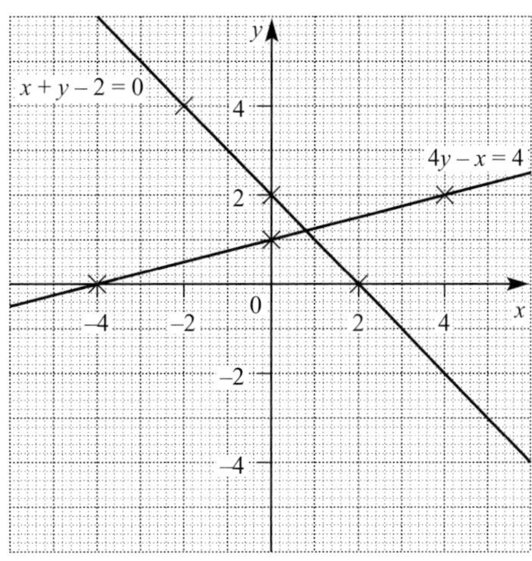

Figure 5.22

(ii) The point of intersection is (0.8, 1.2), so the solution to the simultaneous equations is

$$x = 0.8, y = 1.2.$$

? Why should you plot three points for each line?

? Two lines may not intersect. When is this the case?

EXERCISE 5D

You will need graph paper for this exercise.

1 Solve these pairs of simultaneous equations by plotting their graphs. In each case you are given a suitable range of values of x.

(i) $x = 3y + 1$ $\quad 0 \leqslant x \leqslant 3$
$y = x - 1$

(ii) $3x + 2y = 5$ $\quad -2 \leqslant x \leqslant 2$
$x + y = 3$

(iii) $y = 2x - 4$ $\quad 0 \leqslant x \leqslant 6$
$3x + 4y = 17$

(iv) $6x + y = 1$ $\quad 0 \leqslant x \leqslant 2$
$4x - y = 4$

2 **(i)** Plot the lines $x = 4$, $y = x + 4$ and $4x + 3y = 12$ on the same axes for $-1 \leqslant x \leqslant 5$.

(ii) State the co-ordinates of the three points of intersection, and for each point give the pair of simultaneous equations that are satisfied there.

(iii) Find the area of the triangle enclosed by the three lines.

3 **(i)** Using the same scale for both axes, plot the lines $2y + x = 4$ and $2y + x = 10$ on the same axes, for $0 \leqslant x \leqslant 6$, and say what you notice about them. Why is this the case?

(ii) Add the line $y = 2x$ to your graph. What do you notice now? Can you justify what you see?

(iii) State the co-ordinates of the two points of intersection, and for each point give the pair of simultaneous equations that are satisfied there.

The circle

When you draw a circle, you open your compasses to a fixed distance (the radius) and choose a position (the centre) for the point of your compasses. These facts are used to derive the *equation* of the circle.

Circles with centre (0, 0)

Figure 5.23 shows a circle with centre O (0, 0) and radius 4. P (x, y) is a general point on the circle.

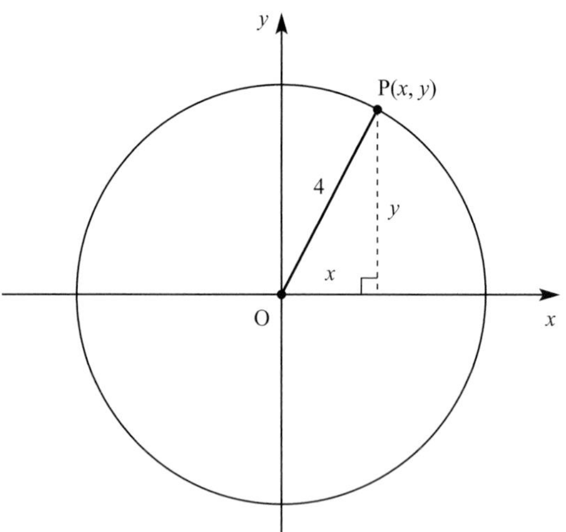

Figure 5.23

$$\text{OP} = \sqrt{(x-0)^2 + (y-0)^2} = 4$$

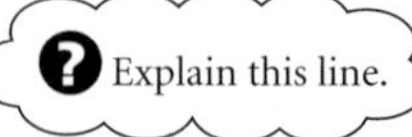

This simplifies to $x^2 + y^2 = 16$, which is the equation of the circle.

This can be generalised. A circle with centre (0, 0), radius r has equation

$$x^2 + y^2 = r^2.$$

Circles with centre (*a*, *b*)

Figure 5.24 shows a circle with centre C (4, 5) and radius 3. P (x, y) is a general point on the circle.

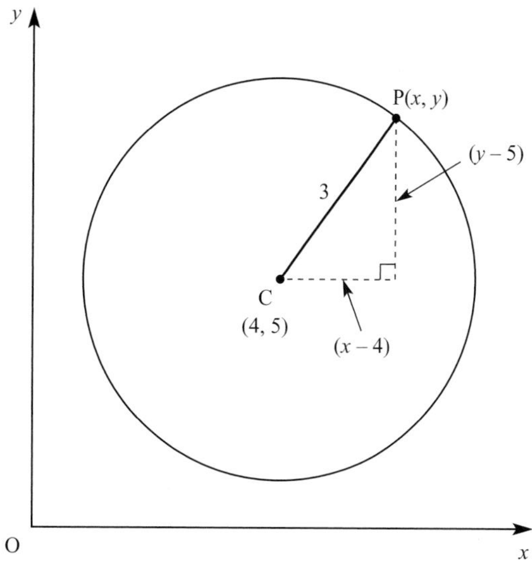

Figure 5.24

$$\text{CP} = \sqrt{(x-4)^2 + (y-5)^2} = 3$$

This simplifies to $(x-4)^2 + (y-5)^2 = 9$, which is the equation of the circle.

This can be generalised. A circle with centre (a, b), radius r has equation

$$(x-a)^2 + (y-b)^2 = r^2.$$

NOTE

Multiplying out this equation gives

$$x^2 - 2ax + a^2 + y^2 - 2by + b^2 = r^2.$$

This rearranges to

$$x^2 + y^2 - 2ax - 2by + (a^2 + b^2 - r^2) = 0.$$

This form of the equation highlights some of the important characteristics of the equation of a circle. In particular:

- the coefficients of x^2 and y^2 are equal,
- there is no xy term.

EXAMPLE 5.13

Find the centre and radius of the circle

$$x^2 + (y + 3)^2 = 25.$$

SOLUTION

Comparing with the general equation for a circle with radius r and centre (a, b),

$$(x - a)^2 + (y - b)^2 = r^2$$

gives $a = 0$, $b = -3$ and $r = 5$

$\Rightarrow$ the centre is $(0, -3)$, the radius is 5.

EXAMPLE 5.14

Figure 5.25 shows a circle with centre $(1, -2)$, which passes through the point $(4, 2)$.

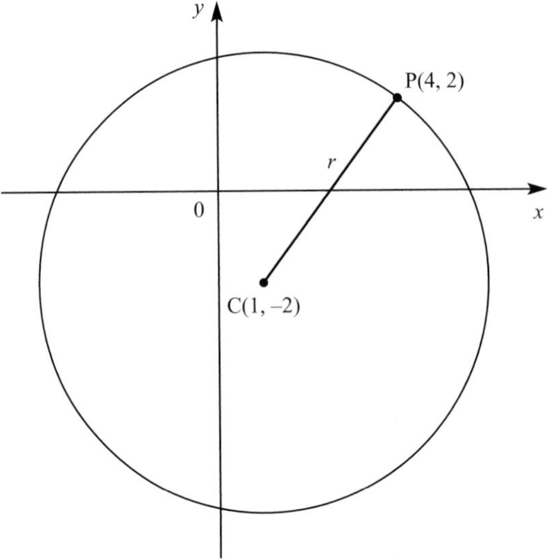

Figure 5.25

(i) Find the radius of the circle.

(ii) Find the equation of the circle.

SOLUTION

(i) Use the two points you are given to find the radius of the circle.

$$r^2 = (4 - 1)^2 + (2 - (-2))^2$$
$$= 25$$
$$\Rightarrow \quad \text{radius} = 5$$

(ii) Now using $(x - a)^2 + (y - b)^2 = r^2$

$$\Rightarrow \quad (x - 1)^2 + (y + 2)^2 = 25$$

is the equation of the circle.

EXAMPLE 5.15

Show that the equation $x^2 + y^2 + 4x - 6y - 3 = 0$ represents a circle.

Hence give the co-ordinates of the centre and the radius of the circle.

SOLUTION

By completing the square, $x^2 + y^2 + 4x - 6y - 3 = 0$ can be written as

$$(x + 2)^2 - 4 + (y - 3)^2 - 9 - 3 = 0$$

$$\Rightarrow \quad (x + 2)^2 + (y - 3)^2 = 16.$$

This represents a circle with centre $(-2, 3)$, radius 4.

EXERCISE 5E

1 Find the equations of these circles.

(i) centre $(1, 2)$, radius 3

(ii) centre $(4, -3)$, radius 4

(iii) centre $(1, 0)$, radius 5

(iv) centre $(-2, -2)$, radius 2

(v) centre $(-4, 3)$, radius 1

2 For each of the circles given below

(a) state the co-ordinates of the centre

(b) state the radius

(c) sketch the circle, paying particular attention to its position in relation to the origin and the co-ordinate axes.

(i) $x^2 + y^2 = 25$

(ii) $(x - 3)^2 + y^2 = 9$

(iii) $(x + 4)^2 + (y - 3)^2 = 25$

(iv) $(x + 1)^2 + (y + 6)^2 = 36$

(v) $(x - 4)^2 + (y - 4)^2 = 16$

3 Find the equation of the circle with centre $(2, -3)$ which passes through $(1, -1)$.

4 A and B are $(4, -4)$ and $(2, 6)$ respectively.
Find

(i) the midpoint C of AB

(ii) the distance AC

(iii) the equation of the circle that has AB as its diameter.

5 Show that the equation $x^2 + y^2 - 4x - 8y + 4 = 0$ represents a circle.

Hence give the co-ordinates of the centre and the radius of the circle, and sketch the circle.

KEY POINTS

1. The gradient of the straight line joining the points (x_1, y_1) and (x_2, y_2) is given by

$$\text{gradient} = \frac{y_2 - y_1}{x_2 - x_1}$$

2. Two lines are parallel when their gradients are equal.

3. Two lines are perpendicular when the product of their gradients is -1.

4. When the points A and B have co-ordinates (x_1, y_1) and (x_2, y_2) respectively then

$$\text{distance AB} = \sqrt{(x_2 - x_1)^2 + (y_2 - y_1)^2}$$

$$\text{midpoint of AB is } \left(\frac{x_1 + x_2}{2}, \frac{y_1 + y_2}{2}\right).$$

5. The equation of a straight line may take any of these forms.
   - line parallel to the y axis: $x = a$
   - line parallel to the x axis: $y = b$
   - line through the origin with gradient m: $y = mx$
   - line through $(0, c)$ with gradient m: $y = mx + c$
   - line through (x_1, y_1) with gradient m: $y - y_1 = m(x - x_1)$
   - line through (x_1, y_1) and (x_2, y_2):

$$\frac{y - y_1}{y_2 - y_1} = \frac{x - x_1}{x_2 - x_1} \quad \text{or} \quad \frac{y - y_1}{x - x_1} = \frac{y_2 - y_1}{x_2 - x_1}$$

6. The co-ordinates of the point of intersection of two lines are found by solving their equations simultaneously.

7. The equation of a circle with centre (h, k) and radius r is

$$(x - h)^2 + (y - k)^2 = r^2.$$

   When the centre is at the origin $(0, 0)$ this simplifies to

$$x^2 + y^2 = r^2.$$

6 Co-ordinate geometry II – applications

Most of the fundamental ideas of science are essentially simple, and may, as a rule, be expressed in a language comprehensible to everyone.

Albert Einstein

Amy is visiting the fair. At the hoop-la stall, she wins a prize if she circles it with her hoop.

The prizes are on a table that is 3 metres square, and she is standing 8 metres from the nearest point of the table, as in figure 6.1.

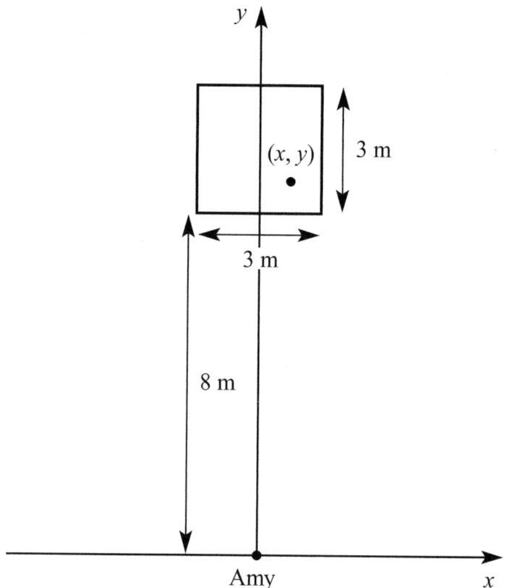

Figure 6.1

❓ The hoop lands with its centre at the point (x, y).
What can you say about the values of x and y if Amy has a chance of winning a prize?

Inequalities

You have met linear inequalities in one variable in Chapter 2, and solved these algebraically. In this chapter you will be *illustrating* linear inequalities in one or two variables and applying this theory to the solution of some practical problems.

When the linear inequality $x \leqslant 2$ is plotted on a graph, you obtain the *region* that lies to the left of the line $x = 2$, as in figure 6.2.

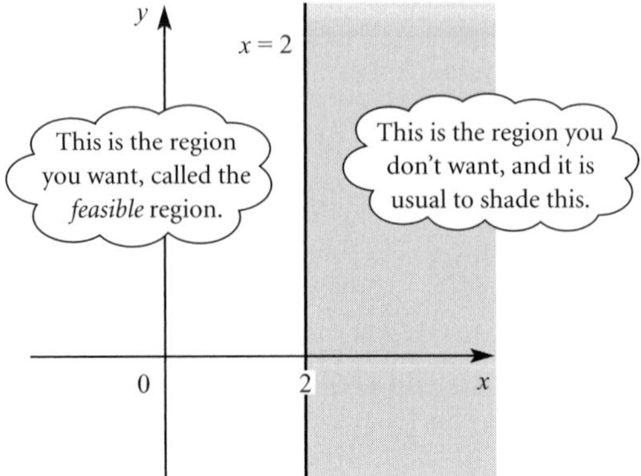

Figure 6.2

Convention

In order to distinguish between the regions defined by, for example, $x < 3$ and $x \leqslant 3$ draw the boundary for $x < 3$ as a *broken* line and that for $x \leqslant 3$ as a *solid* line. Similarly, the boundary for $x + y > 5$ is drawn as a broken line and that for $2y - 3x \geqslant 0$ as a solid line.

Specify the region that you want by shading out the *other* side of the line.

? Why is it more helpful to shade the area you don't want?

When the boundary line is sloping it is not always obvious which side of the line you want. To determine this, use a *test point.*

This can be any point not on the boundary, and (0, 0) is the simplest point to use if the line does not pass through the origin.

EXAMPLE 6.1 Illustrate each of these inequalities on a graph.

(i) $x < 3$
(ii) $y \leqslant 5$
(iii) $x + y > 5$
(iv) $2y - 3x \geqslant 0$

SOLUTION

(i) Draw the line $x = 3$ as a *broken* line.

Test using (0, 0): $0 < 3$ is true so the origin is in the feasible region.

Shade to the right of the line (figure 6.3).

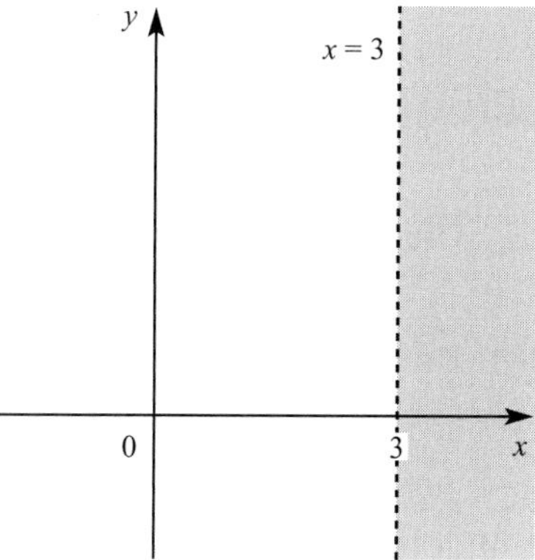

Figure 6.3

(ii) Draw the line $y = 5$ as a *solid* line.

Test using (0, 0): $0 \leqslant 5$ is true so the origin is in the feasible region.

Shade above the line (figure 6.4).

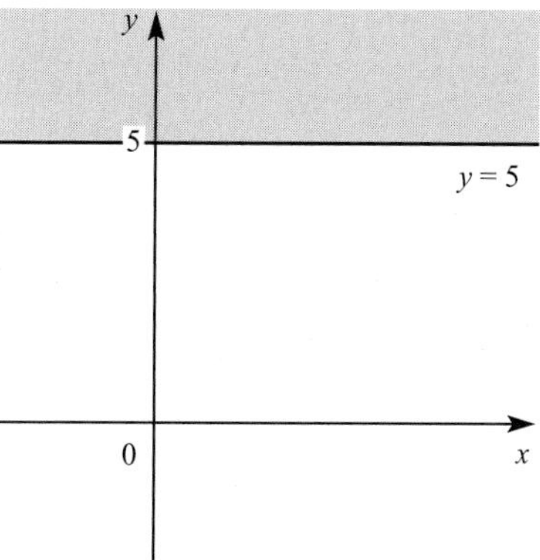

Figure 6.4

(iii) Draw the line $x + y = 5$ as a *broken* line.
(This passes through (0, 5) and (5, 0).)

Test using (0, 0): $0 > 5$ is *not* true, so the origin is not in the feasible region.

Shade the side of the line that includes the origin (figure 6.5).

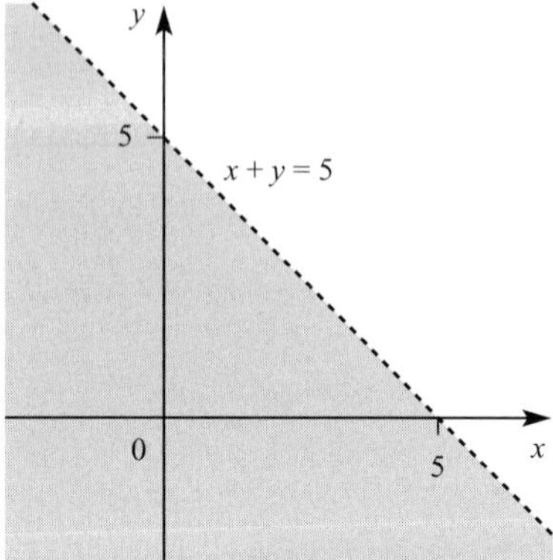

Figure 6.5

(iv) Draw the line $2y - 3x = 0$ as a *solid* line.
(This passes through (0, 0) and (2, 3).)

Since the line passes through the origin, choose a test point that you are sure is not on the line, for example (2, 0).

Using these values, $-6 \geqslant 0$ is *not* true, so the point (2, 0) is not in the feasible region.

Shade the side of the line that includes (2, 0) (figure 6.6).

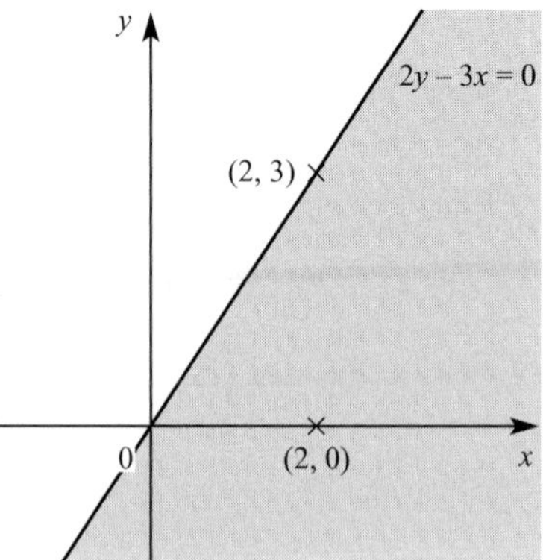

Figure 6.6

EXAMPLE 6.2

Illustrate the region defined by $-2 \leqslant x < 3$.

SOLUTION

The boundaries are $x = -2$ (solid line) and $x = 3$ (broken line). Using (0, 0) as a test point shows that the region required is between the two lines (figure 6.7).

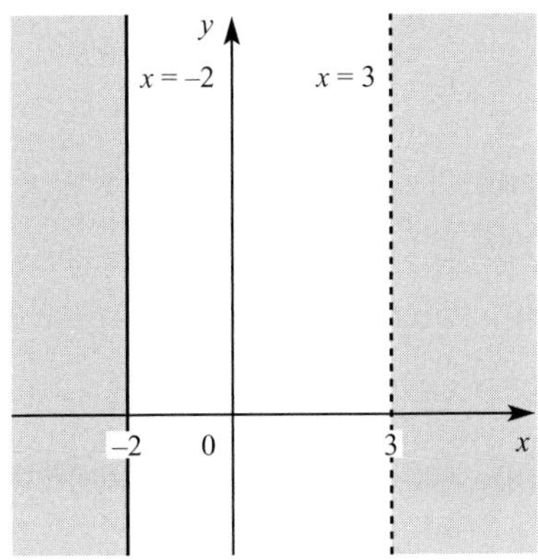

Figure 6.7

EXAMPLE 6.3

Illustrate the region which satisfies the three inequalities $y > 0$, $y < 3x$, $3x + 5y \leqslant 15$. List the points (x, y), where x and y are both integers, that lie in this region.

SOLUTION

The boundaries are $y = 0$ (broken line, the x axis), $y = 3x$ (broken line), through (0, 0) and (1, 3), and $3x + 5y = 15$ (solid line) through (0, 3) and (5, 0).

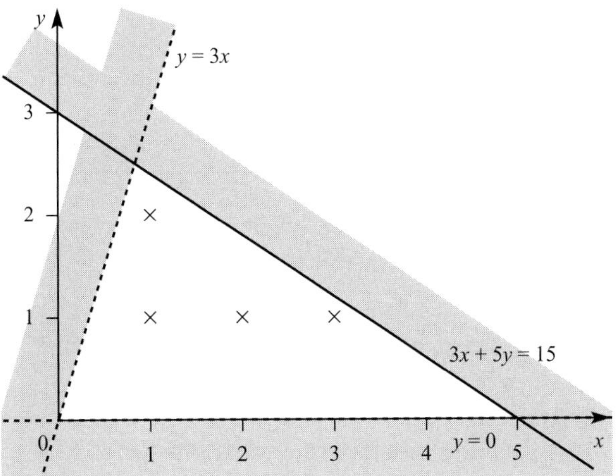

Figure 6.8

From figure 6.8 the points that are in the region are (1, 1), (2, 1), (3, 1), (1, 2).

Why is (5, 0) not in the feasible region?

EXERCISE 6A

You will need graph paper for this exercise.

1 On separate diagrams use shading to denote each of these inequalities.

(i) $x \geqslant -2$

(ii) $y < 3$

(iii) $y > x - 3$

(iv) $y \leqslant 2x + 1$

(v) $x - 2y \leqslant 4$

(vi) $3x + 4y > 12$

(vii) $-1 < x < 5$

(viii) $2 \leqslant y < 7$

(ix) $2x - 3y \leqslant 6$

(x) $x > 3y - 9$

2 **(i)** On the same axes, illustrate the regions defined by these inequalities.

$$x > 2 \qquad 2x + 3y \leqslant 12 \qquad y > -2$$

(ii) Are the following points in the region defined by all three inequalities?

(a) $(3, -1)$

(b) $(4, 2)$

(c) $(2, 1)$

3 Write down the inequalities that define the region illustrated.

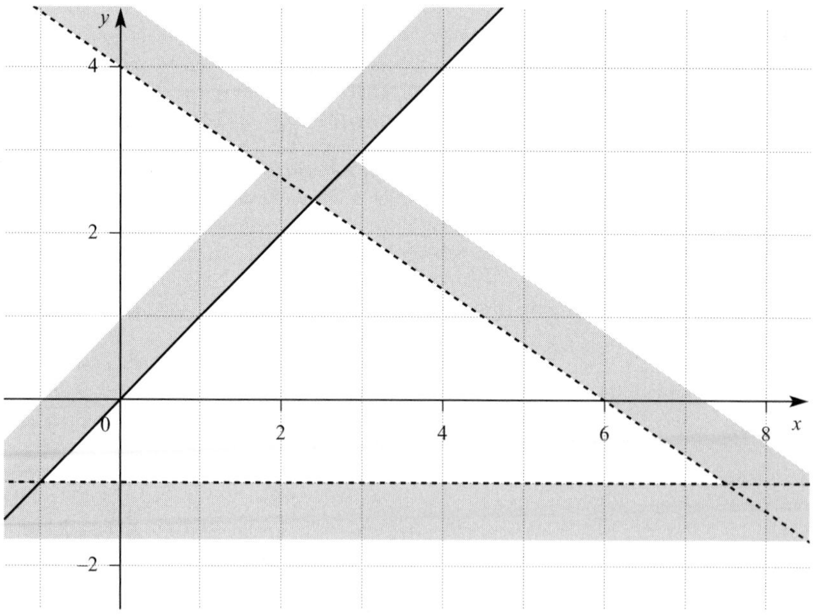

4 Illustrate the region that satisfies all of these inequalities.

$$-2 \leqslant y < 4 \qquad y < x \qquad x + y \leqslant 9$$

Using inequalities for problem solving

Inequalities occur naturally in many everyday situations.

'The journey will take between two and three hours.'
'I spend at least £3 per week on sweets.'
'My test result will be no more than 75%.'

? Write sentences to illustrate the meaning of each of the following inequality statements, and then express each using mathematical symbols.

(i) at least
(ii) at most
(iii) more than
(iv) less than
(v) no more than
(vi) no less than
(vii) under
(viii) over

EXAMPLE 6.4

At the fair, a ride on the hula hoop costs £2 and one on the big spinner costs £1.50. Oliver has £10 to spend. He has H hula hoop rides and S rides on the big spinner.

(i) Use the information above to form five inequalities.
(ii) Illustrate the region satisfied by these inequalities, using the horizontal axis for H and the vertical axis for S.
(iii) Mark with a cross all possible combinations of rides that satisfy these conditions.
(iv) Which combination gives a mixture of rides and spends the whole £10?

SOLUTION

(i) £H is the cost of a ride $\Rightarrow$ $H \geqslant 0$.

Oliver only has £10 to spend, so $2H \leqslant 10$ $\Rightarrow$ $H \leqslant 5$.

£S is the cost of a ride $\Rightarrow$ $S \geqslant 0$.

Oliver only has £10 to spend, so $1.5S \leqslant 10$ $\Rightarrow$ $S \leqslant 6\frac{2}{3}$, but S is a whole number, so $S \leqslant 6$.

When there is a mixture of rides, the cost still cannot exceed £10. This gives $2H + 1.5S \leqslant 10$ $\Rightarrow$ $4H + 3S \leqslant 20$.

(ii) and **(iii)**

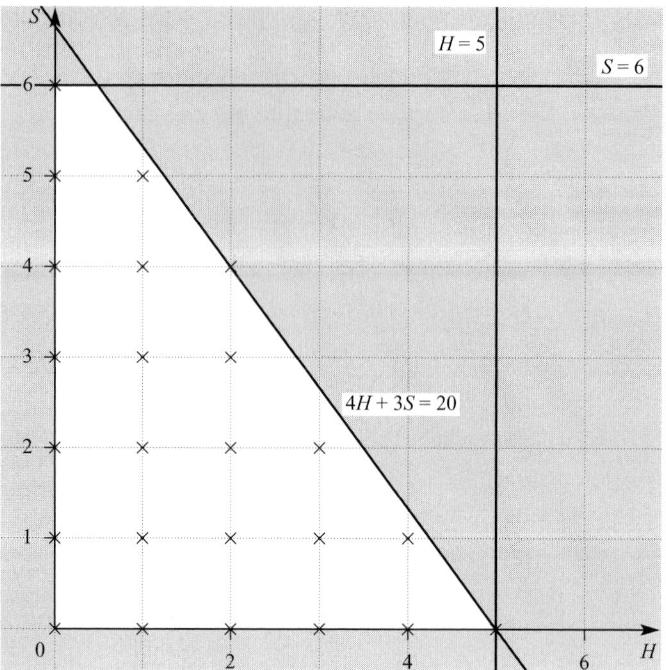

Figure 6.9

(iv) If Oliver spends exactly £10, the option required lies on the line $4H + 3S = 20$. The only points on this line are (0, 6) (which does not give a mixture of rides) and (2, 4).

The combination required is two rides on the hula hoop and four on the big spinner.

EXERCISE 6B

You will need graph paper for this exercise.

1 Anisa is buying apples and bananas for the tennis club picnic and has £4 to spend. Apples cost 24p each and bananas cost 20p each.
She must buy at least 18 pieces of fruit altogether, and would like to buy at least six apples and at least eight bananas.
She buys a apples and b bananas.

(i) Explain why

(a) $a \geqslant 6$

(b) $b \geqslant 8$

(c) $a + b \geqslant 18$

(d) $6a + 5b \leqslant 100$

(ii) Illustrate the region satisfied by these inequalities, using the horizontal axis for a and the vertical axis for b.

(iii) Mark with a cross all possible combinations of fruit.

2 A smallholder keeps s sheep and p pigs.

(i) Make each of these conditions into an inequality.

(a) He has housing for only eight animals.

(b) He must have at least two pigs to keep each other company.

(c) He needs at least three sheep to keep his grass short.

(d) His wife insists that the sheep outnumber the pigs.

(ii) Illustrate the region satisfied by these inequalities, using the horizontal axis for s and the vertical axis for p.

(iii) Mark with a cross all possible combinations of animals.

3 A gardener decides to buy a number of trees: some fruit trees, f, and some trees for landscaping, l.
Fruit trees cost £20 each and others £15 each.
Altogether he has £250 to spend and he would like at least five fruit trees and four others.

(i) Make each of these conditions into an inequality.

(a) He would like at least five fruit trees.

(b) He would like at least four others.

(c) The total cost must be no more than £250.

(ii) Illustrate the region satisfied by these inequalities, using the horizontal axis for f and the vertical axis for l.

(iii) What is the greatest number of each sort of tree that he can have?

4 Lily has decided to buy some daffodil bulbs, d, and some tulip bulbs, t, for her garden.
Daffodil bulbs cost 15p each and tulip bulbs 18p each and she has a total of £18 to spend.
She would like to buy at least as many daffodil bulbs as tulip bulbs but not more that twice as many.

(i) Make each of these conditions into an inequality.

(a) She would like to buy at least as many daffodil bulbs as tulip bulbs.

(b) She doesn't want more than twice as many daffodil bulbs as tulip bulbs.

(c) The total cost must be no more that £18.

(ii) Illustrate the region satisfied by these inequalities, using the horizontal axis for d and the vertical axis for t.

5 The 240 students in Year 7 are going on a geography field trip.
The local coach company has 4 fifty-seat coaches and 6 thirty-seat coaches available.
The school hires f fifty-seat coaches and t thirty-seat coaches.

(i) Explain why

(a) $0 \leqslant f \leqslant 4$

(b) $0 \leqslant t \leqslant 6$

(c) $5f + 3t \geqslant 24$

(ii) Illustrate the region satisfied by these inequalities, using the horizontal axis for f and the vertical axis for t.

(iii) Mark with a cross all possible combinations of coaches.

(iv) Which of these combinations would result in the fewest empty seats?

It is then realised that this calculation does not take into account the members of staff who will be accompanying the trip, and there must be at least one member of staff on each thirty-seat coach and two on each fifty-seat coach.

(v) Explain how **(i)(c)** should be modified to take this into account.

Linear programming

EXAMPLE 6.5

A combined car and lorry park is to be marked into parking spaces.

There are c car spaces and l lorry spaces.

The spaces for cars are each to have an area of $10\,\text{m}^2$ and those for lorries an area of $30\,\text{m}^2$.

The total area available for parking is $2000\,\text{m}^2$.

There must be at least 50 car spaces and 20 lorry spaces.

(i) Write down three inequalities that need to be satisfied.

(ii) Illustrate the region satisfied by these inequalities, using the horizontal axis for c and the vertical axis for l.

(iii) The parking charges are £1.50 per hour for a car and £2.50 per hour for a lorry.
Write down an expression for the hourly revenue, R.

(iv) Find the values of c and l to give the greatest potential hourly revenue.

(v) What is the greatest potential hourly revenue?

SOLUTION

(i) 'at least 50 car bays': $c \geqslant 50$

'at least 20 lorry bays': $l \geqslant 20$

The area constraint: $10c + 30l \leqslant 2000$

$\Rightarrow \quad c + 3l \leqslant 200$

(ii)

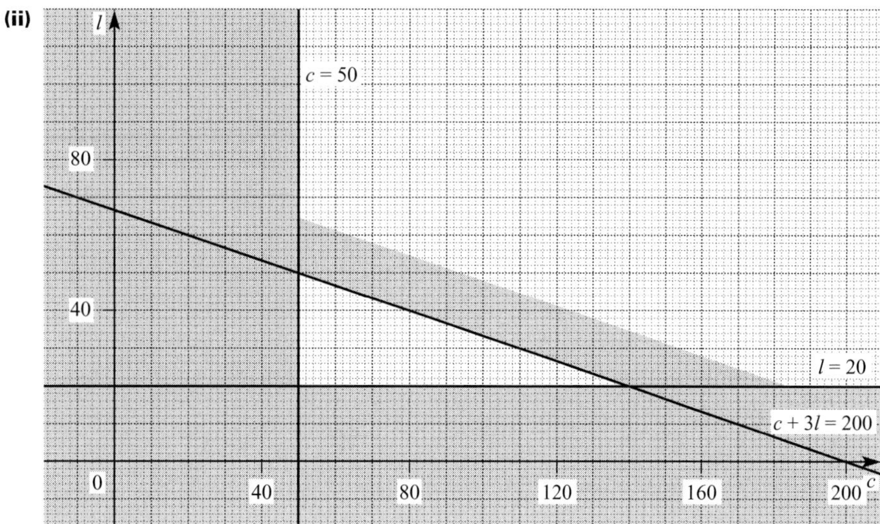

Figure 6.10

(iii) $R = 1.5c + 2.5l$.

This is called the *objective function*. You are looking for the largest numerical value of R so that the line $1.5c + 2.5l = R$ passes through a point in the feasible region.

(iv) For different values of R these lines are all parallel, and each line in the family would represent a particular hourly income (figure 6.11).

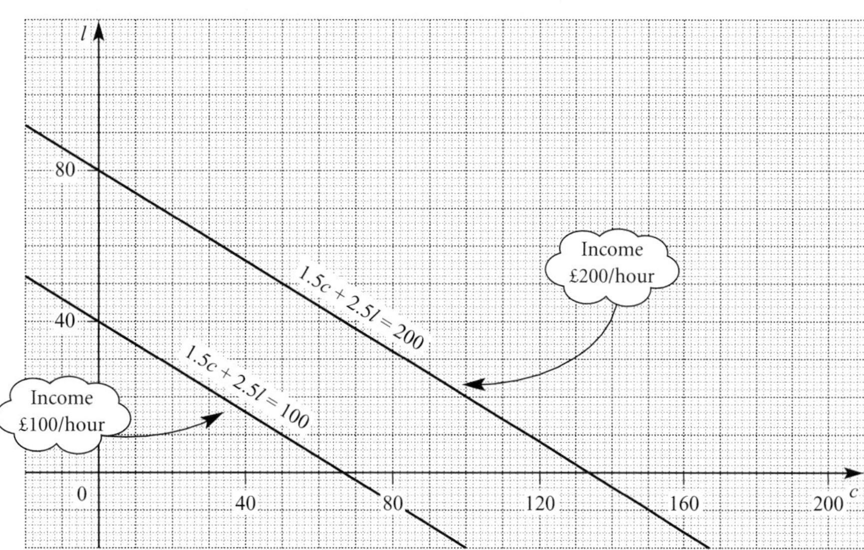

Figure 6.11

In theory you could add all possible lines in this family to the feasible region, but in practice this is not necessary. The best solution will be found at, or near, a vertex of the region.

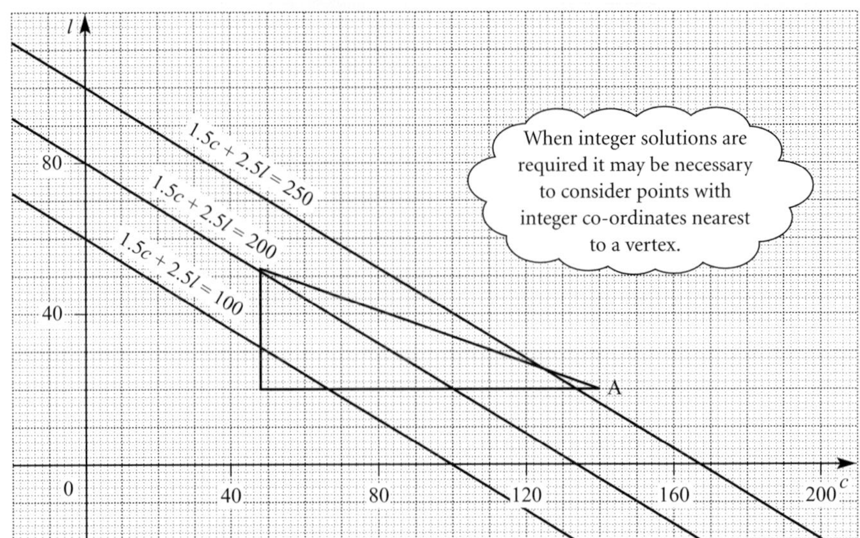

Figure 6.12

Superimposing lines of the form $1.5c + 2.5l = R$ (for different values of R) on the feasible region (figure 6.12) shows that the optimum solution is at (or near) the vertex A.

At A: $l = 20$ and $c + 3l = 200 \quad \Rightarrow \quad c = 140$

(v) When $l = 20$ and $c = 140$, $R = 1.5c + 2.5l = 260$

The greatest potential revenue will be £260 per hour.

Since there are integer solutions for both c and l at the vertex, there is no need to look at any other points.

EXAMPLE 6.6

Mel's band hire a hall that holds 200 people for a concert.

Tickets cost £2 or £3 each.

They need to raise at least £450 from this concert.

They decide that the number of £3 tickets must be not greater than twice the number of £2 tickets.

There are x tickets at £2 each and y tickets at £3 each.

(a) Explain why
  (i) $x + y \leqslant 200$
  (ii) $2x + 3y \geqslant 450$
  (iii) $y \leqslant 2x$

(b) Illustrate the region satisfied by these inequalities.

(c) The profit from the concert is to be maximised.
Write down the objective function.
Find the number of £2 and £3 tickets that must be sold to make the maximum profit and find the profit in this case.

SOLUTION

(a) (i) There are only 200 seats, so $x + y \leqslant 200$.

(ii) £$(2x + 3y)$ is the income raised from selling x tickets at £2 each and y tickets at £3 each. This must be at least £450.

(iii) The number of £3 tickets must not be greater than twice the number of £2 tickets, so $y \leqslant 2x$.

(b) The boundaries are

$x + y = 200$, passing through (200, 0) and (0, 200)
$2x + 3y = 450$, passing through (225, 0) and (0, 150)
$y = 2x$, passing through (0, 0) and (100, 200).

Figure 6.13 shows the region that satisfies all three inequalities.

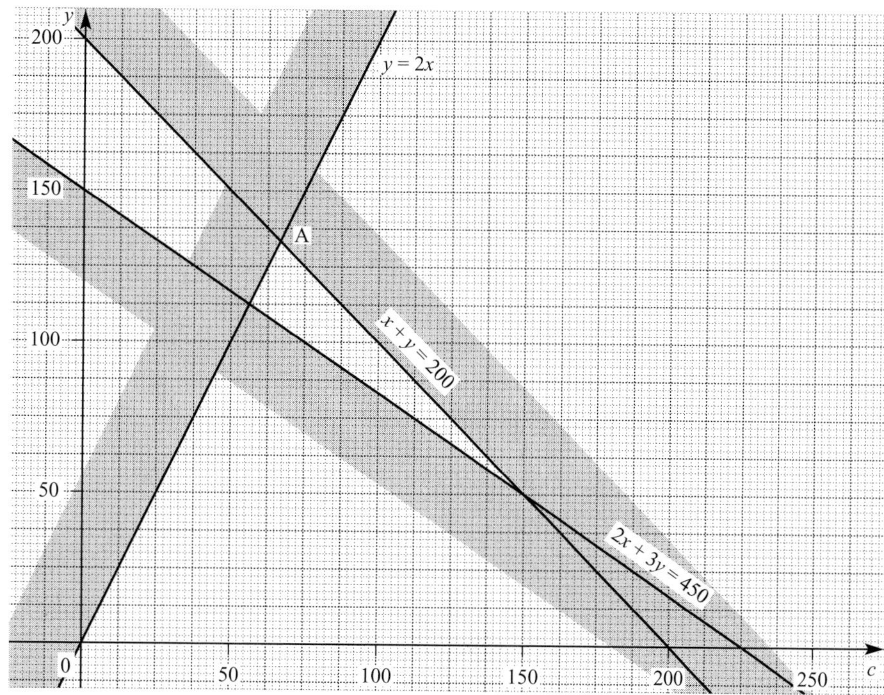

Figure 6.13

(c) The objective function is $P = 2x + 3y$, where P is the profit.
As P varies, all lines describing this function will be parallel to the line $2x + 3y = 450$.

The maximum profit is obtained at the point A, if A has integer co-ordinates. If not, we need the point with integer co-ordinates which is nearest to A and is inside the feasible region.

At A:
$$\left.\begin{array}{r} y = 2x \\ x + y = 200 \end{array}\right\} \Rightarrow x = 66\tfrac{2}{3};\ y = 133\tfrac{1}{3}$$

Points with integer co-ordinates that are nearest to this vertex and inside the feasible region are (66, 132) and (67, 133).

Checking $P = 2x + 3y$ gives a maximum profit of £533 when 67 £2 tickets and 133 £3 tickets are sold.

EXERCISE 6C

You will need graph paper for this exercise.

1 Anna has two after-school jobs, babysitting and working in a local restaurant. She can spend at most 18 hours a week between these two jobs.

She always babysits for her neighbours for three hours after school on Thursday and for four hours on Saturday night, and in order to keep her employment at the restaurant she must work there for at least three hours a week.

(i) In one week she spends b hours babysitting and r hours in the restaurant. Write down three inequalities that need to be satisfied.

(ii) Illustrate the region satisfied by these inequalities, using the horizontal axis for b and the vertical axis for r, and shading the unwanted region.

(iii) Anna can do some of her homework while she is babysitting, so prefers to spend at least twice as much time doing this as working in the restaurant. Write down another inequality and add it to your graph.

Anna charges £3 per hour for babysitting and is paid £4 per hour while working in the restaurant.

(iv) Her weekly earnings are to be maximised.
Write down the objective function and find the maximum amount she can earn in a week.

2 The diagram shows a rectangular grating made of wire, all angles being right angles.

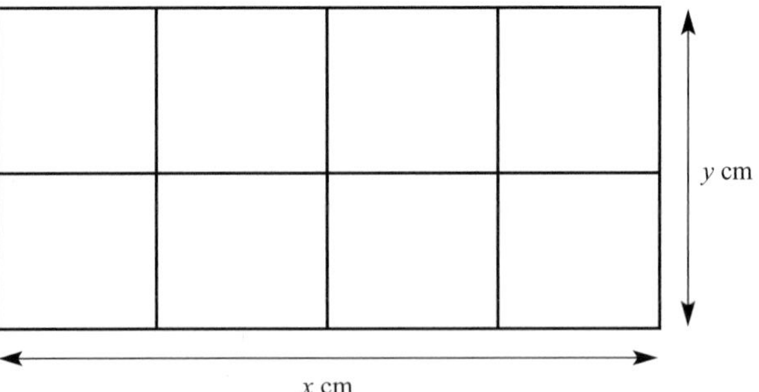

(i) Taking its length and width as x cm and y cm respectively, write down expressions for:

(a) the perimeter of the rectangular grating

(b) the total length of wire used.

(ii) Manufacture of these gratings is subject to a number of conditions:

(a) the perimeter must be no more than 100 cm

(b) the total length of wire must be at most 200 cm

(c) the length of the grating must not be less than twice its width

(d) the length of the grating must not be more than four times its width.

Write down four inequalities that need to be satisfied.

(iii) Illustrate the region satisfied by these inequalities, using the horizontal axis for x and the vertical axis for y, and shading the unwanted region.

(iv) The length of the grating is to be maximised.
Write down the objective function and find the dimensions of the grating with maximum length.

3 St Andrew's Church is putting on a Nativity play in order to raise money for repairs.
The church hall seats a maximum of 400 people for performances and tickets are to be sold for £4 each or £3 for children and concessions.
The church needs to raise at least £1000 and it is decided that the number of £4 tickets must not be greater than twice the number of £3 tickets so that children can be given every opportunity to attend.

(i) There are t tickets at £4 each and c tickets at £3 each. Write down three inequalities that need to be satisfied.

(ii) Illustrate the region satisfied by these inequalities, using the horizontal axis for t and the vertical axis for c, and shading the unwanted region.

(iii) The income is to be maximised. Write down the objective function and find:

(a) the number of tickets of each category which need to be sold to raise the greatest amount of money

(b) the maximum income.

4 A farmer intends to buy some pullets of two different breeds.
Each pullet of breed A costs £1.20 and requires 0.4 m^2 of floor space, while each of breed B costs £1.60 but requires only 0.3 m^2 of floor space.
The farmer cannot spend more than £1000 and the area of the hen-house floor is 240 m^2.

(i) Taking a and b as the numbers of breeds A and B respectively which he could buy, write down two inequalities connecting a and b.

(ii) Illustrate the region satisfied by these inequalities, using the horizontal axis for a and the vertical axis for b, and shading the unwanted region.

(iii) From your graph find the maximum number which can be bought.

5 A manufacturer is to market a new compost which is a mixture of two types, X and Y, that are already on the market.

The details of the two types are in the table below (percentages are by volume).

	peat	**rotted vegetable matter**	**other ingredients**	**cost per litre**
type X	60%	20%	20%	3p
type Y	84%	10%	6%	4.5p

It has been decided that the new type must contain at least 75% peat and 12% rotted vegetable matter.

(i) x litres of type X are to be combined with y litres of type Y to make 100 litres of the new compost.
Write down one equation and two inequalities that need to be satisfied.

(ii) Illustrate these on a graph and identify the line segment that represents all possible values of x and y.

(iii) The manufacturer would like to make this compost up as cheaply as possible.
Write down the objective function and determine the amounts of each of X and Y in 100 litres of the new compost.

6 Mia is going to make some small cakes to sell at school and raise money for charity.
She has decided to make some chocolate muffins and some munchkin cakes.
She would like to make as many cakes as possible but discovers that she only has 2 kg of flour and 750 g of butter. She has plenty of the other ingredients.

The cakes must be made in batches; for 12 muffins she needs 300 g of flour and 50 g of butter, and for 16 munchkins she needs 200 g of flour and 125 g of butter.

(i) Using c to represent the number of batches of chocolate muffins and m to represent the number of batches of munchkin cakes, write down and simplify two inequalities relating to the available ingredients.

(ii) Illustrate the region satisfied by these inequalities, using the horizontal axis for c and the vertical axis for m, and shading the unwanted region.

(iii) Write down the objective function for the total number of cakes and find the greatest number of cakes that she can make.

KEY POINTS

1 When illustrating linear inequalities:
- represent the boundaries for $<$ and $>$ as a *broken* line
- represent the boundaries for $\leqslant$ and $\geqslant$ as a *solid* line
- specify the region you want by shading out the *other* side of the line.

2 The region where a number of inequalities are satisfied simultaneously is called the *feasible* region.

3 In linear programming, the *objective function* is the algebraic expression describing the quantity that you are required to maximise or minimise.

4 The maximum and minimum values of the objective function will lie at, or near, a vertex of the feasible region.

SECTION 3
Trigonometry

Trigonometry I

The difficulty lies, not in the new ideas, but in escaping the old ones, which ramify, for those brought up as most of us have been, into every corner of our minds.

John Maynard Keynes

The London Eye has 32 cars equally spaced around a circle.
What information would you need in order to work out:

(i) the distance of a car above or below the centre
(ii) the distance of a car to the right or left the centre?

Using trigonometry in right-angled triangles

You have met definitions of the three trigonometrical functions, sin, cos and tan, using the sides of a right-angled triangle.

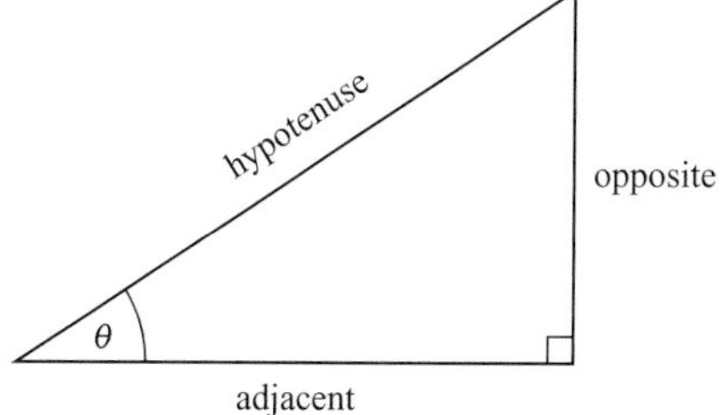

Figure 7.1

In figure 7.1

$$\sin\theta = \frac{\text{opposite}}{\text{hypotenuse}} \qquad \cos\theta = \frac{\text{adjacent}}{\text{hypotenuse}} \qquad \tan\theta = \frac{\text{opposite}}{\text{adjacent}}$$

Do these definitions work for angles of any size?

sin is an abbreviation of sine, cos of cosine and tan of tangent.

EXAMPLE 7.1

Find the length of the side marked a in the triangle in figure 7.2.

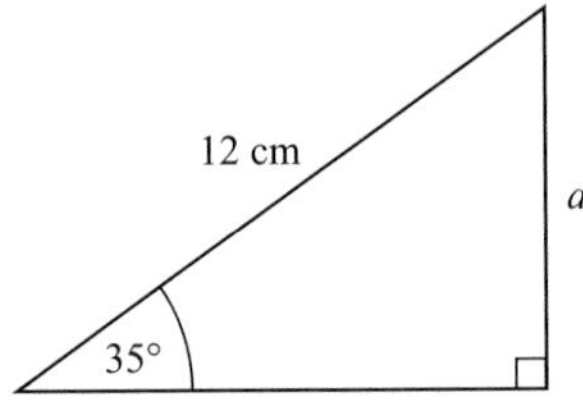

Figure 7.2

SOLUTION

Side a is *opposite* the angle of 35°, and the *hypotenuse* is 12 cm, so we use sin 35°.

$$\sin 35° = \frac{\text{opposite}}{\text{hypotenuse}}$$

$$= \frac{a}{12}$$

$$\Rightarrow \quad a = 12 \sin 35°$$

$$\Rightarrow \quad a = 6.9\,\text{cm (1 d.p.)}$$

EXAMPLE 7.2

Find the length of the hypotenuse in the triangle in figure 7.3.

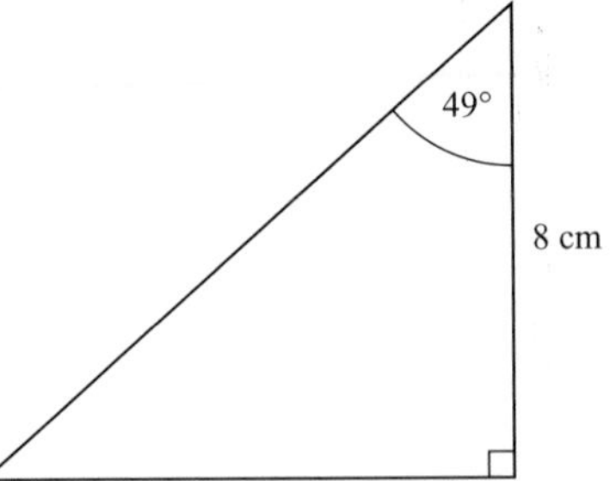

Figure 7.3

SOLUTION

The side of length 8 cm is *adjacent* to the angle of 49°, and you want the *hypotenuse* so you use cos 49°.

$$\cos 49° = \frac{\text{adjacent}}{\text{hypotenuse}}$$
$$= \frac{8}{\text{hypotenuse}}$$
$$\Rightarrow \quad \text{hypotenuse} = \frac{8}{\cos 49°}$$
$$\Rightarrow \quad \text{hypotenuse} = 12.2\,\text{cm (1 d.p.)}$$

EXAMPLE 7.3 Find the size of the angle marked θ in the triangle in figure 7.4.

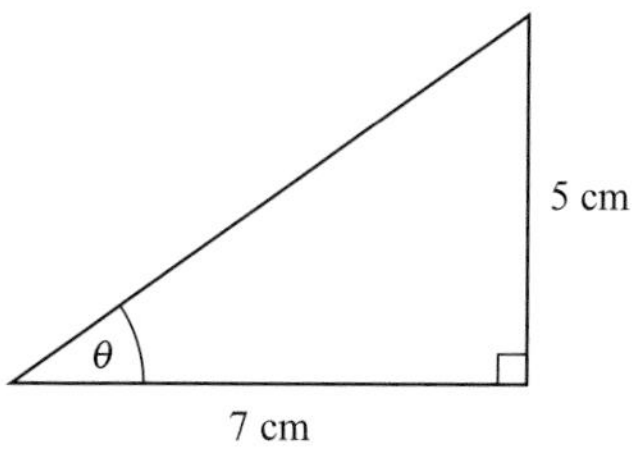

Figure 7.4

SOLUTION

The sides whose lengths are known are those *opposite* and *adjacent* to θ so we use $\tan\theta$.

$$\tan\theta = \frac{\text{opposite}}{\text{adjacent}} = \frac{5}{7} = 0.714\,285\,714$$
$$\Rightarrow \quad \theta = 35.5° \text{ (1 d.p.)}$$

? The full calculator value for $\frac{5}{7}$ has been used to find the value of θ. What is the least number of decimal places that you could use to give the same value for the angle (to 1 d.p.) in this example?

Trigonometry may often be applied to practical situations as in the following example.

EXAMPLE 7.4 A bird flies straight from the top of a 15 m tall tree, at an angle of depression of 27°, to catch a worm on the ground.

(i) How far does the bird fly?

(ii) How far was the worm from the bottom of the tree?

SOLUTION

First draw a sketch, labelling the information given and using letters to mark what you want to find.

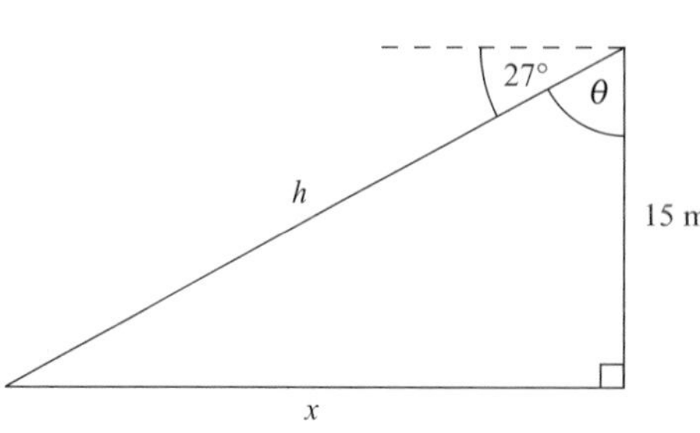

Remember, *angles of depression* are measured down from the horizontal, and *angles of elevation* are measured up from the horizontal.

Figure 7.5

(i) $\theta + 27° = 90°$

$\Rightarrow \quad \theta = 63°$

$$\cos 63° = \frac{15}{h}$$

$$\Rightarrow \quad h = \frac{15}{\cos 63°} = 33.040\,338\,97$$

$\Rightarrow$ the bird flies 33 m.

(ii) Using Pythagoras' theorem

$$h^2 = x^2 + 15^2$$

$$\Rightarrow \quad x^2 = 33.040\,338\,97^2 - 15^2 = 866.663\,999^2$$

$$\Rightarrow \quad x = 29.439\,157\,58$$

$\Rightarrow$ The worm is 29.4 m from the bottom of the tree.

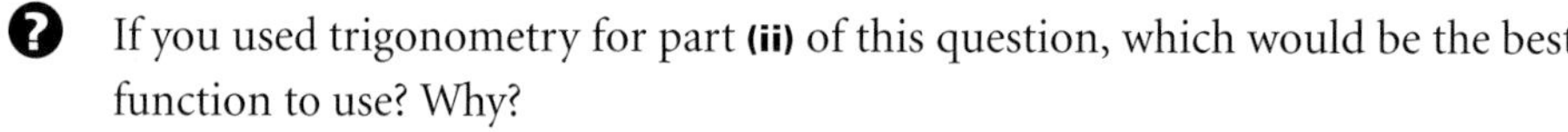

? If you used trigonometry for part **(ii)** of this question, which would be the best function to use? Why?

Historical note

The word for trigonometry is derived from three Greek words.

Tria: *three*	gonia: *angle*	metron *measure*
$(\tau\rho\iota\alpha)$	$(\gamma o\nu\iota\alpha)$	$(\mu\varepsilon\tau\rho o\nu)$

This shows how trigonometry developed from studying angles, often in connection with astronomy, although the subject was probably discovered independently by a number of people. Hipparchus (150 BC) is believed to have produced the first trigonometrical tables which gave lengths of chords of a circle of unit radius. His work was further developed by Ptolemy in AD 100.

EXERCISE 7A

1 Find the length marked x in each of these triangles. Give your answers correct to 1 decimal place.

(i)

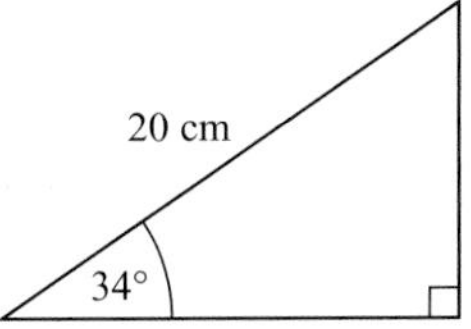

(ii)

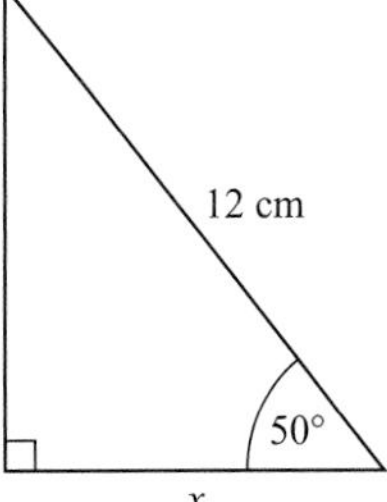

(iii)

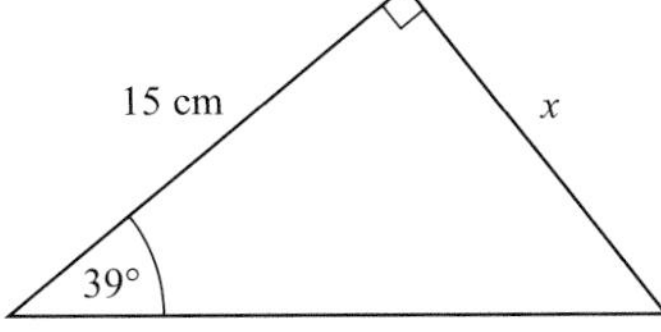

(iv)

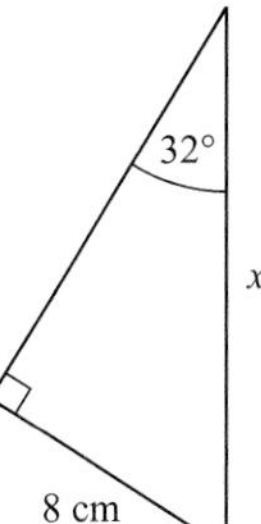

(v)

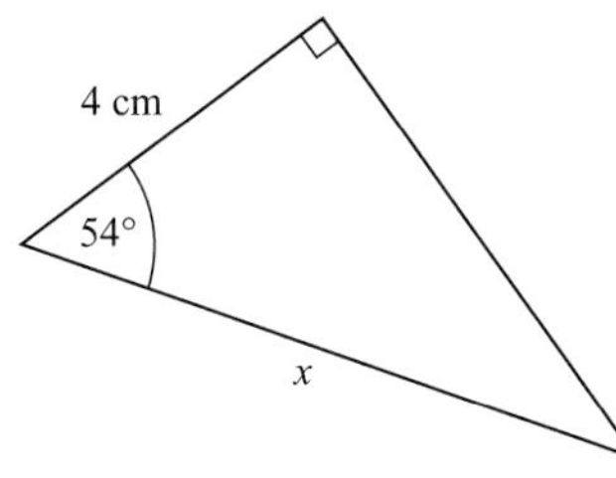

(vi)

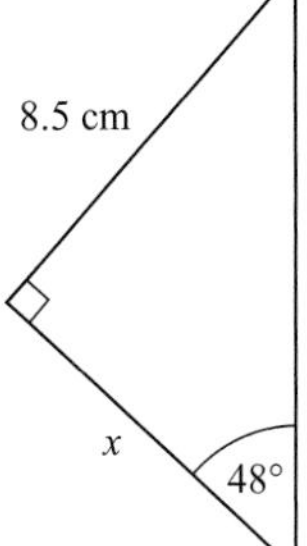

2 Find the angle marked θ in each of these triangles. Give your answers correct to 1 decimal place.

(i)

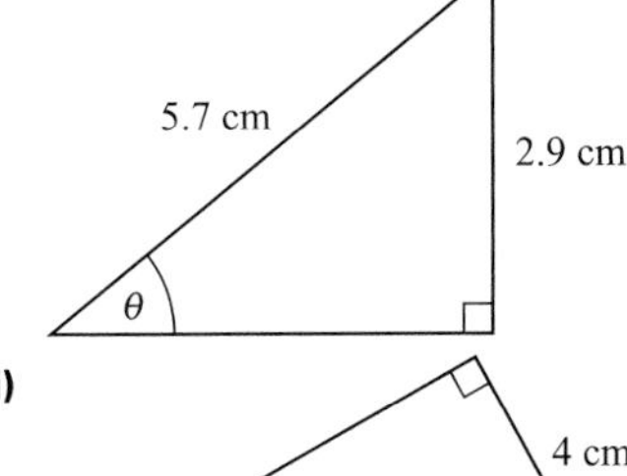

(ii)

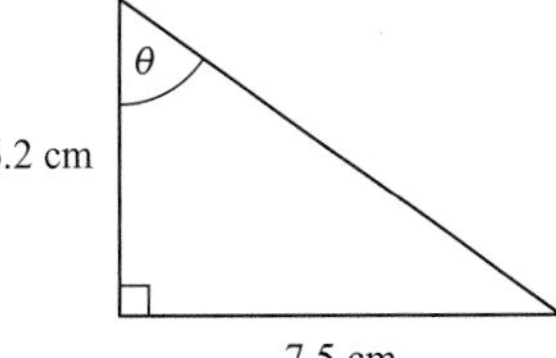

(iii)

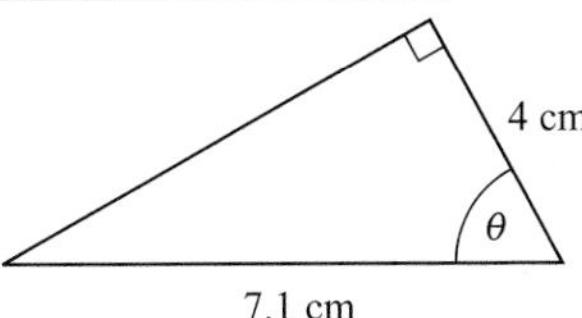

(iv)

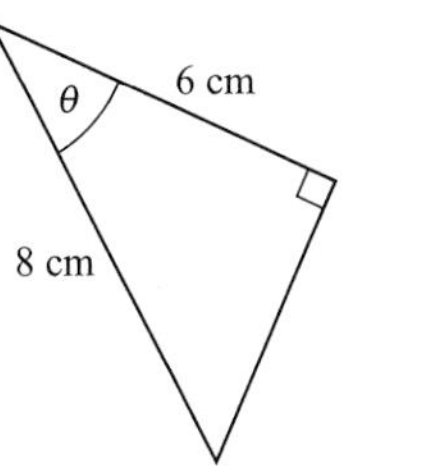

(v)

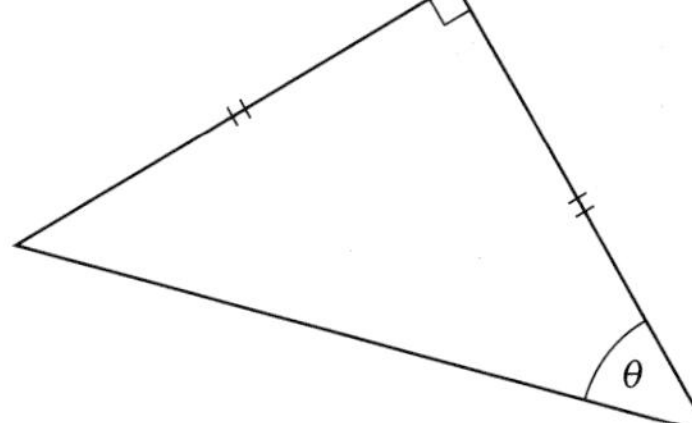

(vi)

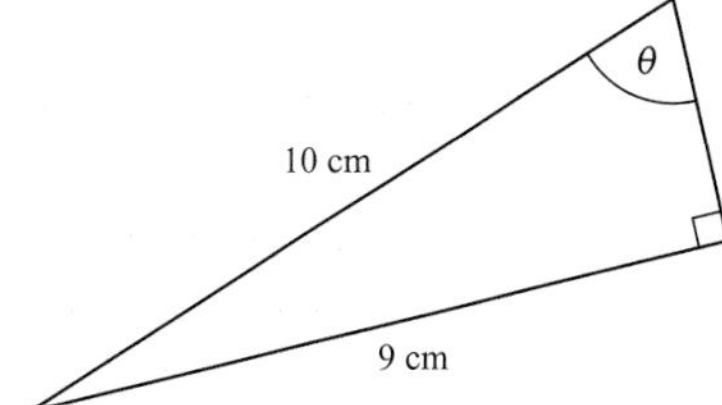

3 In an isosceles triangle, the line of symmetry bisects the base of the triangle. Use this fact to find the angle θ and the lengths x and y in these diagrams.

(i) **(ii)**

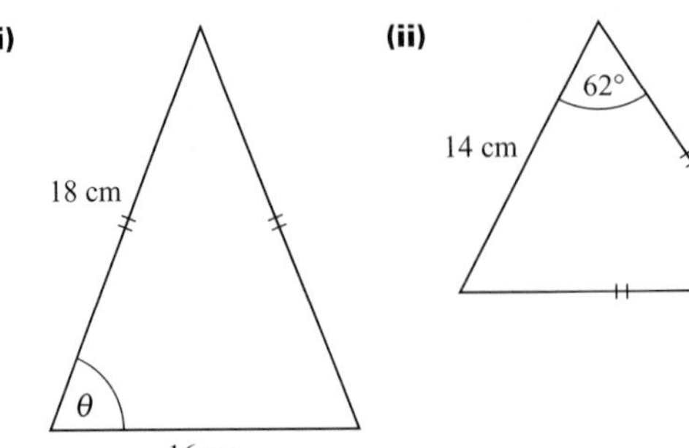

(iii)

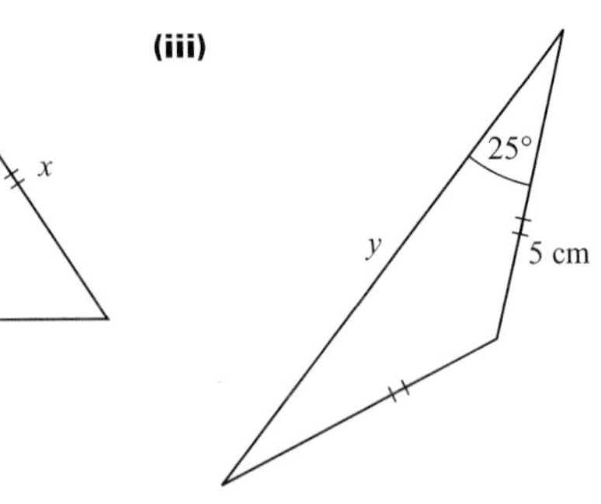

4 A ladder 5 m long rests against a wall. The foot of the ladder makes an angle of 65° with the ground.
How far up the wall does the ladder reach?

5 From the top of a vertical cliff 30 m high, the angle of depression of a boat at sea is 21°.
How far is the boat from the bottom of the cliff?

6 From a point 120 m from the base of an office block, the angle of elevation of the top of the block is 67°.
How tall is the block?

7 A rectangle has sides of length 12 cm and 8 cm.
What angle does the diagonal make with the longest side?

8 The diagram shows the positions of three airports:
E (East Midlands), M (Manchester) and L (Leeds).
The distance from M to L is 65 km on a bearing of 060°.
Angle LME = 90° and ME = 100 km.

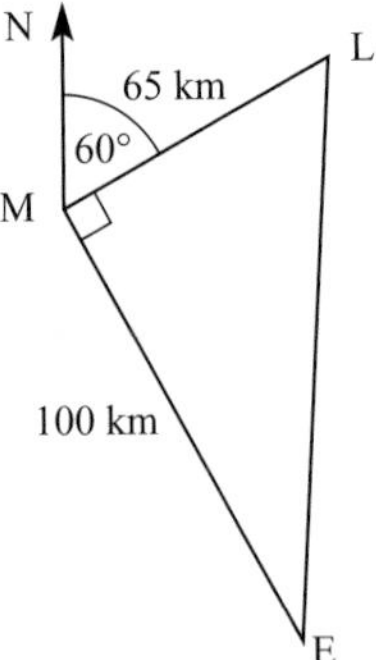

(i) Calculate, correct to three significant figures, the distance LE.

(ii) Calculate, correct to the nearest degree, the size of angle MEL.

(iii) An aircraft leaves M at 10.45 am and flies direct to E, arriving at 11.03 am. Calculate, correct to three significant figures, the average speed of the aircraft in kilometres per hour.

[MEG]

Trigonometrical functions for angles of any size

Positive and negative angles

By convention, angles are measured from the positive x axis (figure 7.6). Anticlockwise is taken to be positive and clockwise to be negative.

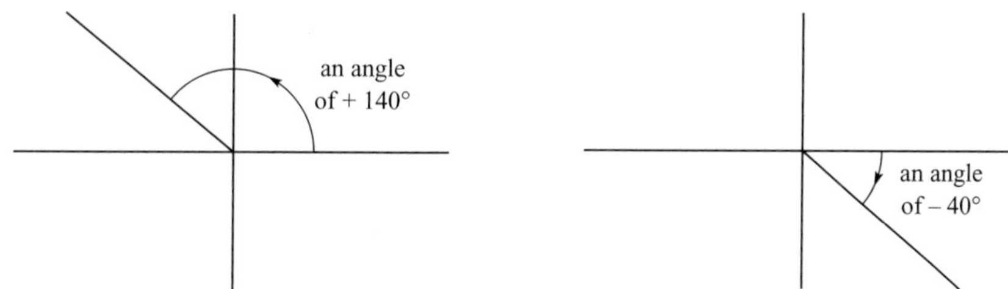

Figure 7.6

The only exception is for compass bearings, which are measured clockwise from the north.

? Is it possible to extend the definitions of trigonometrical functions to angles greater than 90°, like sin 156°, cos 202° or tan 320°?

It is not difficult to extend these definitions, as follows.

First look at the right-angled triangle in figure 7.7 which has a hypotenuse of unit length.

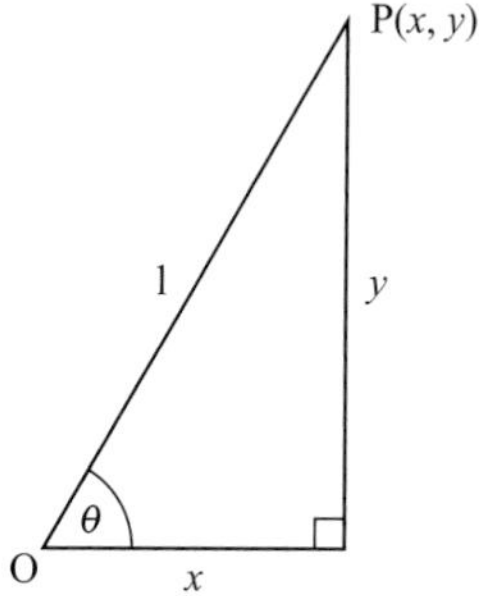

Figure 7.7

This gives the definitions:

$$\sin\theta = \frac{y}{1} = y \qquad \cos\theta = \frac{x}{1} = x \qquad \tan\theta = \frac{y}{x}$$

Now imagine the angle θ situated at the origin, as in figure 7.8, and allow θ to take any value. The vertex marked P has co-ordinates $(\cos\theta, \sin\theta)$ and can now be anywhere on the unit circle.

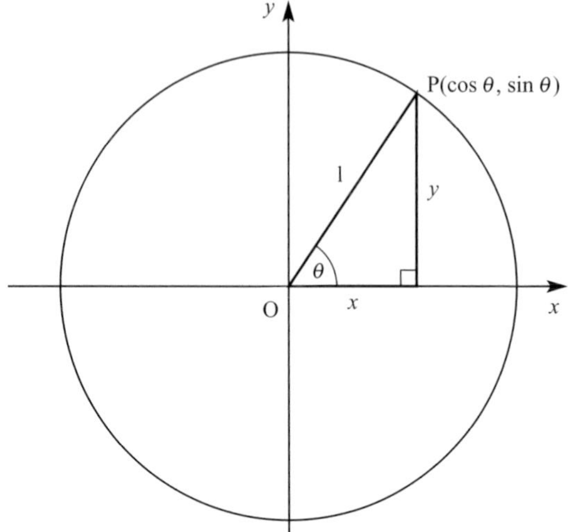

Figure 7.8

You can now see that the definitions on the previous page can be applied to *any* angle θ, whether it is positive or negative, and whether it is less than or greater than 90°.

$$\sin\theta = y \qquad \cos\theta = x \qquad \tan\theta = \frac{y}{x}$$

For some angles, x or y (or both) will take a negative value, so the signs of $\sin\theta$, $\cos\theta$ and $\tan\theta$ will vary accordingly.

The sine and cosine graphs

Look at figure 7.9. There is a unit circle and angles have been drawn at intervals of 30°. The resulting y co-ordinates are plotted relative to the axes on the right. They have been joined with a continuous curve to give the graph of $\sin\theta$ for $0 \leqslant \theta \leqslant 360°$.

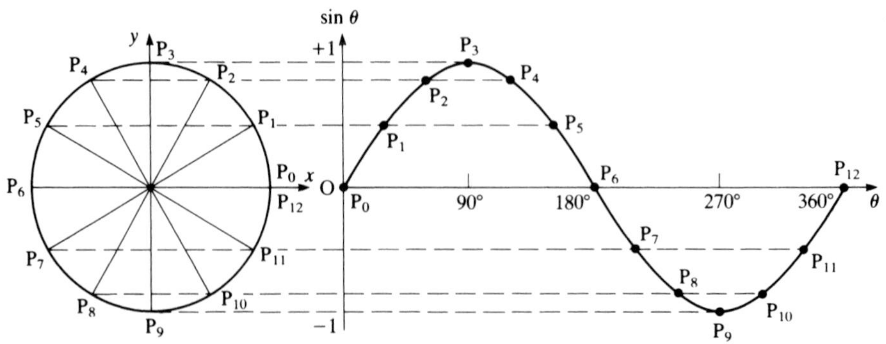

Figure 7.9

Continue this process for angles 390°, 420°, … and angles –30°, –60°, ….

What do you notice?

Since the curve repeats itself every 360°, as shown in figure 7.10, the sine function is described as *periodic* with *period* 360°.

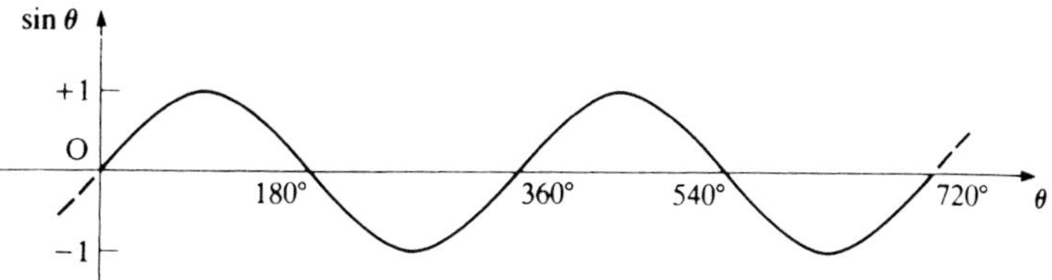

Figure 7.10

In a similar way you can transfer the x co-ordinates onto a set of axes to obtain the graph of $\cos\theta$. This is most easily illustrated if you first rotate the circle through 90° anticlockwise.

Figure 7.11 shows this new orientation, together with the resulting graph.

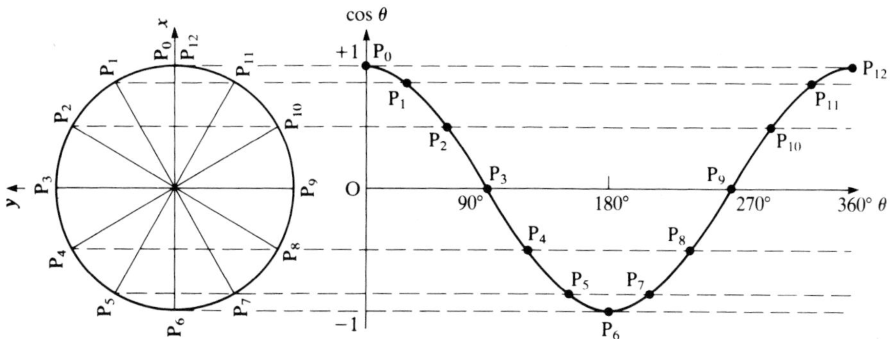

Figure 7.11

For angles in the interval $360° \leqslant \theta \leqslant 720°$, the cosine curve will repeat itself. You can see that the cosine function is also periodic with a period of 360°.

Notice that the graphs of $\sin\theta$ and $\cos\theta$ have exactly the same shape. The cosine graph can be obtained by translating the sine graph 90° to the left, as shown in figure 7.12.

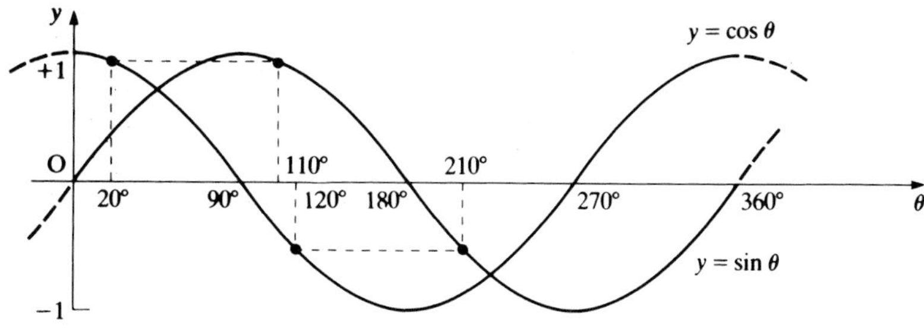

Figure 7.12

? Look at the graphs. How do they show you that $\cos\theta = \sin(\theta + 90°)$?

From the graphs it can be seen that, for example.

$\cos 20° = \sin 110°$ $\quad\cos 90° = \sin 180°$ $\quad\cos 120° = \sin 210°$ and so on....

In general

$$\cos\theta \equiv \sin(\theta + 90°)$$

ACTIVITY 7.2

(i) Draw the graph of $y = \sin\theta$ for $0° \leqslant \theta \leqslant 90°$.
Show how the rest of the curve for $-180° \leqslant \theta \leqslant 360°$ can be obtained from this using only reflections, rotations and translations.
Where is there a line of symmetry?

(ii) Draw the graph of $y = \cos\theta$ for $0° \leqslant \theta \leqslant 90°$.
Show how the rest of the curve for $-360° \leqslant \theta \leqslant 360°$ can be obtained from this using only reflections, rotations and translations.
Where is there a line of symmetry?

The tangent graph

The value of $\tan\theta$ can be worked out from the definition $\tan\theta = \frac{y}{x}$ or by using

$$\tan\theta = \frac{\sin\theta}{\cos\theta}$$

? The function $\tan\theta$ is undefined for $\theta = 90°$. What does *undefined* mean?
How can you tell that tan 90° is undefined?
For which other values of θ is $\tan\theta$ undefined?

The graph of $\tan\theta$ is shown in figure 7.13. The dotted lines $\theta = \pm 90°$ and $\theta =$ 270° are *asymptotes*, they are not actually part of the curve.

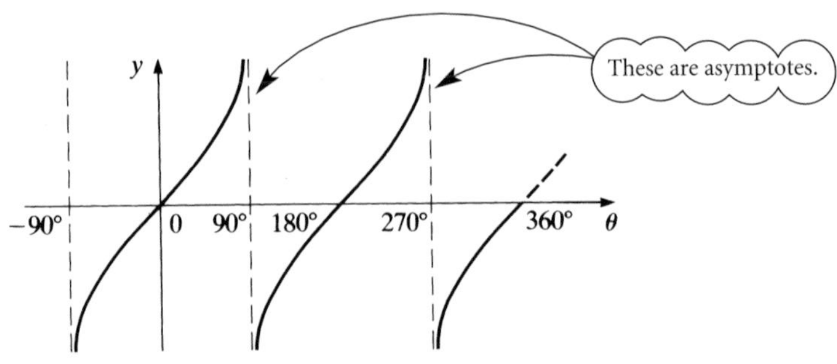

Figure 7.13

How would you describe an asymptote to a friend ?

The graph of $\tan\theta$ is periodic, like those for $\sin\theta$ and $\cos\theta$. What is the period of this graph?

Show how the part of the curve for $0° \leqslant \theta < 90°$ can be used to generate the rest of the curve using rotations and translations.

ACTIVITY 7.3 Draw the graphs of $y = \sin\theta$, $y = \cos\theta$ and $y = \tan\theta$ for values of θ between $-180°$ and $360°$.

These graphs will be useful for solving trigonometrical equations, so keep them handy. It is also a good idea to learn them at this stage.

Solution of equations using graphs of trigonometrical functions

Suppose that you want to solve the equation

$$\sin\theta = 0.5.$$

You start by pressing the calculator keys

and the answer comes up as 30°.

NOTE

The $\sin^{-1}$ key may also be labelled invsin or arcsin.

However, look at the graph of $y = \sin\theta$ (figure 7.14). You can see that there are other roots as well.

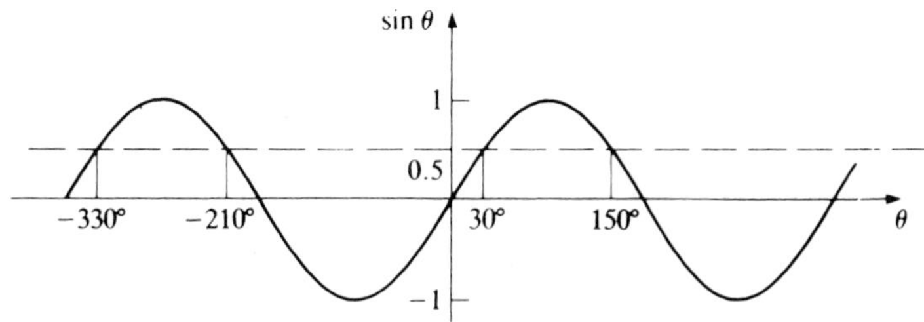

Figure 7.14

How many roots does the equation have?

The root 30° is called the *principal value.*

Other roots can be found by looking at the graph. The roots for $\sin\theta = 0.5$ are seen (figure 7.14) to be:

$$\theta = \ldots, -330°, -210°, 30°, 150°, \ldots.$$

NOTE

A calculator always gives the principal value of the solution. These values are in the range

$0° \leqslant \theta \leqslant 180°$ (cos)

$-90° \leqslant \theta \leqslant 90°$ (sin)

$-90° < \theta < 90°$ (tan)

EXAMPLE 7.5

Find values of θ in the interval $-360° \leqslant \theta \leqslant 360°$ for which $\cos\theta = 0.4$.

SOLUTION

$\cos\theta = 0.4 \Rightarrow \quad \theta = 66.4°$ (principal value).

Figure 7.15 shows the graph of $y = \cos\theta$.

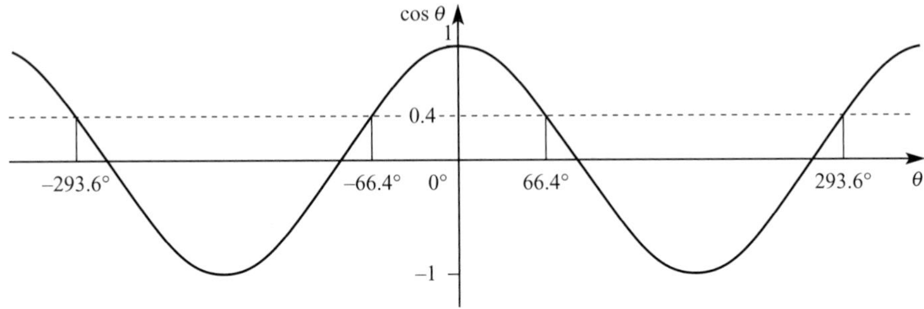

Figure 7.15

The values of θ for which $\cos\theta = 0.4$ are

$$-293.6°, -66.4°, 66.4°, 293.6°.$$

Here all values are given rounded to 1 decimal place.

? How do you arrive at 293.6°?

EXAMPLE 7.6

Find values of θ in the interval $-360° \leqslant \theta \leqslant 360°$ for which $\tan\theta = -0.7$.

SOLUTION

$$\tan\theta = -0.7 \quad \Rightarrow \quad \theta = -35.0° \text{ (principal value)}.$$

Figure 7.16 shows the graph of $y = \tan\theta$.

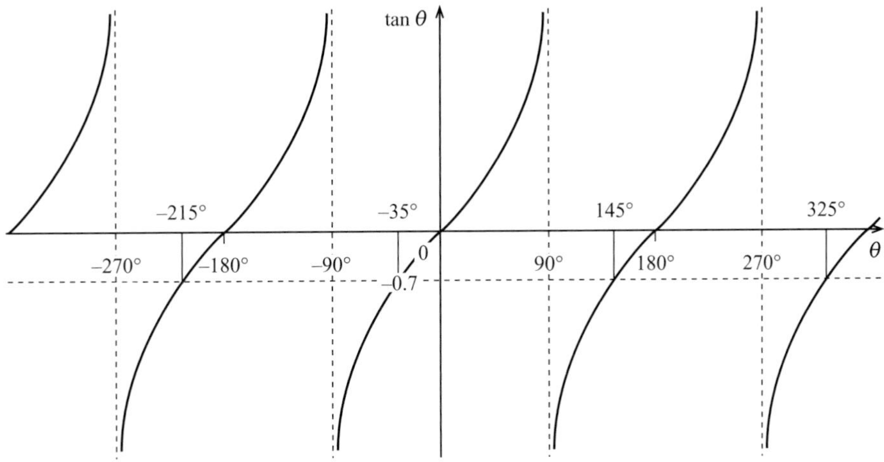

Figure 7.16

The values of θ for which $\tan\theta = -0.7$ are

$$-215.0°, -35.0°, 145.0°, 325.0°.$$

ACTIVITY 7.4

Draw x and y axes. For each of the four quadrants formed, work out the sign of $\sin\theta$, $\cos\theta$ and $\tan\theta$ from the definitions

$$\sin\theta = y, \;\; \cos\theta = x, \;\; \tan\theta = \frac{y}{x}.$$

EXERCISE 7B

1 Solve the following equations for $0° \leqslant \theta \leqslant 360°$. Give answers to 1 decimal place where necessary.

(i) $\cos\theta = 0.5$ **(ii)** $\tan\theta = 1$ **(iii)** $\sin\theta = \frac{\sqrt{3}}{2}$

(iv) $\sin\theta = -0.5$ **(v)** $\cos\theta = 0$ **(vi)** $\tan\theta = -5$

(vii) $\tan\theta = 0$ **(viii)** $\cos\theta = -0.54$ **(ix)** $\sin\theta = 1$

2 In this question all the angles are in the interval 0° to 360°. Give your answers to 1 decimal place.

(i) Find α (alpha) if $\cos\alpha = 0.5469$ and $\sin\alpha < 0$.

(ii) Find β (beta) if $\tan\beta = -5.76$ and $\cos\beta < 0$.

(iii) Find γ (gamma) if $\sin\gamma = 0.5432$ and $\tan\gamma > 0$.

(iv) Find δ (delta) if $\cos\delta = -0.75$ and $\tan\delta < 0$.

(v) Find ε (epsilon) if $\sin\varepsilon = 0$ and $\cos\varepsilon < 0$.

(vi) Find φ (phi) if $\tan\varphi = 0.5$ and $\sin\varphi < 0$.

3 Find all possible angles θ for which the following conditions are true.

(i) $\cos\theta = \cos 65°$ and $0° \leqslant \theta \leqslant 360°$

(ii) $\sin\theta = \sin 120°$ and $0° \leqslant \theta \leqslant 360°$

(iii) $\tan\theta = \tan 45°$ and $-180° \leqslant \theta \leqslant 180°$

(iv) $\sin\theta = \sin(-50°)$ and $-180° \leqslant \theta \leqslant 180°$

(v) $\cos\theta = \cos(-90°)$ and $-180° \leqslant \theta \leqslant 180°$

(vi) $\tan\theta = \tan(-75°)$ and $0° \leqslant \theta \leqslant 360°$

4 (i) Sketch the curve $y = \sin x$ for $-90° \leqslant x \leqslant 450°$.

(ii) Solve the equation $\sin x = \frac{\sqrt{3}}{2}$ for $-90° \leqslant x \leqslant 450°$, and illustrate all the roots on your sketch.

(iii) Sketch the curve $y = \cos x$ for $-90° \leqslant x \leqslant 450°$.

(iv) Solve the equation $\cos x = 0.5$ for $-90° \leqslant x \leqslant 450°$, and illustrate all the roots on your sketch.

(v) Explain why some of the roots of $\sin x = \frac{\sqrt{3}}{2}$ are the same as those for $\cos x = 0.5$ and some are different.

5 Solve the following equations for $0° \leqslant \theta \leqslant 360°$. Give your answers to 1 decimal place where necessary.

(i) $3\cos\theta = 2$ **(ii)** $7\sin\theta = 5$ **(iii)** $3\tan\theta = 8$

(iv) $6\sin\theta + 5 = 0$ **(v)** $5\cos\theta + 2 = 0$ **(vi)** $5 - 9\tan\theta = 10$

6 (i) Solve the equation $\sin\theta = 0.2$ for $0° \leqslant \theta \leqslant 180°$.

(ii) Solve the equation $\sin(\varphi - 20°) = 0.2$ for $0° \leqslant \varphi \leqslant 180°$. (You may find the substitution $\varphi - 20° = \theta$ useful.)

7 Solve the following equations for $0° \leqslant \theta \leqslant 360°$.

(i) $\sin^2\theta = 0.75$ **(ii)** $\cos^2\theta = 0.5$ **(iii)** $\tan^2\theta = 1$

8 (i) Factorise $2x^2 + x - 1$

(ii) Hence solve $2x^2 + x - 1 = 0$

(iii) Use your results to solve these equations for $0° \leqslant \theta \leqslant 360°$.

(a) $2\sin^2\theta + \sin\theta - 1 = 0$

(b) $2\cos^2\theta + \cos\theta - 1 = 0$

(c) $2\tan^2\theta + \tan\theta - 1 = 0$

ACTIVITY 7.5 Use the right-angled triangle in figure 7.17 to show that $\sin\theta = \cos(90° - \theta)$.

Investigate whether this result is still true when θ takes values that are greater than 90° or negative. Can you justify your answer?

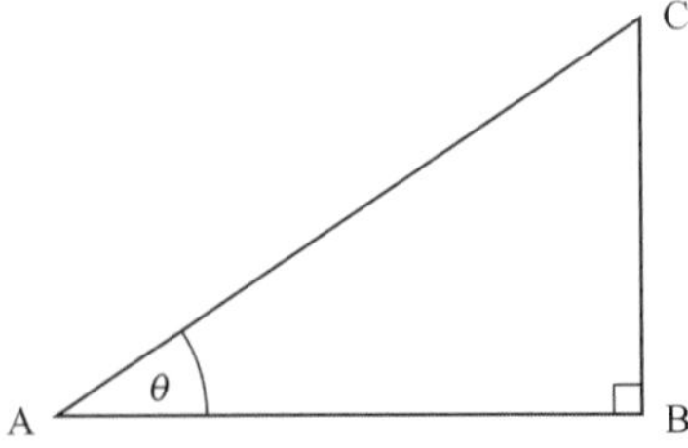

Figure 7.17

? Identify any symmetries in the graphs of the following.

(i) $y = \sin\theta$

(ii) $y = \cos\theta$

(iii) $y = \tan\theta$

INVESTIGATION

Here is an alternative way of finding the additional roots to the equation $\sin\theta = 0.5$.

1 Look at the line of symmetry marked on the graph of $y = \sin\theta$ in figure 7.18.

2 The principal value for $\sin\theta = 0.5$ is $\theta = 30°$. Reflect this in the line $\theta = 90°$, to give the value $\theta = 150°$. Notice that this is another root of $\sin\theta = 0.5$.

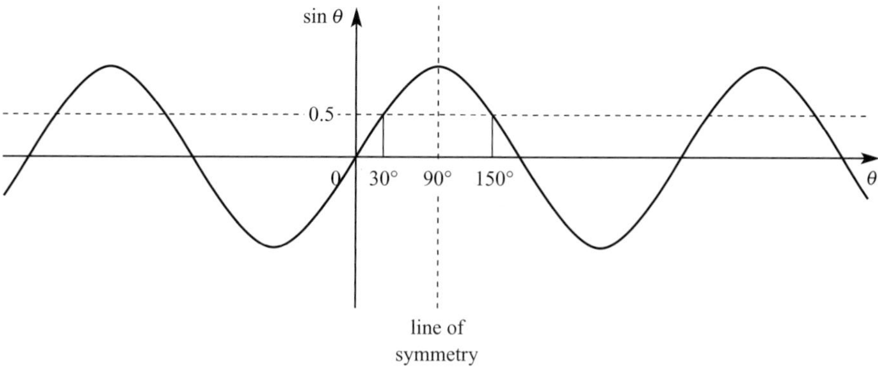

Figure 7.18

3 These two roots are shown differently in figure 7.19.

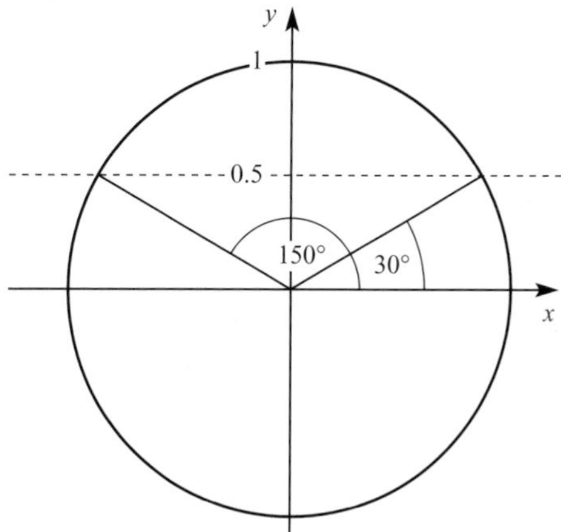

Figure 7.19

4 By measuring the angles anticlockwise, rename $\theta = 150°$ as $\theta = -210°$, and $\theta = 30°$ as $\theta = -330°$. This is illustrated in figure 7.20.

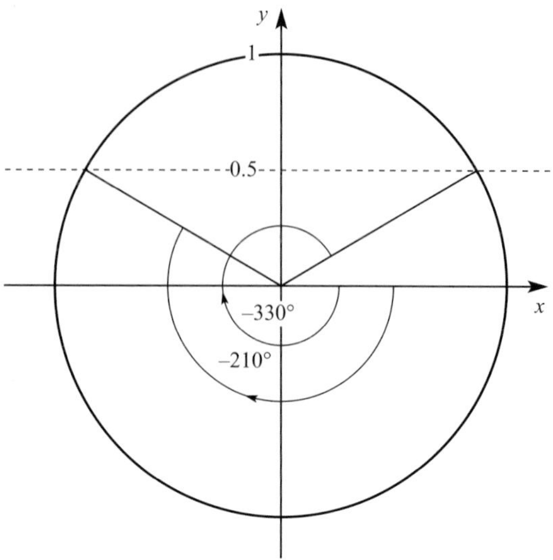

Figure 7.20

You now have all the roots of $\sin\theta = 0.5$ in the range $-360° \leqslant \theta \leqslant 360°$, as found earlier in this chapter.

Now look at the equations solved in Examples 7.5 and 7.6 using a method similar to this.

(i) For $\cos\theta = 0.4$, follow the instructions above, but use the line of symmetry $\theta = 0$ for the cosine graph.

(ii) For $\tan\theta = -0.7$, notice how the curve repeats itself every 180°. This means that the principal root of the equation is translated by multiples of 180° (which may be negative multiples) to obtain the other values. How would you illustrate this on a diagram of the four quadrants?

Identities involving $\sin\theta$, $\cos\theta$ and $\tan\theta$

Remember the earlier definitions for trigonometrical functions of angles of any magnitude

$$\sin\theta = y \qquad \cos\theta = x \qquad \tan\theta = \frac{y}{x}$$

where the angle θ was defined by a point P(x, y) on a circle of unit radius (figure 7.21).

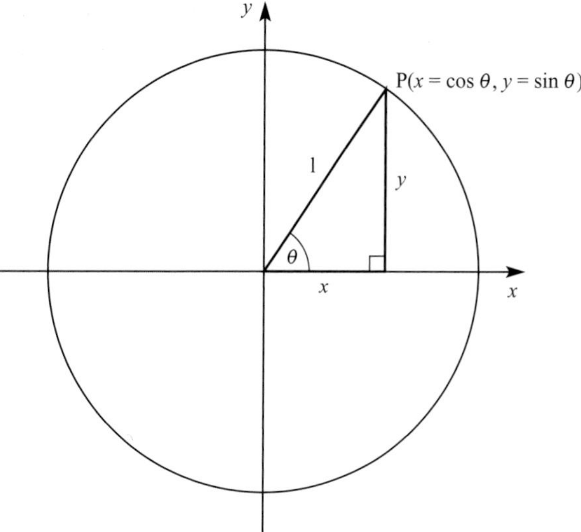

Figure 7.21

From these it follows that

$$\tan\theta = \frac{y}{x}$$

so $$\tan\theta = \frac{\sin\theta}{\cos\theta}$$

NOTE

It would be more accurate here to use the identity sign, ≡, since the relationship is true for all values of θ.

$$\tan\theta \equiv \frac{\sin\theta}{\cos\theta}$$

However, in this book, as in mathematics generally, we often use the equals sign where it would be more correct to use an identity sign. The identity sign is kept for situations where we really want to emphasise that the relationship is an identity and not an equation.

Another useful identity can be found by applying Pythagoras' theorem to any point P(x, y) on the unit circle. Look at figure 7.22.

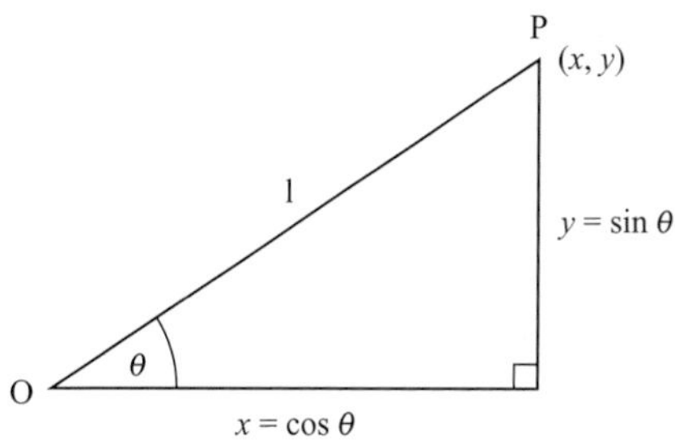

Figure 7.22

$$y^2 + x^2 \equiv \text{OP}^2$$
$$\Rightarrow \quad (\sin\theta)^2 + (\cos\theta)^2 \equiv 1$$

This is written as

$$\sin^2\theta + \cos^2\theta \equiv 1$$

This notation is more economical since it doesn't need brackets. It tells you that it is the trigonometrical function that is being squared, not the angle θ.

What is the difference between an identity and an equation?

Using trigonometrical identities to solve equations

The identity

$$\sin^2\theta + \cos^2\theta \; = 1$$

is particularly useful when solving equations which contain both $\sin\theta$ and $\cos\theta$ with one of the functions squared. You can rearrange it to get

$$\sin^2\theta = 1 - \cos^2\theta \quad \text{or} \quad \cos^2\theta = 1 - \sin^2\theta.$$

EXAMPLE 7.7

Solve the equation $2\cos^2\theta - \sin\theta - 1 = 0$ for values of θ in the range 0° to 360°.

SOLUTION

$$2\cos^2\theta - \sin\theta - 1 = 0$$
$$\Rightarrow \quad 2(1 - \sin^2\theta) - \sin\theta - 1 = 0$$
$$\Rightarrow \quad 2 - 2\sin^2\theta - \sin\theta - 1 = 0$$
$$\Rightarrow \quad 0 = 2\sin^2\theta + \sin\theta - 1$$

This is now a quadratic in $\sin\theta$.

Factorising

$$\Rightarrow \quad (\sin\theta + 1)(2\sin\theta - 1) = 0$$
$$\Rightarrow \quad \sin\theta = \; -1 \text{ or } \sin\theta = 0.5$$

$$\sin\theta = -1 \Rightarrow \theta = -\,90° \text{ or } \theta = 270°$$

$\theta = -90°$ is the principal value but it is outside the required range. $\theta = 270°$ is inside the required range.

$$\sin\theta = 0.5 \Rightarrow \theta = 30° \text{ (principal value) or } \theta = 150°$$

$\theta = 30°$ and $\theta = 150°$ are both inside the required range. The next value, 390°, is too big.

The solution of the equation is therefore $\theta = 30°$, 150°, or 270°.

NOTE

For your answer it is usual to list all the values of θ that you have found in increasing order.

? Sometimes one of the brackets may be of the form $(\sin\theta - 2)$ for example. How would you solve

$(2\sin\theta - 1)(\sin\theta - 2) = 0$?

? You may meet a quadratic equation that does not factorise. What would you do then?

EXERCISE 7C

1 For each of the equations **(i)–(v)**:

(a) use the identity $\sin^2\theta + \cos^2\theta = 1$ to rewrite the equation in a form involving only one trigonometrical function

(b) factorise, and hence solve, the resulting equation for $0° \leqslant \theta \leqslant 360°$.

(i) $2\cos^2\theta + \sin\theta - 1 = 0$

(ii) $\sin^2\theta + \cos\theta + 1 = 0$

(iii) $2\sin^2\theta - \cos\theta - 1 = 0$

(iv) $\cos^2\theta + \sin\theta = 1$

(v) $1 + \sin\theta - 2\cos^2\theta = 0$

2 For each of the equations **(i)–(iii)**:

(a) use the identity $\sin^2\theta + \cos^2\theta = 1$ to rewrite the equation in a form involving only one trigonometrical function

(b) use the quadratic formula to solve the resulting equation for $-180° \leqslant \theta \leqslant 180°$.

(i) $\sin^2\theta - 2\cos\theta + 1 = 0$

(ii) $\cos^2\theta - \sin\theta = 0$

(iii) $\sin^2\theta - 3\cos\theta = 0$

3 (i) Use the identity

$$\tan\theta = \frac{\sin\theta}{\cos\theta}$$

to rewrite the equation $\sin\theta = 2\cos\theta$ in terms of $\tan\theta$.

(ii) Hence solve the equation $\sin\theta = 2\cos\theta$ for $-180° \leqslant \theta \leqslant 180°$.

4 Use the identity

$$\tan\theta = \frac{\sin\theta}{\cos\theta}$$

to solve the following equations for $0 \leqslant \theta \leqslant 360°$.

(i) $2\sin\theta + \cos\theta = 0$

(ii) $\sqrt{3}\tan\theta = 2\sin\theta$

(iii) $4\cos\theta\,\tan\theta = 1$

The area of a triangle

You are familiar with the use of capital letters to label the vertices of a triangle. In a similar way you can use lower case letters to name the sides. To do this you would use a to denote the length of the side opposite angle A, b to denote the length of the side opposite angle B, etc. as in figure 7.23.

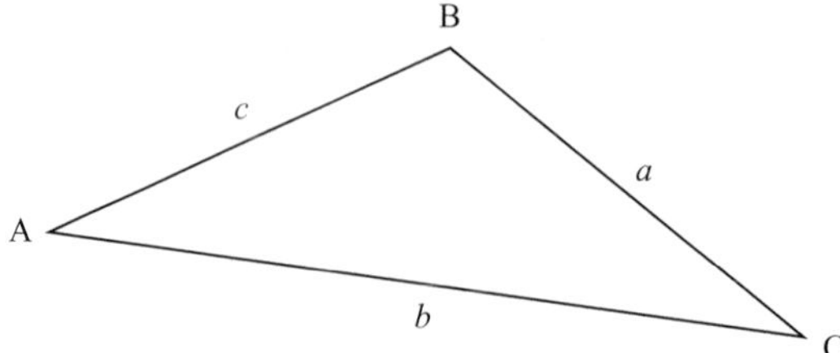

Figure 7.23

Using this notation, for any triangle ABC the area is given by the formula

$$\text{area} = \tfrac{1}{2}bc\sin A.$$

Proof

Figure 7.24 shows a triangle ABC. The perpendicular CD is the height h corresponding to AB as base.

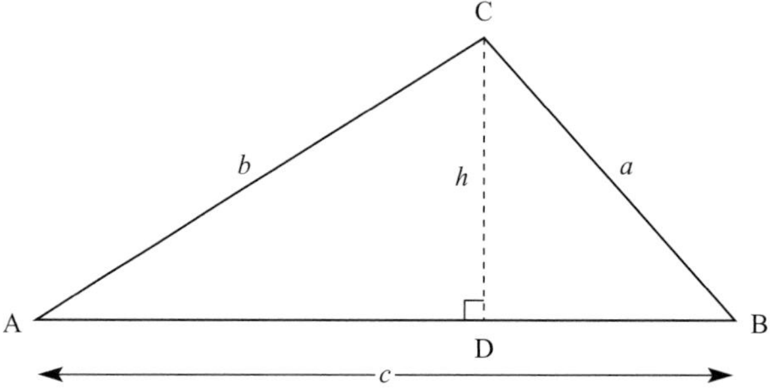

Figure 7.24

Using area of a triangle equals half its base times its height,

$$\text{area} = \tfrac{1}{2}ch. \qquad ①$$

In triangle ACD

$$\sin A = \frac{h}{b}$$

$$\Rightarrow \qquad h = b\sin A.$$

Substituting in ① gives

$$\text{area} = \tfrac{1}{2}bc\sin A.$$

NOTE

Taking the other two points in turn as the top of the triangle gives equivalent results:

$$\text{area} = \frac{1}{2}ca\sin B$$

and

$$\text{area} = \frac{1}{2}ab\sin C.$$

This can be remembered as *half the product of two sides times the sine of the angle between them.*

EXAMPLE 7.8

Figure 7.25 shows a regular pentagon, PQRST, inscribed in a circle, centre C, radius 8 cm. Calculate the area of

(i) triangle CPQ

(ii) the pentagon.

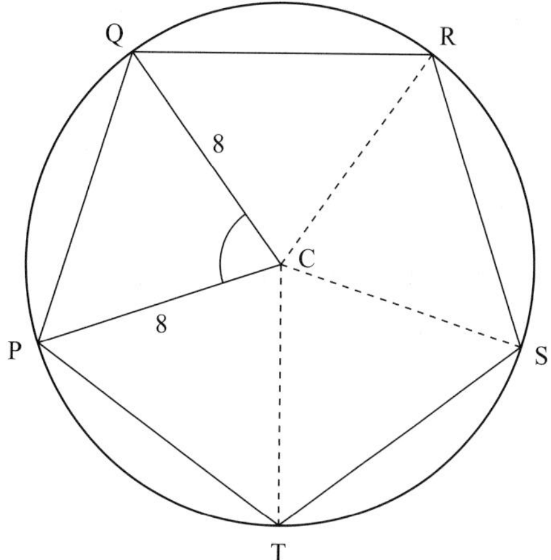

Figure 7.25

SOLUTION

$$\text{angle PCQ} = 360° \div 5$$

$$= 72°$$

$$\text{area PCQ} = \frac{1}{2} \times 8 \times 8 \times \sin 72°$$

$$= 30.4338\ \ldots$$

$$\Rightarrow \quad \text{area PCQ} = 30.4\,\text{cm}^2 \text{ (1 d.p.)}$$

$$\text{area PQRST} = 5 \times 30.4338\ \ldots$$

$$= 152.169\ldots$$

$$\Rightarrow \quad \text{area PQRST} = 152.2\,\text{cm}^2 \text{ (1 d.p.)}$$

EXAMPLE 7.9

Figure 7.26 shows an isosceles triangle with an area of 24 cm^2 and one angle of 40°.

Calculate the lengths of the two equal sides.

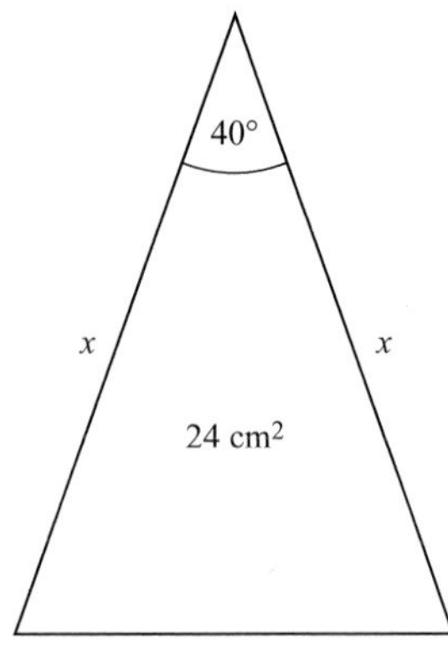

Figure 7.26

SOLUTION

Let the equal sides be of length x cm.

This gives

$$24 = \frac{1}{2} \times x \times x \times \sin 40°$$

$$\Rightarrow \quad x^2 = \frac{48}{\sin 40°}$$

$$\Rightarrow \quad x = 8.64 \text{ cm (3 s.f.)}$$

EXERCISE 7D

1 Find the area of each of the following triangles.

(i)

B
100°
4 cm
5 cm
C
A

(ii)

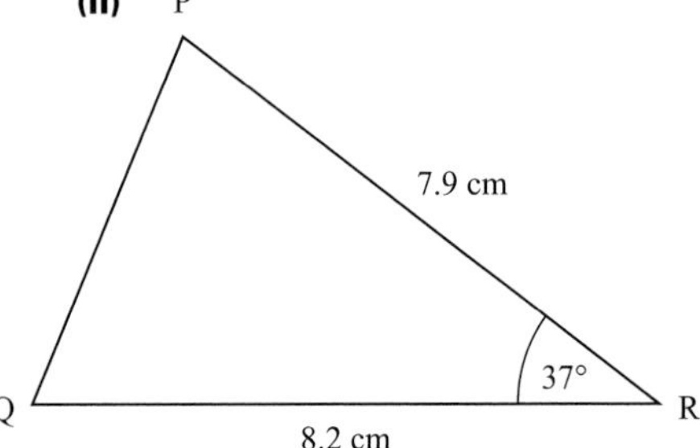

(iii)

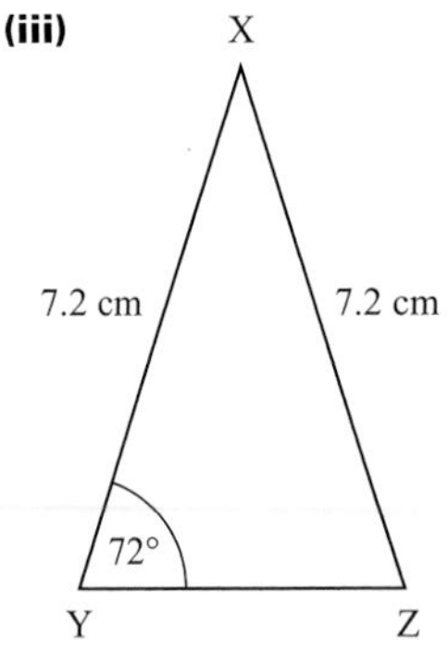

(iv)

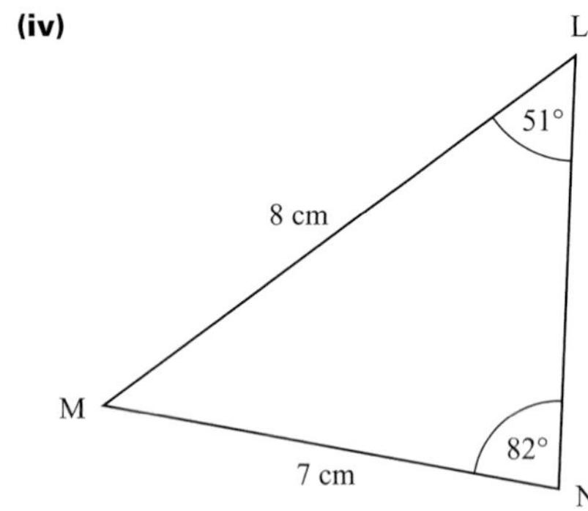

2 A regular hexagon is made up of six equilateral triangles. Find the area of a regular hexagon of side 7 cm.

3 A pyramid on a square base has four identical triangular faces which are isosceles triangles with equal sides 9 cm and equal angles 72°.

(i) Find the area of a triangular face.

(ii) Find the length of a side of the base.

(iii) Hence find the total surface area of the pyramid.

4 A tiler wishes to estimate the number of triangular tiles needed to tile an area of 10 m². The dimensions of each tile are shown in the diagram.

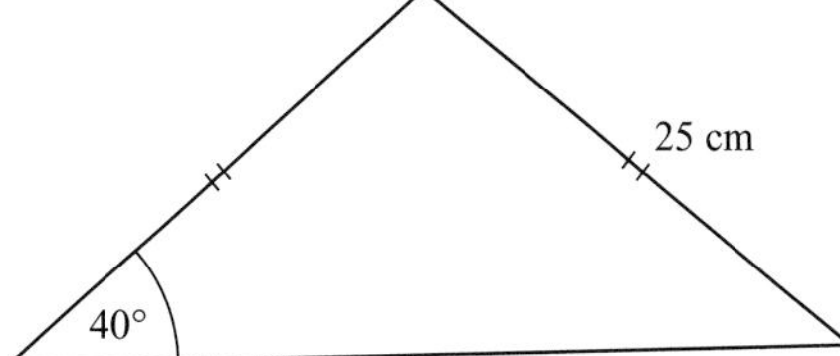

(i) Find the area of a tile.

The tiler then divides 10 m² by this area and rounds to the next whole number.

(ii) What result would this give?

(iii) Explain what is wrong with this estimate.

5 A regular tetrahedron has four sides, each of which is an equilateral triangle of side 10 cm. Find the total surface area of the tetrahedron.

You can use trigonometry to find sides and angles in non-right-angled triangles. This involves two important rules, the sine rule and the cosine rule.

The sine rule

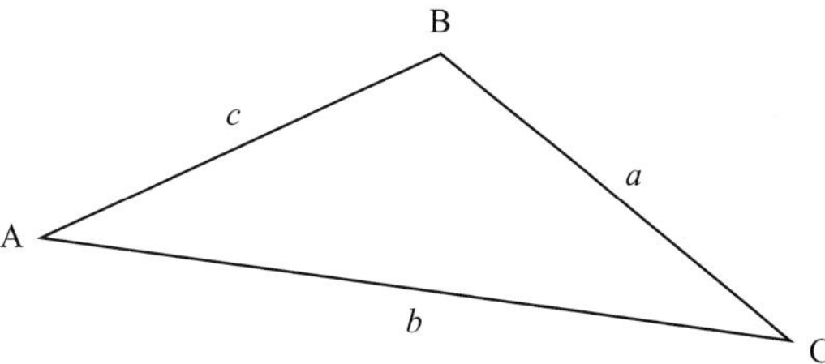

Figure 7.27

You have already seen that for any triangle ABC

$$\text{area} = \tfrac{1}{2}\, bc \sin \mathrm{A} = \tfrac{1}{2}\, ca \sin \mathrm{B} = \tfrac{1}{2}\, ab \sin \mathrm{C}$$

$$\Rightarrow \quad \frac{bc \sin \mathrm{A}}{abc} = \frac{ca \sin \mathrm{B}}{abc} = \frac{ab \sin \mathrm{C}}{abc}$$

$$\Rightarrow \quad \frac{\sin \mathrm{A}}{a} = \frac{\sin \mathrm{B}}{b} = \frac{\sin \mathrm{C}}{c}.$$

This is one form of the *sine rule* and is the version that is easier to use if you want to find an angle.

Inverting this gives

$$\frac{a}{\sin A} = \frac{b}{\sin B} = \frac{c}{\sin C}$$

which is better when you need to find a side.

Why is the inverted form of the sine rule better when you want to find a side?

EXAMPLE 7.10

Find the side BC in the triangle shown in figure 7.28.

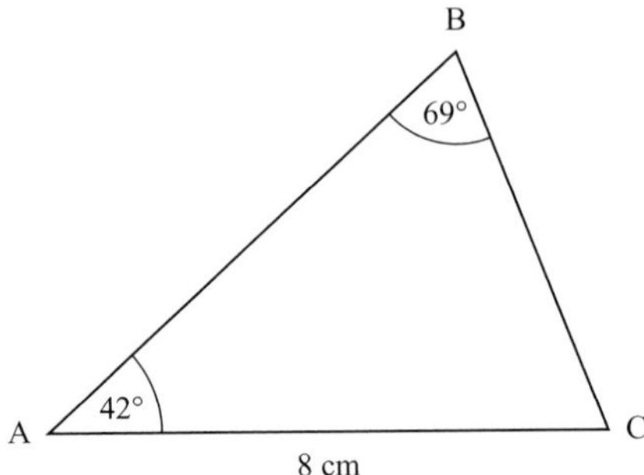

Figure 7.28

SOLUTION

Using the sine rule

$$\frac{a}{\sin A} = \frac{b}{\sin B} = \frac{c}{\sin C}$$

$$\Rightarrow \quad \frac{a}{\sin 42°} = \frac{8}{\sin 69°}$$

$$\Rightarrow \quad a = \frac{8\sin 42°}{\sin 69°}$$

$$= 5.733\,887\ldots$$

$\Rightarrow$ side BC = 5.7 cm (1 d.p.).

Do the calculation entirely on your calculator, and round only the final answer.

When using the sine rule to find an angle, you need to be careful because sometimes there are two possible answers, as in the next example.

EXAMPLE 7.11 Find the angle P in the triangle PQR, given that R = 32°, $r = 4$ cm and $p = 7$ cm.

SOLUTION

The sine rule for $\triangle$ PQR is

$$\frac{\sin P}{p} = \frac{\sin Q}{q} = \frac{\sin R}{r}$$

$$\Rightarrow \quad \frac{\sin P}{7} = \frac{\sin 32°}{4}$$

$$\Rightarrow \quad \sin P = 0.927\,358\,712$$

$$\Rightarrow \quad P = 68.0° \text{ (1 d.p.) or } P = (180° - 68.0°) = 112° \text{ (1 d.p.).}$$

Both solutions are possible as indicated in figure 7.29.

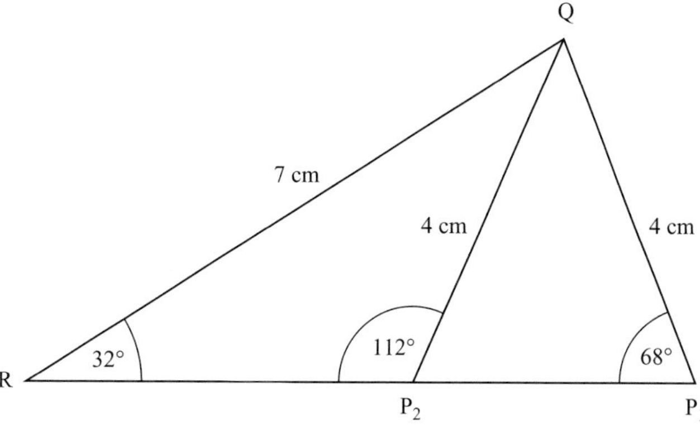

Figure 7.29

You should always check to see if there is a second solution, but sometimes only one solution is possible since the second would give an angle sum greater than 180°.

Figure 7.30 shows triangle XYZ with XY = 6 cm, XZ = 8 cm and $\angle$XYZ = 78°. What happens when you use the sine rule to calculate the remaining angles?

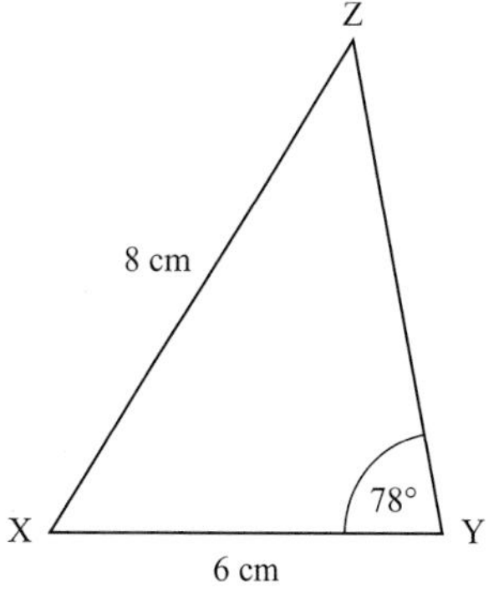

Figure 7.30

EXERCISE 7E

1 Find the length x in each of these triangles.

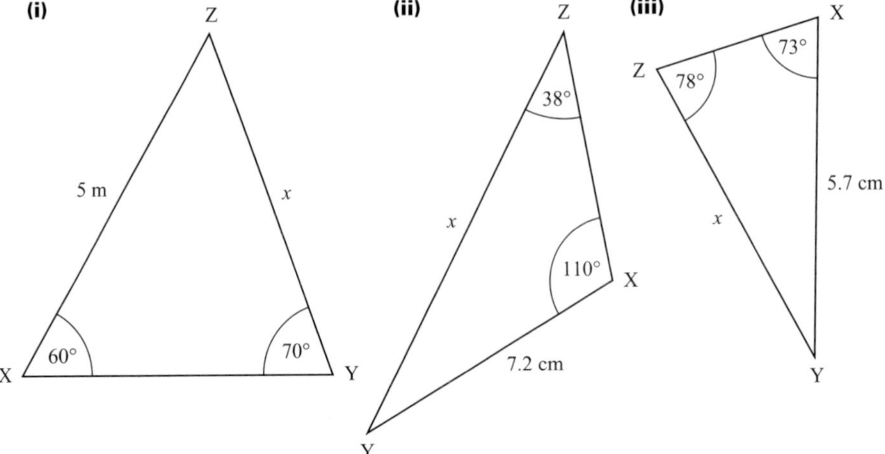

2 Find the angle θ in each of these triangles.

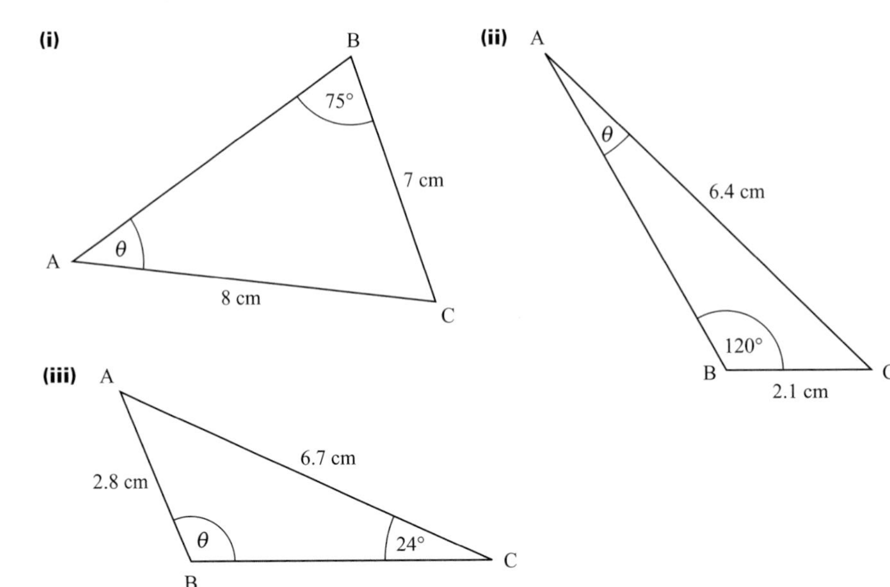

The cosine rule

Sometimes it is not possible to use the sine rule with the information you have about a triangle. For example, you know all three sides but none of the angles.

Like the sine rule, the cosine rule can be applied to any triangle, and again there are equivalent versions.

$$a^2 = b^2 + c^2 - 2bc\cos A$$

When you want to find a side.

$$\cos A = \frac{b^2 + c^2 - a^2}{2bc}$$

When you want to find an angle.

Proof

For the $\triangle$ABC, CD is the perpendicular from C to AB as shown in figure 7.31.

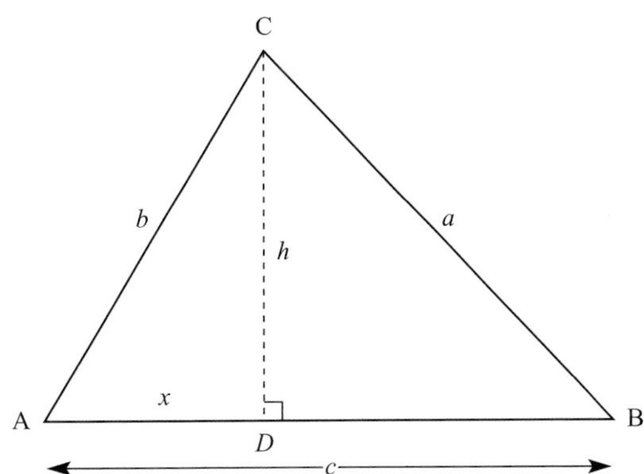

Figure 7.31

In $\triangle$ ACD

$$b^2 = x^2 + h^2 \qquad \text{(Pythagoras' theorem)} \qquad ①$$

$$\cos \text{A} = \frac{x}{b}, \text{ so } x = b\cos\text{A}. \qquad ②$$

In $\triangle$BCD

$$a^2 = (c - x)^2 + h^2 \qquad \text{(Pythagoras' theorem)}$$

$\Rightarrow \quad a^2 = c^2 - 2cx + x^2 + h^2$

$\Rightarrow \quad a^2 = c^2 - 2cx + b^2$ using ①

$\Rightarrow \quad a^2 = c^2 - 2cb\cos\text{A} + b^2$ using ②

$\Rightarrow \quad a^2 = b^2 + c^2 - 2bc\cos\text{A}$ as required.

Rearranging this gives

$$2bc\cos\text{A} = b^2 + c^2 - a^2$$

$$\Rightarrow \quad \cos\text{A} = \frac{b^2 + c^2 - a^2}{2bc}$$

which is the second form of the cosine rule.

NOTE

Starting with a perpendicular from a different vertex would give the following similar results.

$$b^2 = a^2 + c^2 - 2ac\cos\text{B}$$

$$\cos\text{B} = \frac{a^2 + c^2 - b^2}{2ac}$$

$$c^2 = a^2 + b^2 - 2ab\cos\text{C}$$

$$\cos\text{C} = \frac{a^2 + b^2 - c^2}{2ac}$$

EXAMPLE 7.12 Find the side AB in the triangle shown in figure 7.32.

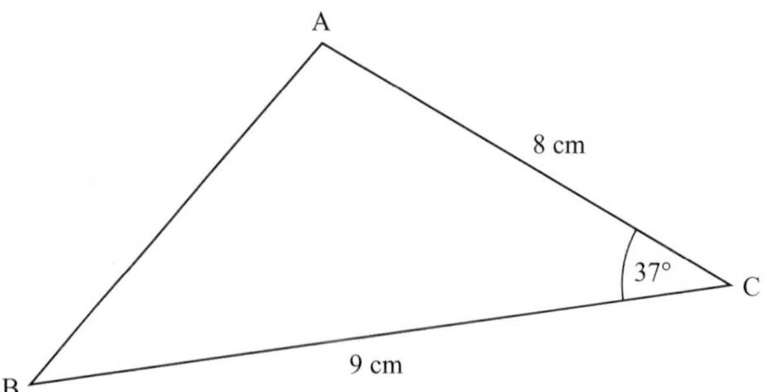

Figure 7.32

SOLUTION

$$c^2 = a^2 + b^2 - 2ab\cos C$$
$$= 9^2 + 8^2 - 2 \times 9 \times 8\cos 37°$$
$$AB = 5.5\,\text{cm (1 d.p.)}$$

EXAMPLE 7.13 Find the angle P in the triangle shown in figure 7.33.

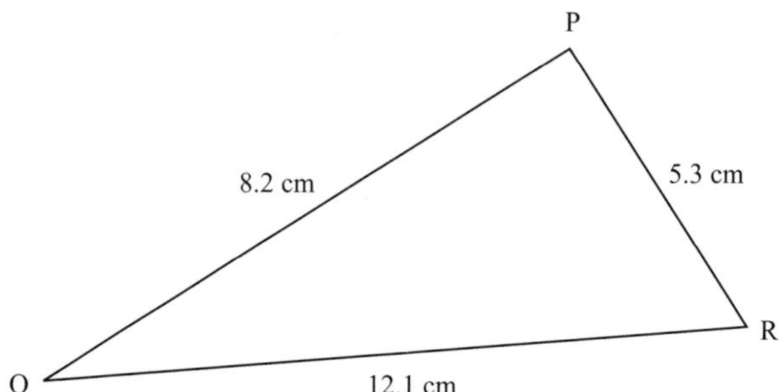

Figure 7.33

SOLUTION

The cosine rule for this triangle can be written as

$$\cos P = \frac{q^2 + r^2 - p^2}{2qr}$$
$$= \frac{5.3^2 + 8.2^2 - 12.1^2}{2 \times 5.3 \times 8.2}$$
$$P = 126.0° \text{ (1 d.p.)}$$

EXERCISE 7F

1 Find the length x in each of these triangles.

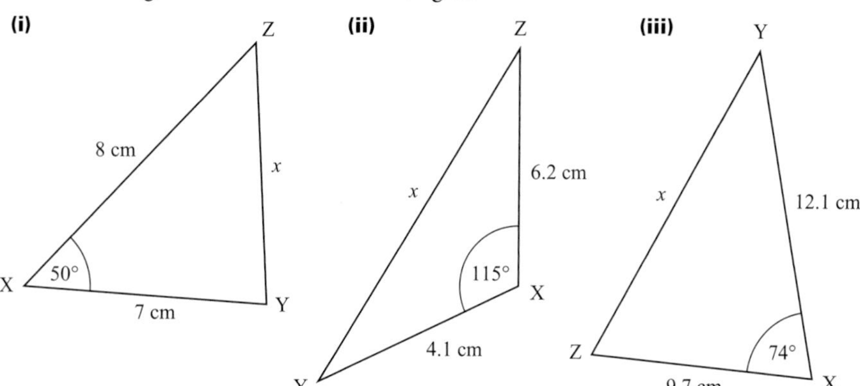

2 Find the angle θ in each of the following triangles.

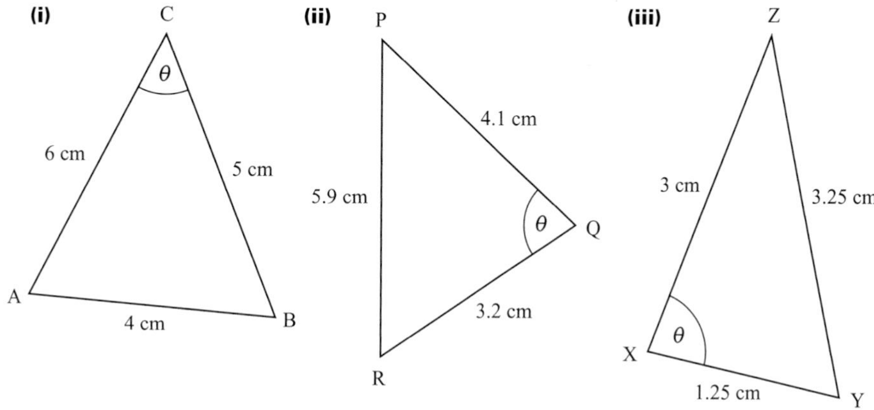

3 The diagonals of a parallelogram have lengths of 12 cm and 18 cm and the angle between them is 72°. Find the lengths of the sides of the parallelogram.

4 The diagram shows a quadrilateral ABCD with AB = 8 cm, BC = 6 cm, CD = 7 cm, DA = 5 cm and $\angle$ABC = 90°. Calculate

(i) AC

(ii) $\angle$ADC.

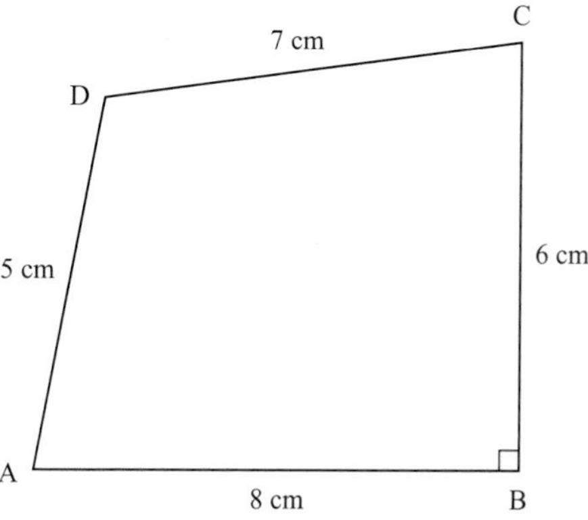

5 The diagram shows two circles. One has centre A and a radius of 8 cm. The other has centre B and a radius of 10 cm. AB = 12 cm and the circles intersect at P and Q. Calculate ∠PAB.

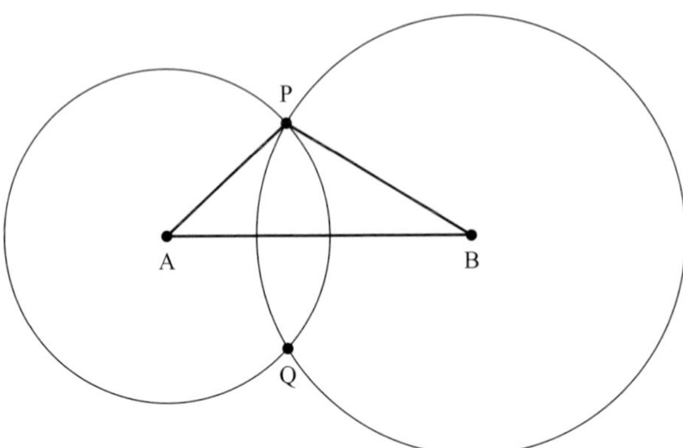

Using the sine and cosine rules together

Sometimes you need to use both the sine and cosine rules in the same problem, as in the next example.

EXAMPLE 7.14

Figure 7.34 shows the positions of three towns, Aldbury, Bentham and Chorton. Bentham is 8 km from Aldbury on a bearing of 037° and Chorton is 9 km from Bentham on a bearing of 150°. Find

(i) the angle ABC
(ii) the distance of Chorton from Aldbury (to 0.1 km)
(iii) the bearing of Chorton from Aldbury (to 1°).

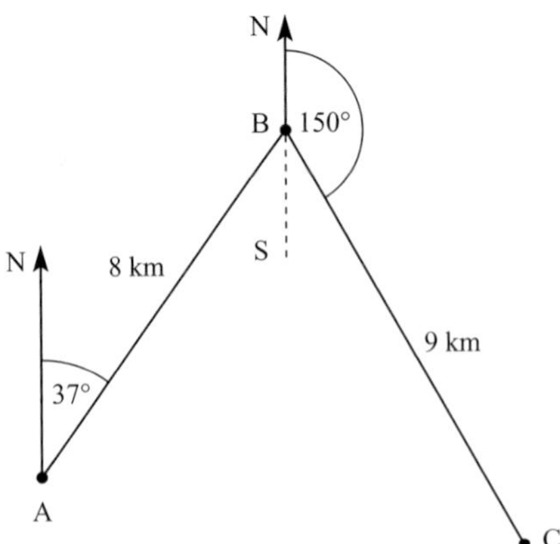

Figure 7.34

SOLUTION

(i) $\angle ABS = 37°$ (alternate angles)
and $\angle SBC = 30°$ (adjacent angles)
so $\angle ABC = 67°$

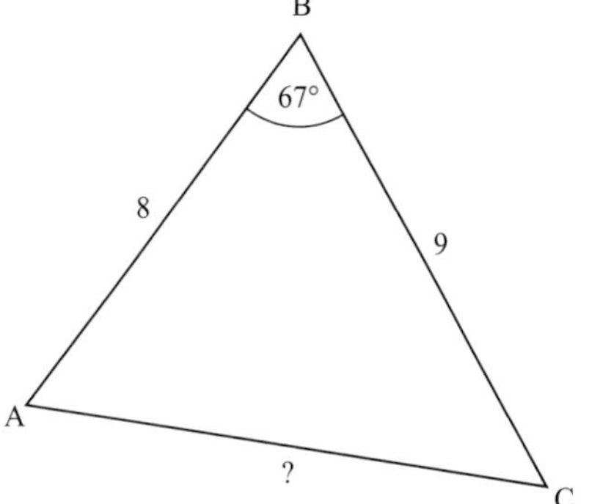

Figure 7.35

(ii) Using the cosine rule

$$b^2 = a^2 + c^2 - 2ac\cos B$$
$$= 9^2 + 8^2 - 2 \times 9 \times 8 \cos 67$$
$$= 88.7347\ldots$$
$$b = 9.4199\ldots$$

Store this result in your calculator memory so that you can use it later.

Chorton is 9.4 km from Aldbury (1 d.p.).

(iii) Using the sine rule

$$\frac{\sin A}{a} = \frac{\sin B}{b}$$
$$\frac{\sin A}{9} = \frac{\sin 67°}{9.4199\ldots}$$
$$\sin A = 0.879\,47\ldots$$
$$A = 61.57\ldots°$$

To find the bearing of Chorton from Aldbury, add 37° to give 099°.

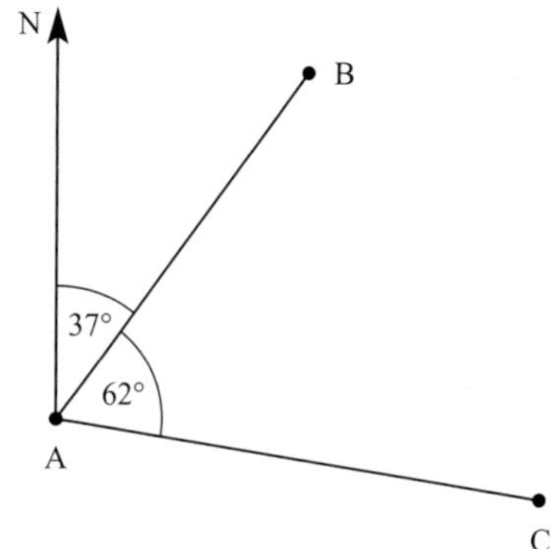

Figure 7.36

? The other value of A that gives sin A = 0.879 47… is 118.42…°
Why does this not give an alternative solution to this problem?

EXAMPLE 7.15

A triangular plot of land has sides of length 70 m, 80 m and 95 m.
Find its area in hectares. (1 hectare (ha) is 1000 m^2)

SOLUTION

First draw a sketch and label the sides.

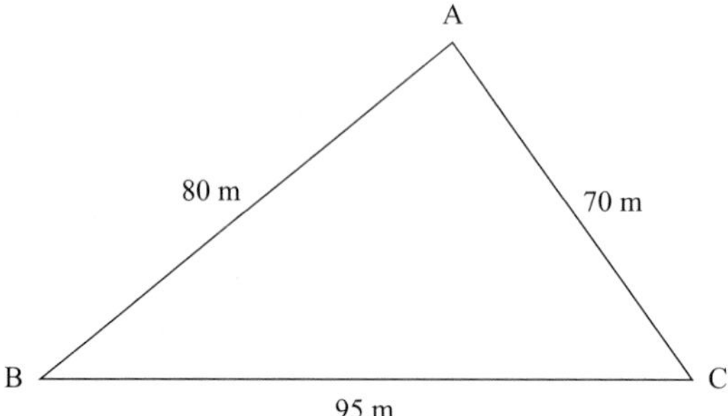

Figure 7.37

You can now see that the first step is to find one of the angles, and this will need the cosine rule.

❓ How do you decide that it is the cosine rule you want to use?

Using $\cos A = \dfrac{b^2 + c^2 - a^2}{2bc}$

gives $\cos A = \dfrac{70^2 + 80^2 - 95^2}{2 \times 70 \times 80} = 0.2031\ldots$

$\Rightarrow \qquad A = 78.28\ldots°.$

Now using area $= \frac{1}{2}bc\sin A$

$\Rightarrow \qquad \text{area} = \frac{1}{2} \times 70 \times 80 \sin 78.28\ldots°$

$\qquad\qquad = 2741.625\ldots \text{ m}^2$

$\Rightarrow \qquad \text{area} = 2.74 \text{ ha} \qquad (2 \text{ d.p.}).$

EXERCISE 7G

1 The hands of a clock have lengths 6 cm and 8 cm.
Find the distance between the tips of the hands at 8 pm.

2 From a lighthouse L, a ship A is 4 km away on a bearing of 340° and a ship B is 5 km away on a bearing of 065°.
Find the distance AB.

3 When I am at a point X, the angle of elevation of the top of a tree T is 27°, but if I walk 20 m towards the tree, to point Y, the angle of elevation is then 47°.

(i) Find the distance TY.

(ii) Find the height of the tree.

4 Two adjacent sides of a parallelogram are of lengths 9.3 cm and 7.2 cm, and the shorter diagonal is of length 8.1 cm.

(i) Find the angles of the parallelogram.

(ii) Find the length of the other diagonal of the parallelogram.

5 A yacht sets off from A and sails for 5 km on a bearing of 067° to a point B so that it can clear the headland before it turns onto a bearing of 146°. It then stays on that course for 8 km until it reaches a point C.

(i) Find the distance AC.

(ii) Find the bearing of C from A.

6 Two ships leave the docks, D, at the same time. *Princess Pearl*, P, sails on a bearing of 160° at a speed of 18 km/h, and *Regal Rose*, R, sails on a bearing of 105°. After two hours the angle DRP is 80°.

(i) Find the distance between the ships at this time.

(ii) Find the speed of the *Regal Rose.*

7 The diagram represents a simplified drawing of the timber cross-section of a roof. Find the lengths of the struts BD and EG.

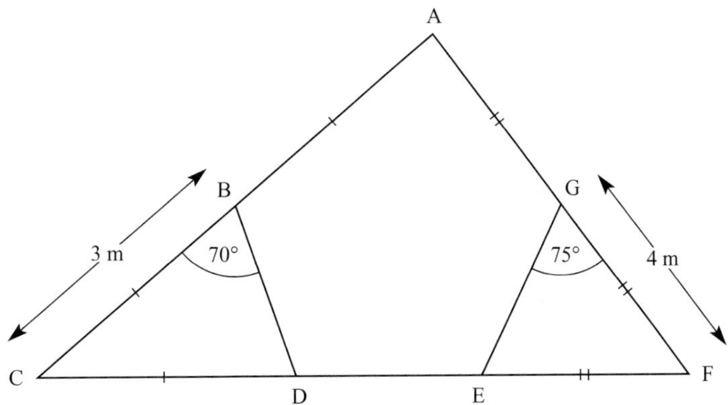

8 Sam and Aziz cycle home from school.
Sam cycles due east for 4 km, and Aziz cycles due south for 3 km and then for 2 km on a bearing of 125°.
How far apart are their homes?

KEY POINTS

1 For an angle θ in a right-angled triangle

$$\sin\theta = \frac{\text{opposite}}{\text{hypotenuse}} \qquad \cos\theta = \frac{\text{adjacent}}{\text{hypotenuse}} \qquad \tan\theta = \frac{\text{opposite}}{\text{adjacent}}$$

2 The point (x, y) at angle θ on the unit circle with centre $(0, 0)$ has co-ordinates $(\cos\theta, \sin\theta)$ for all θ, i.e. $\cos\theta = x$ and $\sin\theta = y$.

This also gives $\tan\theta = \dfrac{y}{x}$.

3 The graphs of $\sin\theta$, $\cos\theta$ and $\tan\theta$ are as shown below.

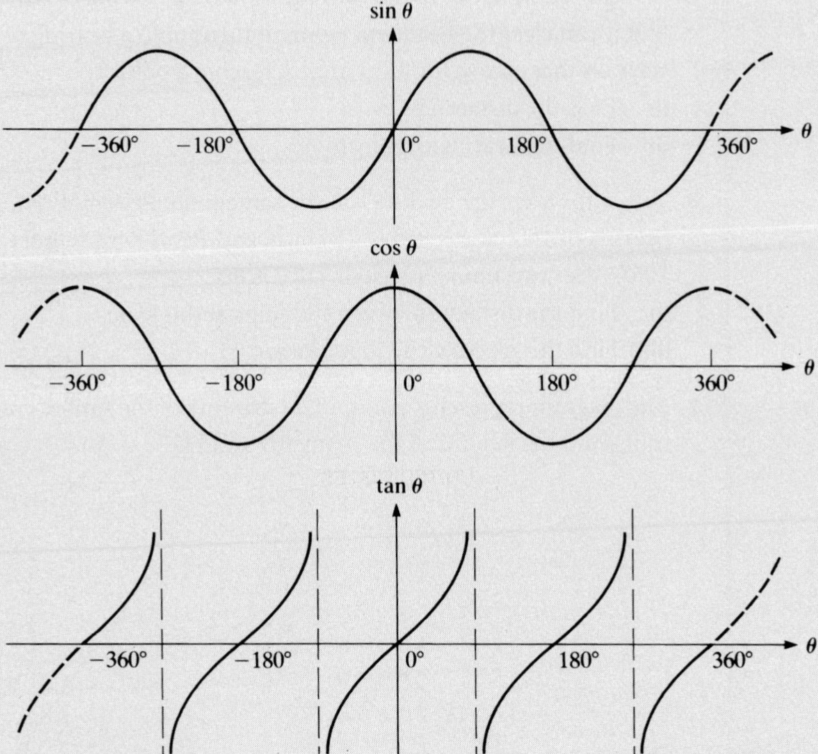

4 $\tan\theta = \dfrac{\sin\theta}{\cos\theta}$

5 $\sin^2\theta + \cos^2\theta = 1$

6 For a triangle ABC

area $= \frac{1}{2}bc\sin A$

$$\left.\begin{aligned} &\frac{a}{\sin A} = \frac{b}{\sin B} = \frac{c}{\sin C} \\ &\frac{\sin A}{a} = \frac{\sin B}{b} = \frac{\sin C}{c} \end{aligned}\right\}$$

the sine rule

$$\left.\begin{aligned} &a^2 = b^2 + c^2 - 2bc\cos A \\ &\cos A = \frac{b^2 + c^2 - a^2}{2bc} \end{aligned}\right\}$$

8 Trigonometry II – applications

What is it that breathes fire into the equations and makes a universe for them to describe?

Stephen W Hawking

Figure 8.1

An aircraft flying between two places at the same latitude doesn't usually follow a route along the line of latitude. Why?

Working in three dimensions

When you are solving three-dimensional problems it is extremely important to draw good diagrams. There are two types:

- representations of three-dimensional objects
- true shape diagrams of two-dimensional sections within a three-dimensional object.

Representations of three-dimensional objects

Figures 8.2 and 8.3 illustrate ways in which you can draw a clear diagram.

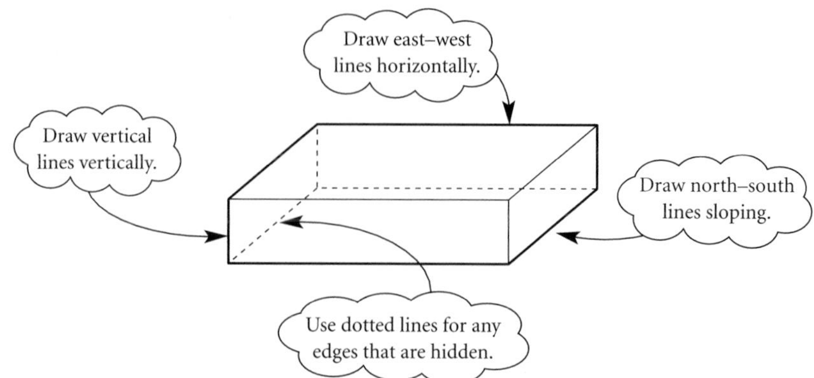

Figure 8.2

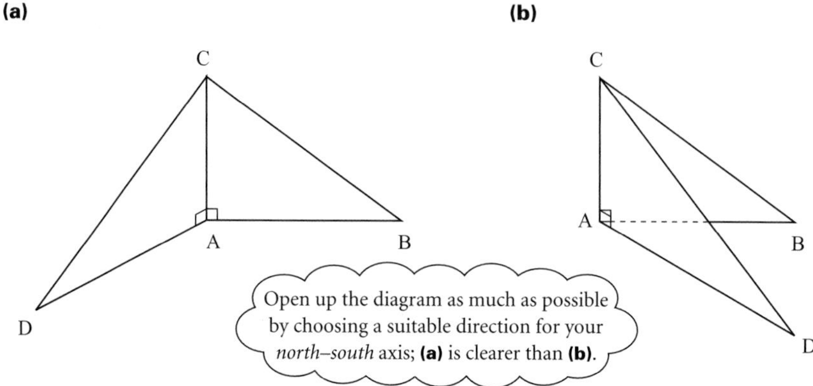

Figure 8.3

True shape diagrams

In a two-dimensional representation of a three-dimensional object, right angles do not always appear to be 90°, so draw as many true shape diagrams as necessary.

For example, if you need to do calculations on the triangular cross-section BCD in figure 8.4(a), you should draw the triangle so that the right angle really does look 90° as in figure 8.4(b).

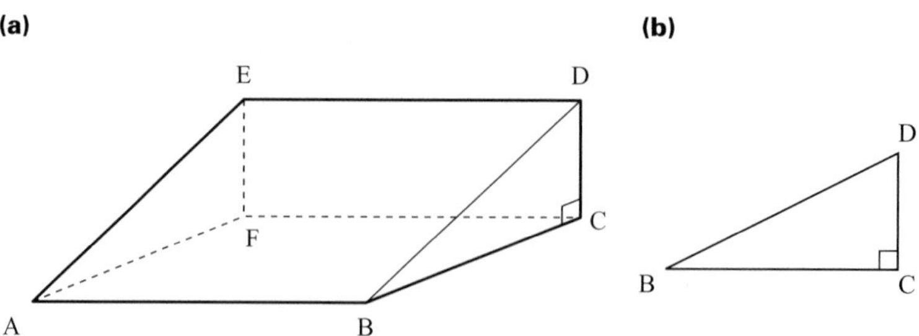

Figure 8.4

Lines and planes in three dimensions

A *plane* is a flat surface (not necessarily horizontal).

A *line of greatest slope* of a sloping plane is a line of greatest gradient, i.e. the line that a ball would follow if allowed to roll down it. This is shown in figure 8.5.

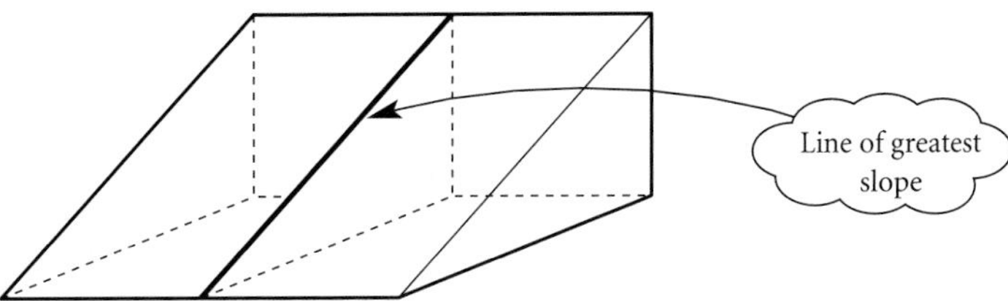

Figure 8.5

? Give an example of a sloping plane from everyday life.

In three-dimensional problems you need to be aware of the relationships between lines and planes.

Two lines

In two dimensions, two lines either meet (when extended if necessary), or they are parallel.

In three dimensions, there is a third option: they are *skew*, as in figure 8.6.

Figure 8.6

Two planes

In three dimensions there are two options.

1 The two planes are *parallel*. Opposite walls of a room are *parallel*.

2 The two planes meet *in a line*. The ceiling meets each wall of a room *in a line*. An open gate and a wall meet *in a line*.

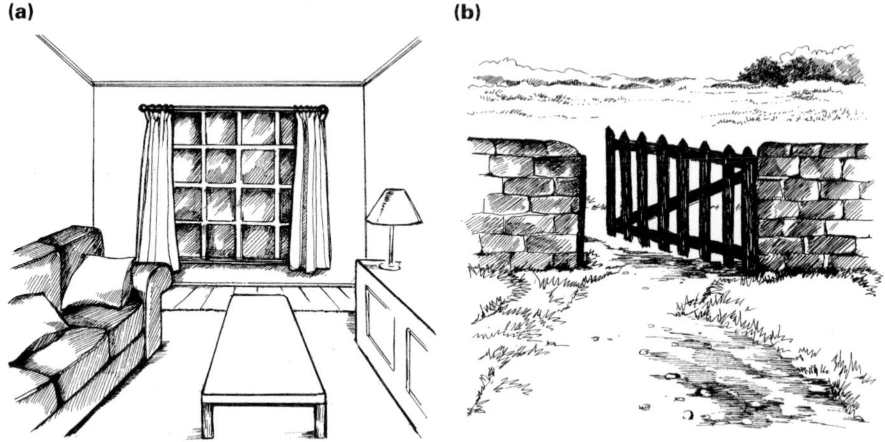

Figure 8.7

Give other examples of these cases.

A line and a plane

In three dimensions there are three options.

1 The line and the plane are *parallel*. A curtain rail is *parallel* to the floor.

2 The line meets the plane at a *single point*. When you are writing, your pen meets the paper at a *single point*.

3 The line *lies in* the plane. When you put your pen down, your pen *lies in* the plane of the paper.

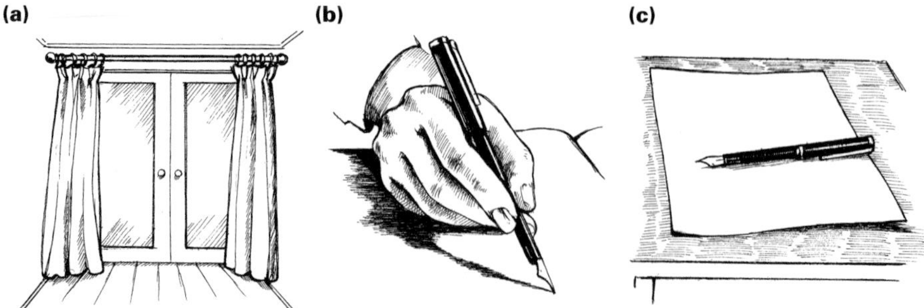

Figure 8.8

For option 2, you can calculate the angle between the line and the plane.

When a book is opened and positioned on a table, as in figure 8.9, the spine is perpendicular to the table. The bottom edge of each page defines a line on the table that is perpendicular to the spine.

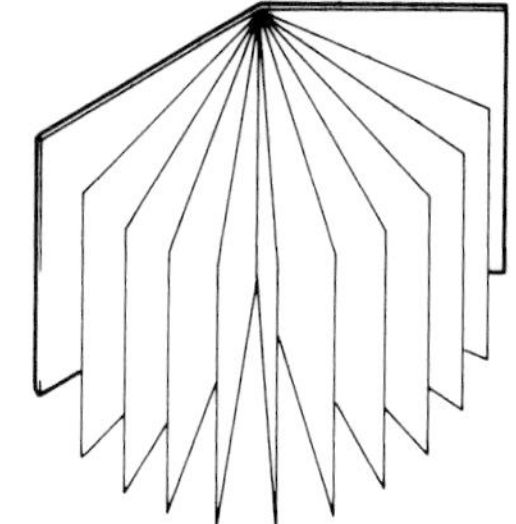

Figure 8.9

EXAMPLE 8.1

Figure 8.10 shows a wedge ABCDEF with AB = 8 cm, BC = 6 cm and CD = 2 cm. The angle BCD is 90°.

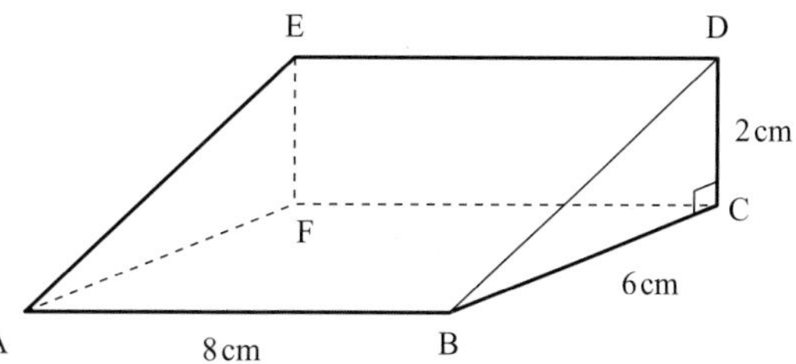

Figure 8.10

Find

(i) AC **(ii)** AD **(iii)** ∠DAC **(iv)** ∠DBC

SOLUTION

(i) From figure 8.11a

$$AC^2 = 8^2 + 6^2 \quad \text{(Pythagoras)}$$

$$\Rightarrow \quad AC = 10\,\text{cm}$$

(ii) From figure 8.11b

$$AD^2 = AC^2 + 2^2 \quad \text{(Pythagoras)}$$

$$\Rightarrow \quad AD = 10.2\,\text{cm} \quad \text{(1 d.p.)}$$

(iii) From figure 8.11b

$$\tan \angle DAC = \frac{2}{10}$$

$$\Rightarrow \quad \angle DAC = 11.3° \quad \text{(1 d.p.)}$$

(iv) From figure 8.11c

$$\tan \angle DBC = \frac{2}{6}$$

$$\Rightarrow \quad \angle DBC = 18.4° \quad \text{(1 d.p.)}$$

(a)

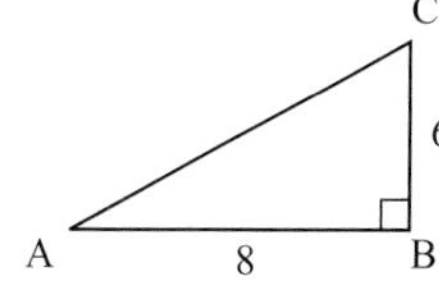

(b)

D
2
A
10
C

(c)

D
2
B
6
C

Figure 8.11

EXAMPLE 8.2

Figure 8.12 shows a straight level road AB, 400 m long. A vertical radio mast XY stands some distance from the road and the bottom of the mast, X, is on the same level as the road. The angle of elevation of Y from A is 30°, ∠XAB = 25° and ∠AXB = 90°. Calculate

(i) the distance AX

(ii) the height of the mast

(iii) the distance of X from the road.

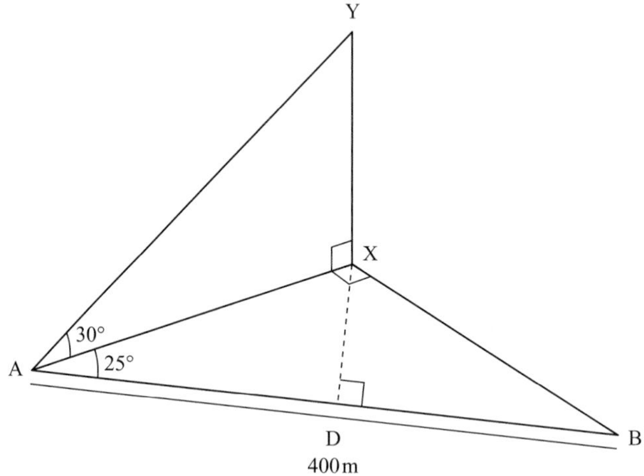

Figure 8.12

SOLUTION

(i) From figure 8.13(a)

$$\frac{AX}{400} = \cos 25°$$

$\Rightarrow$ $AX = 362.523\ldots$

$\Rightarrow$ The distance AX = 363 m.

(ii) From figure 8.13(b)

$$\frac{XY}{362.523\ldots} = \tan 30°$$

$\Rightarrow$ $XY = 209.302\ldots$

$\Rightarrow$ The height of the mast XY = 209 m.

(iii) From figure 8.13(c)

$$\frac{DX}{362.523\ldots} = \sin 25°$$

$\Rightarrow$ $DX = 153.208\ldots$

$\Rightarrow$ The distance of X from the road = 153 m.

(a)

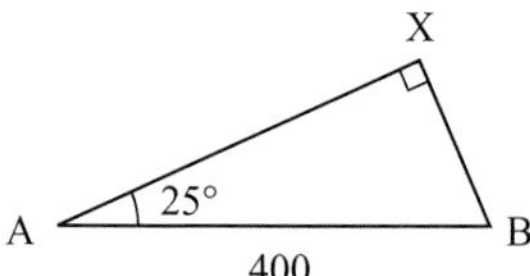

(b)

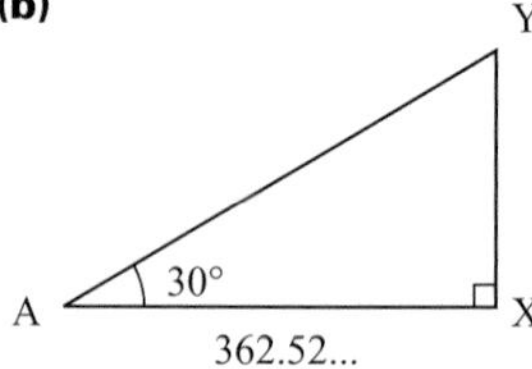

(c)

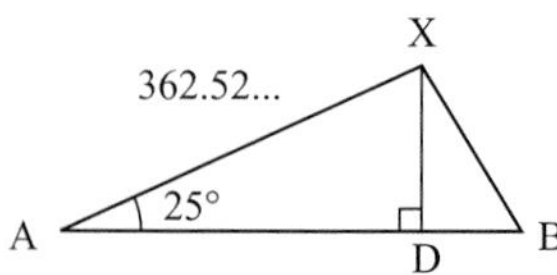

Figure 8.13

EXERCISE 8A

1 The cube ABCDEFGH shown in the diagram has sides of length 10 cm.

Find

(i) the length AC

(ii) the length AG

(iii) the angle GAC

(iv) the angle between GA and EC.

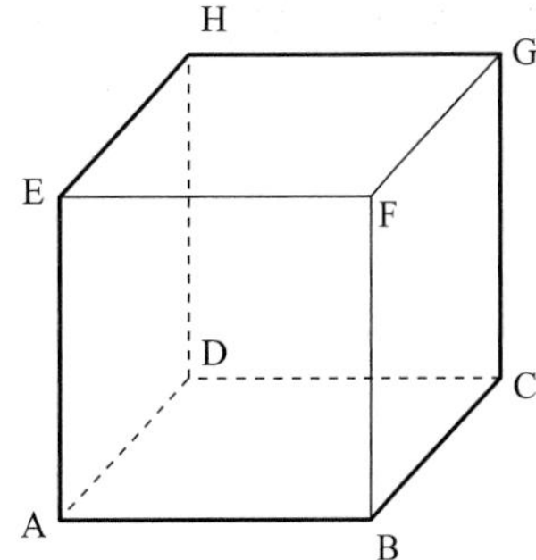

2 The diagram represents a pyramid ABCD.
ABC is an isosceles triangle with AB = AC = 5 cm and BC = 8 cm.
BCD is an isosceles triangle with BD = CD = 13 cm.
D is vertically above A and ∠BAD = ∠CAD = 90°.
M is the midpoint of BC.

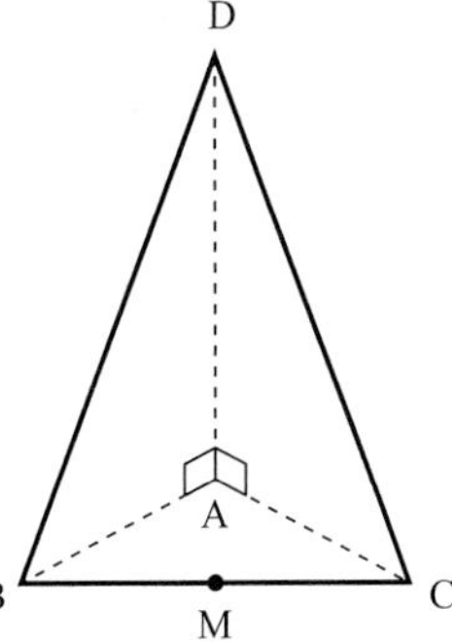

Calculate

(i) the length AM

(ii) the angle BCD

(iii) the angle DMA.

3 The diagram shows a wedge ABCDEF which has been made to hold a door open.
AB = 5 cm, BC = 12 cm and FC = 4 cm.
Find

(i) the angle FBC

(ii) the length AC

(iii) the angle FAC.

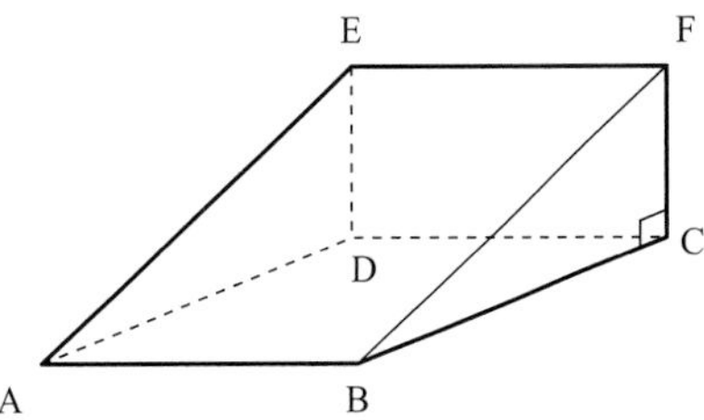

There is a gap of 2 cm between the door and the floor.

(iv) How far along BF will the base of the door meet the wedge?

4 A, B and C are points on a horizontal plane.
A is 75 m from C on a bearing of 210° and the bearing of B from C is 120°. The bearing of B from A is 075°.
From A the angle of elevation of the top T of a vertical tower at C is 42°.
Find
(i) the distance BC
(ii) the height of the tower
(iii) the angle of elevation of T from B.

5 C is the foot of a vertical tower CT 28 m high.
A and B are points in the same horizontal plane as C and CA = CB.
P is the point on AB that is nearest to C.
The angle of elevation of the top of the tower from P is 40° and $\angle ACB = 120°$.
Calculate
(i) the length CP
(ii) the length CB
(iii) the length AB
(iv) the angle of elevation of the top of the tower from B.

6 The waste-paper basket shown in the diagram has a top ABCD that is a square of side 30 cm and a base PQRS that is a square of side 20 cm.
The line joining the centres of the top and base is perpendicular to both and is 40 cm long.
Find
(i) the length PR
(ii) the length AC
(iii) the length AP.

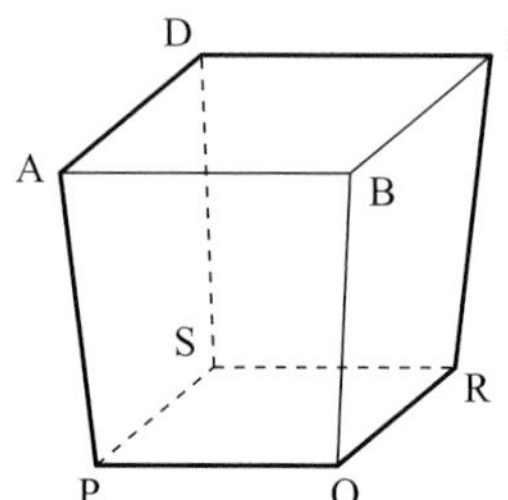

7 In Egypt, pyramids were used as burial chambers for the Pharaohs.
The largest of these, shown in the diagram and built about 2500 BC for Cheops, is 146 m high and has a square base of side 231 m.
E is the centre of the base and VE = 146 m.

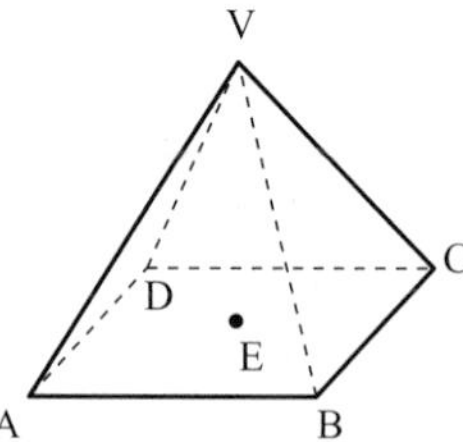

Find
(i) the angle VAE
(ii) the length VA
(iii) the length VM where M is the midpoint of AB.
The pyramid is to be treated against erosion by applying a sealant, 10 litres of which covers 48 m^2.
(iv) How many litres would be needed to protect the whole pyramid?

8 The tent shown in the diagram has a base that is 2.2 m wide and 3.6 m long. The ends are isosceles triangles, inclined at an angle of 80° to the base. $\angle AEB = \angle DFC = 70°$ and M is the midpoint of AB.

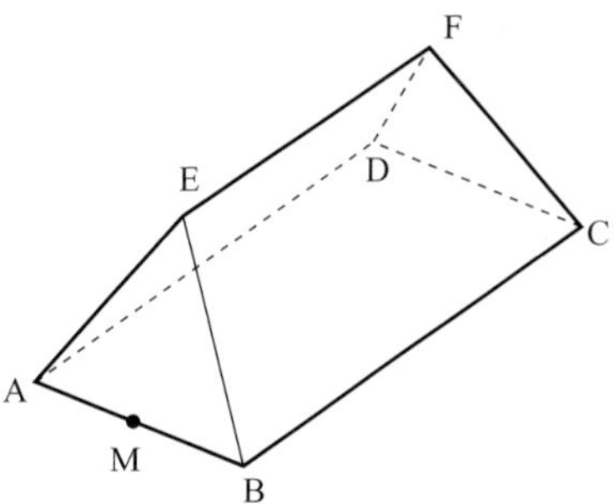

Find

(i) the length of EM

(ii) the height of EF above the base

(iii) the length of EF.

9 A, B and C are three points on the ground. AB = 200 km, BC = 120 km and $\angle ABC = 90°$.
A satellite is observed from A and C at the instant when it passes directly over B. The angle of elevation of the satellite from A is 60° and the surface of the Earth may be treated as flat.
Calculate

(i) the height of the satellite to the nearest km

(ii) the angle of elevation of the satellite from C to the nearest degree.

10 A new perfume is to be packaged in a box that is in the shape of a regular tetrahedron VABC of side 6 cm standing on a triangular prism ABCDEF as shown in the diagram.
The height of the prism is 12 cm.
M is the midpoint of BC.
Find

(i) the length AM

(ii) the length VM

(iii) the angle VAM

(iv) the total height of the box.

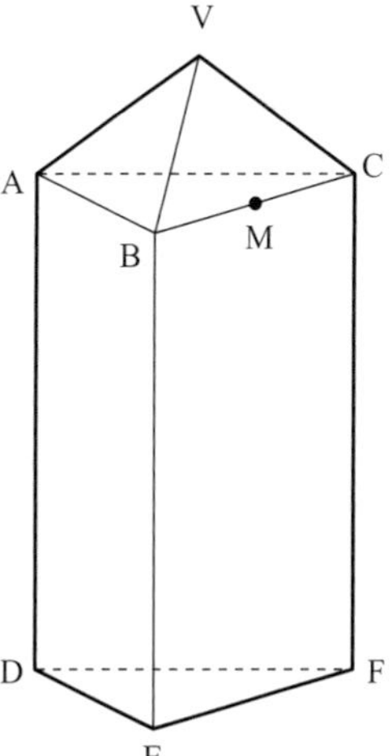

INVESTIGATION

Points on the surface of the Earth are described by giving their latitude and longitude.

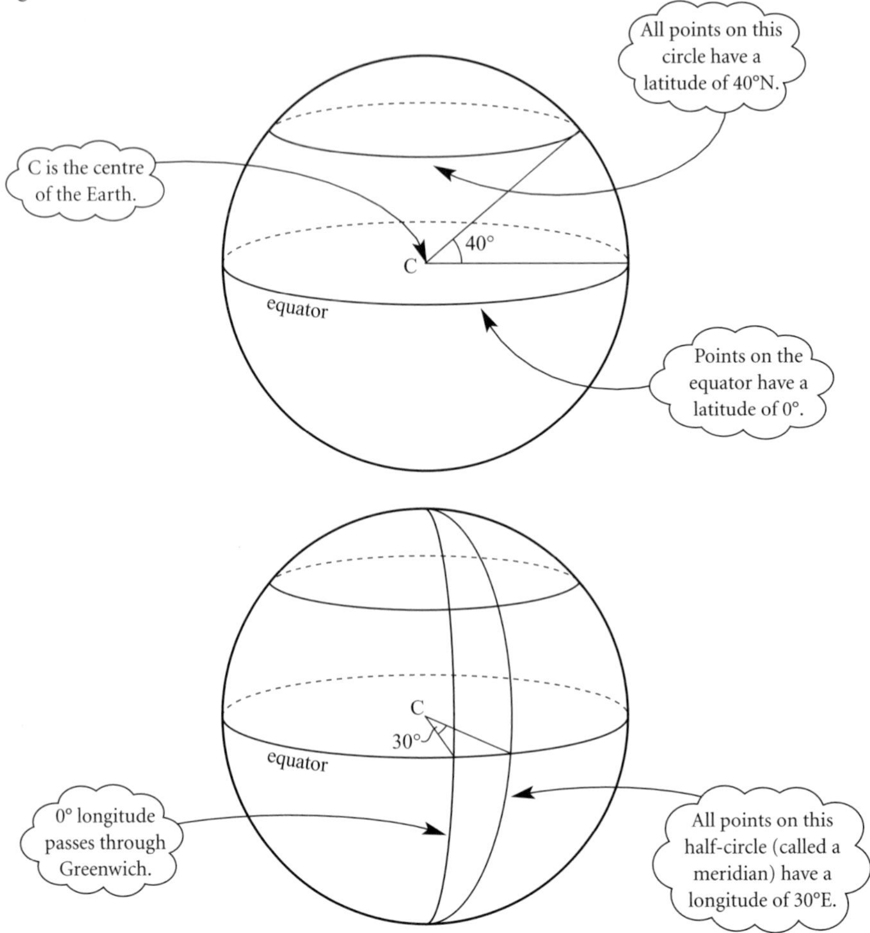

Figure 8.14

Similarly, circles of latitude south of the equator are denoted by S, and half-circles to the west of the Greenwich meridian are denoted by W.

At the beginning of this chapter you were asked about flights between London and Vancouver.

Taking the radius of the Earth as 6370 km, find:

(i) the radius of the circle of latitude 50°N

(ii) the distance along this circle from London (0°, 50°N) to Vancouver (123°W, 50°N)

(iii) the straight line distance LV (through the Earth).

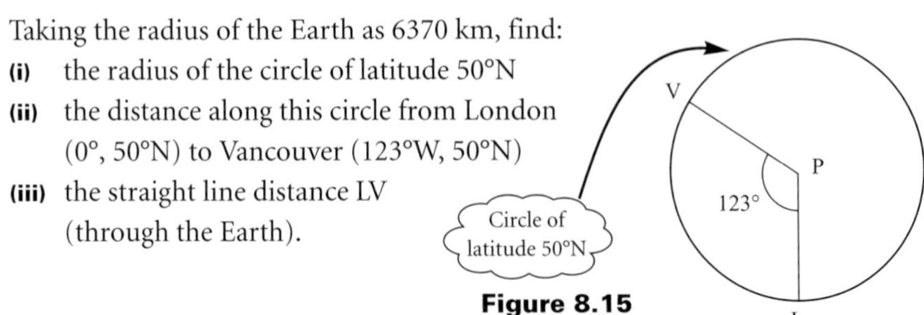

Figure 8.15

Now look at the circle on the surface of the Earth that passes through L and V and has its centre at C.

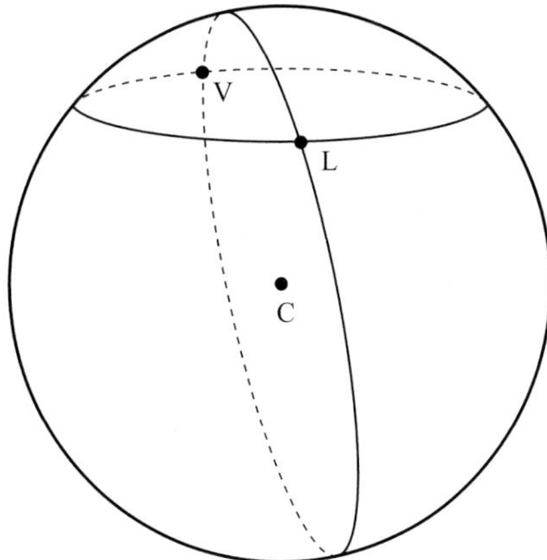

Figure 8.16

(iv) Write down the radius of this circle.
(v) Use triangle VCL to find $\angle VCL$.
(vi) Find the arc length VL on this circle.

? This arc represents the most economical flight path. Why?

KEY POINTS

1 In three dimensions:
- a plane is a flat surface
- the line of greatest slope of a plane is the steepest line contained in the plane
- two lines may meet, be parallel or be skew
- two planes are either parallel or meet in a line
- a line and a plane may be parallel, meet in a single point or the line may lie in the plane

2 When solving three-dimensional problems always draw a clear diagram where:
- vertical lines are drawn vertically
- east–west lines are drawn horizontally
- north–south lines are drawn sloping
- edges that are hidden are drawn as dotted lines.

SECTION 4
Calculus

Calculus I – differentiation

I do not know what I may appear to the world; but to myself I seem to have been only like a boy playing on the seashore, and diverting myself in now and then finding a smoother pebble or a prettier shell than ordinary, whilst the great ocean of truth lay all undiscovered before me.

Isaac Newton (1642 – 1727)

Figure 9.1

Imagine that you are skiing the mogul field in figure 9.1. As you travel, the direction of your skis is always changing. The direction in which you are travelling at any moment is the gradient of your path at that point. This is represented by the direction of your skis and is the gradient of the tangent to the curve at that point.

The gradient of a curve

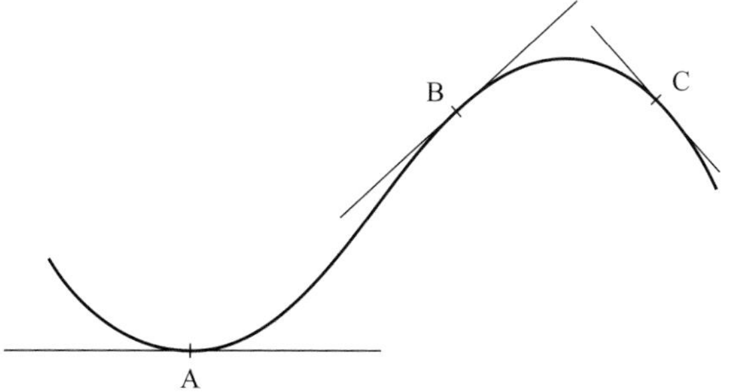

Figure 9.2

In figure 9.2 the curve has a zero gradient at A, a positive gradient at B and a negative gradient at C.

One way of finding these gradients is to draw the tangents and use two points on each one to calculate its gradient. This is time-consuming and the results depend on the accuracy of your drawing and measuring. If you know the equation of the curve, then *differentiation* provides another method of calculating the gradient.

Finding the gradient of a curve

Instead of trying to draw an accurate tangent, this method starts by calculating the gradients of chords PQ_1, PQ_2,.... As the different positions of Q get closer to P, the values of the gradient of PQ get closer to the gradient of the tangent at P. The first few positions of Q are shown in figure 9.3.

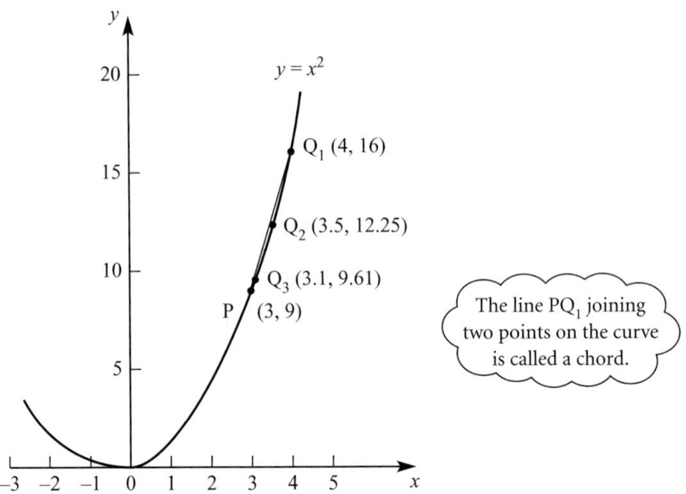

Figure 9.3

For P at (3, 9)

chord	co-ordinate of Q	gradient of PQ
PQ_1	(4, 16)	$\frac{16-9}{4-3}=7$
PQ_2	(3.5, 12.25)	$\frac{12.25-9}{3.5-3}=6.5$
PQ_3	(3.1, 9.61)	$\frac{9.61-9}{3.1-3}=6.1$
PQ_4	(3.01, 9.0601)	$\frac{9.0601-9}{3.01-3}=6.01$
PQ_5	(3.001, 9.006001)	$\frac{9.006001-9}{3.001-3}=6.001$

In this process the gradient of the chord PQ gets closer and closer to that of the tangent, and hence the gradient of the curve at (3, 9).

Look at the sequence formed by the gradients of the chords.

$$7, 6.5, 6.1, 6.01, 6.001, \ldots$$

It looks as though this sequence is converging to 6.

? Do you think the limit will still be 6 if the points Q are positioned on the other side of P?

ACTIVITY 9.1

Take points R_1 to R_5 on the curve $y = x^2$ with x co-ordinates 2, 2.5, 2.9, 2.99, and 2.999 respectively and find the gradients of the chords joining each of these points to P(3, 9).

The calculations above show that the gradient of the curve $y = x^2$ at (3, 9) seems to be 6 or about 6 but do not provide conclusive proof of its value. To do that you need to apply the method in more general terms.

Take the point P(3, 9) and another point Q close to (3, 9) on the curve $y = x^2$. Let the x co-ordinate of Q be $(3 + h)$ where h is small. Since $y = x^2$ at all points on the curve, the y co-ordinate of Q will be $(3 + h)^2$.

Figure 9.4 shows Q in a position where h is positive. Negative values of h would put Q to the left of P.

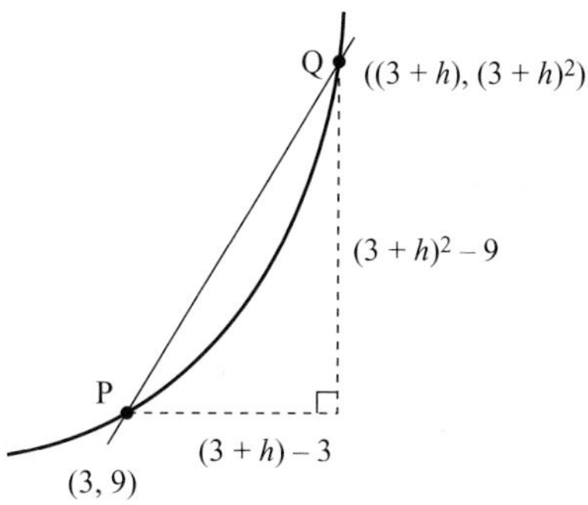

Figure 9.4

From figure 9.4, the gradient of PQ is $\dfrac{(3+h)^2-9}{h}$

$$= \frac{9 + 6h + h^2 - 9}{h}$$

$$= \frac{6h + h^2}{h}$$

$$= \frac{h(6+h)}{h}$$

$$= 6 + h.$$

For example, when $h = 0.001$, the gradient of PQ is 6.001 and when $h = -0.001$, the gradient of PQ is 5.999. The gradient of the tangent at P is between these two values. Similarly the gradient of the tangent at P would be between $6 - h$ and $6 + h$ for all small non-zero values of h.

For this to be true, the gradient of the tangent at (3, 9) must be *exactly* 6.

In this case, 6 was the *limit* of the gradient values, whether you approached P from the right or the left.

ACTIVITY 9.2

Using a similar method, find the gradient of the tangent to the curve at

(i) (2, 4)
(ii) (–1, 1)
(iii) (–3, 9)

What do you notice?

The gradient function

The work so far has involved finding the gradient of the curve $y = x^2$ at just one particular point. It would be very tedious if you had to do this every time and so instead you can consider a general point (x, y) and then substitute the value(s) of x and/or y corresponding to the point(s) of interest.

EXAMPLE 9.1

Find the gradient of the curve $y = x^3$ at the general point (x, y).

SOLUTION

Let P have the general value x as its x co-ordinate, so P is the point (x, x^3) (since it is on the curve $y = x^3$).

Let the x co-ordinate of Q be $(x + h)$ so Q is the point $((x + h), (x + h)^3)$.

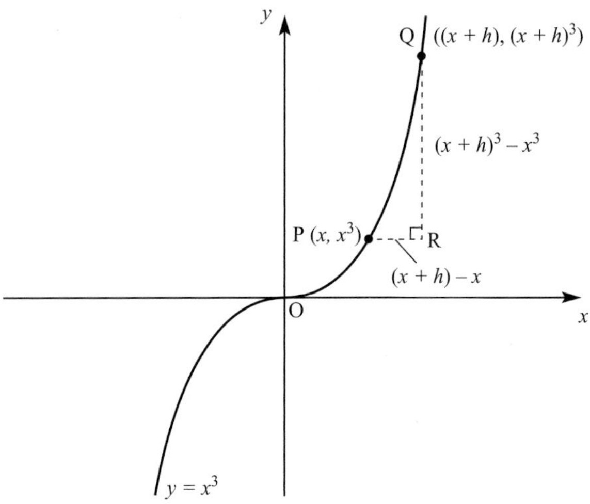

Figure 9.5

The gradient of the chord PQ is given by

$$\begin{aligned}\frac{\text{QR}}{\text{PR}} &= \frac{(x+h)^3 - x^3}{(x+h) - x} \\ &= \frac{x^3 + 3x^2h + 3xh^2 + h^3 - x^3}{h} \\ &= \frac{3x^2h + 3xh^2 + h^3}{h} \\ &= \frac{h(3x^2 + 3xh + h^2)}{h} \\ &= 3x^2 + 3xh + h^2\end{aligned}$$

As Q gets closer to P, h takes smaller and smaller values and the gradient approaches the value of $3x^2$, which is the gradient of the tangent at P.

The gradient of the curve $y = x^3$ at the point (x, y) is equal to $3x^2$.

NOTE

If the equation of the curve is written as $y = \text{f}(x)$ then the *gradient function*, i.e. the gradient of the curve at the general point (x, y), is written as $\text{f}'(x)$. Using this notation, the result in Example 9.1 can be written as

$$\text{f}(x) = x^3 \quad \Rightarrow \quad \text{f}'(x) = 3x^2.$$

EXERCISE 9A

1 **(i)** For the curve $y = x^4$ estimate the gradient at the point P(2, 16) by taking different positions of Q with x co-ordinates 2.1, 2.01 and 2.001 respectively.

(ii) Use a similar method to estimate the gradient at the points (3, 81) and (–1, 1).

2 Use the method in Example 9.1 to prove that the gradient of the curve $y = x^2$ at the point (x, y) is equal to $2x$.

3 Use the binomial theorem to expand $(x + h)^4$ and hence find the gradient of the curve $y = x^4$ at the point (x, y).

4 (i) Copy the table and enter the results you have found already.

(ii) Suggest how the gradient pattern should continue when $f(x) = x^5$, $f(x) = x^6$ and $f(x) = x^n$ (where n is a positive whole number).

$f(x)$	$f'(x)$ (gradient at (x, y))
x^2	
x^3	
x^4	
x^5	
x^6	
⋮	⋮
x^n	

NOTE

The result you should have obtained from question 4 is known as *Wallis's rule* and can be used as a formula.

An alternative notation

So far, h has been used to denote the difference between the x co-ordinates of our points P and Q, where Q is close to P.

h is sometimes replaced by δx. The Greek letter δ(delta) is shorthand for 'a small change in' and so δx represents a small change in x, δy a small change in y and so on.

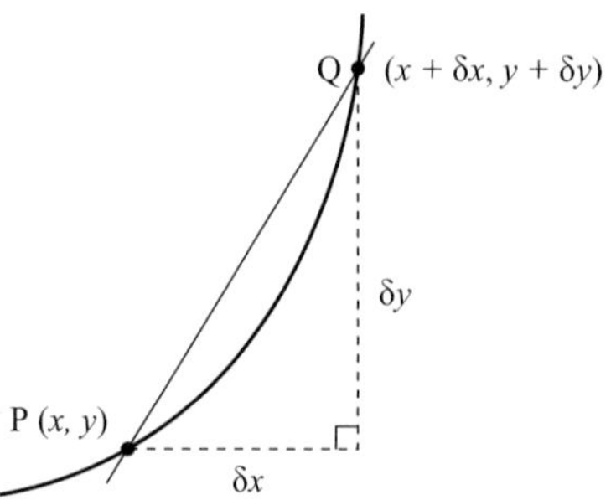

Figure 9.6

In figure 9.6 the gradient of the chord PQ is $\frac{\delta y}{\delta x}$.

In the limit as δx tends towards 0, δx and δy both become infinitesimally small and the value obtained for $\frac{\delta y}{\delta x}$ approaches the gradient of the tangent at P.

Read this as 'the limit as δx tends towards 0'.

$$\lim_{\delta x \to 0} \frac{\delta y}{\delta x} \text{ is written as } \frac{dy}{dx}.$$

Using this notation, you have a rule for differentiation.

$$y = x^n \quad \Rightarrow \quad \frac{dy}{dx} = nx^{n-1}$$

The gradient function, $\frac{dy}{dx}$, is sometimes called the *derivative* of y with respect to x and when you find it you have *differentiated* y with respect to x.

If the curve is written as $y = f(x)$, then the derivative is $f'(x)$.

NOTE

There is nothing special about the letters x, y and f. If, for example, your curve represents time (t) on the horizontal axis and velocity (v) on the vertical axis, then the relationship may be referred to as $v = g(t)$. In this case v is a function of t and the gradient function is given by $\frac{dv}{dt} = g'(t)$.

In this book you are often asked to differentiate a relationship in the form $y = f(x)$. Unless otherwise stated, take this to mean differentiate with respect to x.

ACTIVITY 9.3

(i) Plot the curve with equation $y = x^2 + 2$ for values of x from -2 to $+2$.

(ii) On the same axes and for the same range of values of x, plot the curves $y = x^2 - 2$, $y = x^2$ and $y = x^2 + 5$.

(iii) What do you notice about the gradients of this family of curves when $x = 0$? What about when $x = 1$ and $x = -1$?

(iv) Differentiate the equation $y = x^2 + c$, where c is a constant. How does this result help you to explain your finding in **(iii)**?

Differentiation using standard results

Finding the gradient from first principles establishes a formal basis for differentiation but in practice you would use the differentiation rule. This also includes the results obtained by differentiating (i.e. finding the gradient of) equations which represent straight lines.

The gradient of the line $y = x$ is 1.

The gradient of the line $y = c$ is 0 where c is a constant, since this line is parallel to the x axis.

The rule can be extended further to include functions of the type $y = kx^n$ for any constant k, to give

$$y = kx^n \quad \Rightarrow \quad \frac{dy}{dx} = nkx^{n-1}.$$

You may find it helpful to remember the rule as

multiply by the power of x and reduce the power by 1.

EXAMPLE 9.2

Find the gradient function for each of the following functions.

(i) $y = x^7$ **(ii)** $u = 4x^3$ **(iii)** $v = 5t^2$

SOLUTION

(i) $\frac{dy}{dx} = 7x^6$ **(ii)** $\frac{du}{dx} = 12x^2$ **(iii)** $\frac{dv}{dt} = 10t$

Sums and differences of functions

Many of the functions you will meet are sums or differences of simpler ones. For example, the function $(4x^3 + 3x)$ is the sum of the functions $4x^3$ and $3x$. To differentiate a function such as this you differentiate each part separately and then add the results together.

This may be written in general form as

$$y = f(x) + g(x) \quad \Rightarrow \quad \frac{dy}{dx} = f'(x) + g'(x)$$

EXAMPLE 9.3

Differentiate $y = 4x^3 + 3x$.

SOLUTION

$$\frac{dy}{dx} = 12x^2 + 3$$

EXAMPLE 9.4

Given that $y = 2x^3 - 3x + 4$, find

(i) $\frac{dy}{dx}$

(ii) the gradient of the curve at the point (2, 14).

SOLUTION

(i) $\frac{dy}{dx} = 6x^2 - 3$

(ii) At (2, 14), $x = 2$.

Substituting $x = 2$ in the expression for $\frac{dy}{dx}$ gives

$$\frac{dy}{dx} = 6 \times (2)^2 - 3 = 21.$$

EXERCISE 9B

1 Differentiate the following functions using the rules

$$y = kx^n \quad \Rightarrow \quad \frac{dy}{dx} = nkx^{n-1}$$

and $\quad y = f(x) + g(x) \quad \Rightarrow \quad \frac{dy}{dx} = f'(x) + g'(x).$

(i)	$y = x^4$	**(ii)**	$y = 2x^3$	**(iii)**	$y = 5x^2$
(iv)	$y = 7x^9$	**(v)**	$y = -3x^6$	**(vi)**	$y = 5$
(vii)	$y = 10x$	**(viii)**	$y = 2x^5 + 4x^2$	**(ix)**	$y = 3x^4 + 8x$
(x)	$y = x^3 + 4$	**(xi)**	$y = x - 5x^3$	**(xii)**	$y = 3x^5 + 4x^4 - 3x^2 + 2$
(xiii)	$u = 4x^3 + 2x$	**(xiv)**	$p = 2x + 6$	**(xv)**	$z = x^5 + 12x^3 + 3x$
(xvi)	$v = 3t^5 + 2$	**(xvii)**	$d = \frac{1}{4}p^3$	**(xviii)**	$h = r^3 + 42r^2 - 5r + 24$
(xix)	$C = 2\pi r$	**(xx)**	$A = \pi r^2$		

Tangents and normals

Now that you know how to find the gradient of a curve at any point you can use this to find the equation of the tangent at any particular point on the curve.

EXAMPLE 9.5

(i) Find the equation of the tangent to the curve $y = 3x^2 - 5x - 2$ at the point $(1, -4)$.

(ii) Sketch the curve and show the tangent on your sketch.

SOLUTION

(i) First find the gradient function $\frac{dy}{dx}$.

$$\frac{dy}{dx} = 6x - 5$$

Substitute $x = 1$ into this gradient function to find the gradient, m, of the tangent at $(1, -4)$.

$$\begin{aligned} m &= 6 \times 1 - 5 \\ &= 1 \end{aligned}$$

The equation of the tangent is given by

$$y - y_1 = m(x - x_1)$$

$x_1 = 1$, $y_1 = -4$ and $m = 1$.

$$y - (-4) = 1(x - 1)$$

$$\Rightarrow \quad y = x - 5.$$

This is the equation of the tangent.

(ii) $y = 3x^2 - 5x - 2$ is a ∪-shaped quadratic curve.

It crosses the x axis when $3x^2 - 5x - 2 = 0$.

$\Rightarrow \quad (3x + 1)(x - 2) = 0$

$\Rightarrow \quad x = -\frac{1}{3}$ or $x = 2$

It crosses the y axis when $y = -2$.

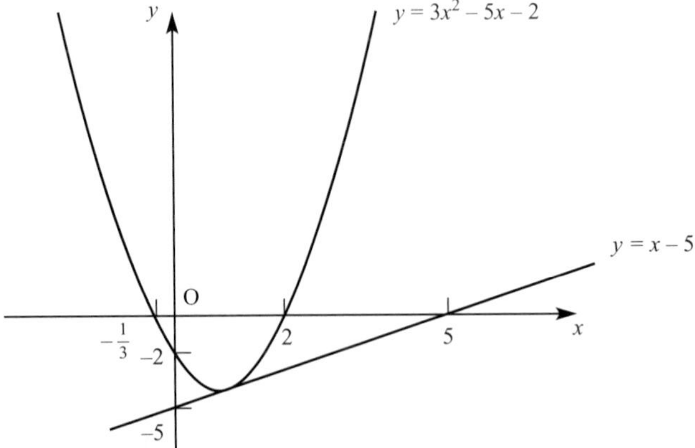

Figure 9.7

The *normal* to a curve at a particular point is the straight line that is at right angles to the tangent at that point (see figure 9.8). Remember that for perpendicular lines $m_1 m_2 = -1$.

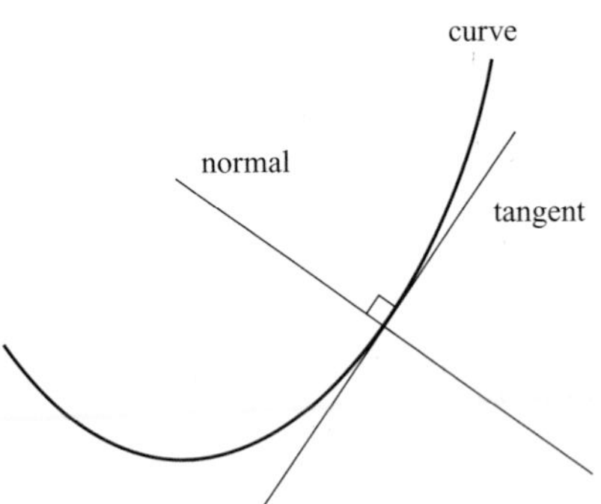

Figure 9.8

EXAMPLE 9.6

Figure 9.9 is a sketch of the curve $y = x^3 - 3x^2 + 2x$ and the point P(3, 6). Find the equation of the normal to the curve $y = x^3 - 3x^2 + 2x$ at P.

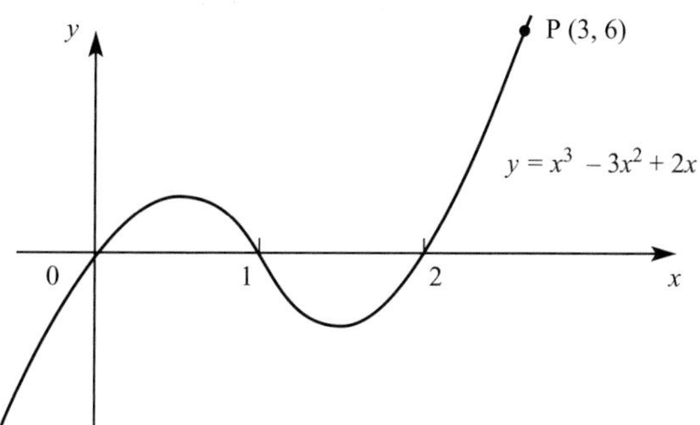

Figure 9.9

SOLUTION

$$y = x^3 - 3x^2 + 2x \quad \Rightarrow \quad \frac{dy}{dx} = 3x^2 - 6x + 2$$

Substitute $x = 3$ to find the gradient, m_1, of the tangent at the point (3, 6).

$$m_1 = 3 \times (3)^2 - 6 \times 3 + 2 = 11$$

The gradient, m_2, of the normal to the curve at this point is given by

$$m_2 = -\frac{1}{m_1} = -\frac{1}{11}.$$

$m_1 m_2 = -1$

The equation of the normal is given by

$$y - y_1 = m_2(x - x_1)$$

(x_1, y_1) is (3, 6).

$$\Rightarrow \quad y - 6 = -\tfrac{1}{11}(x - 3)$$

$$\Rightarrow \quad 11y - 66 = -x + 3$$

$$\Rightarrow \quad x + 11y - 69 = 0$$

Multiply by 11 to eliminate the fraction.

EXERCISE 9C

1 The sketch shows the graph of $y = 5x - x^2$.

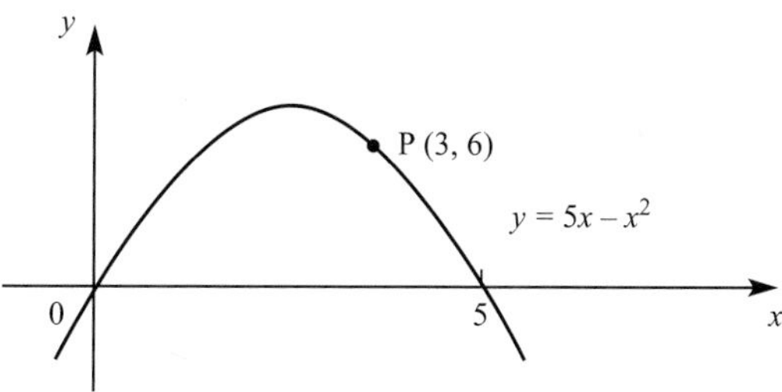

The marked point, P, has co-ordinates (3, 6). Find

(i) the gradient function $\frac{dy}{dx}$

(ii) the gradient of the curve at P

(iii) the equation of the tangent at P

(iv) the equation of the normal at P.

2 The sketch shows the graph of $y = 3x^2 - x^3$.

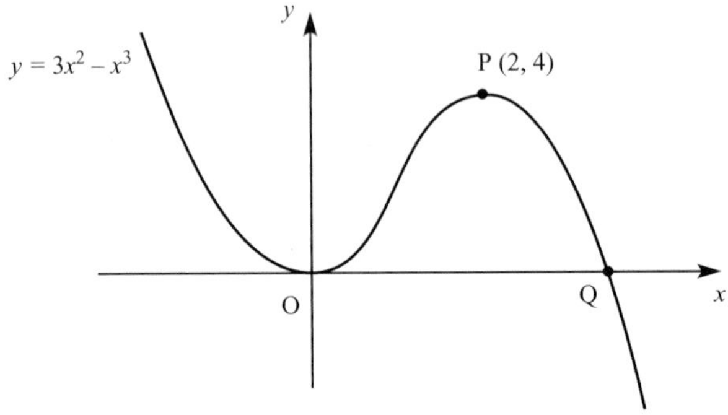

(i) The marked point, P, has co-ordinates (2, 4). Find

(a) the gradient function $\frac{dy}{dx}$

(b) the gradient of the curve at P

(c) the equation of the tangent at P

(d) the equation of the normal at P.

(ii) The graph touches the x axis at the origin O and crosses it at the point Q. Find

(a) the co-ordinates of Q

(b) the gradient of the curve at Q

(c) the equation of the tangent at Q.

(iii) Without further calculation, state the equation of the tangent to the curve at O.

3 The sketch shows the graph of $y = x^5 - x^3$.

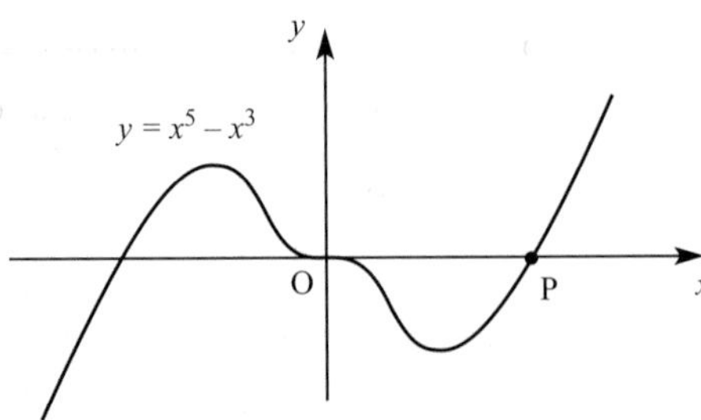

(i) Find the co-ordinates of the point P where the curve crosses the positive x axis.

(ii) Find the equation of the tangent at P.

(iii) Find the equation of the normal at P.

The tangent at P meets the y axis at Q and the normal meets the y axis at R.

(iv) Find the co-ordinates of Q and R and hence find the area of triangle PQR.

4 (i) Given that $f(x) = x^3 - 3x^2 + 4x + 1$, find $f'(x)$.

(ii) The point P is on the curve $y = f(x)$ and its x co-ordinate is 2.

(a) Calculate the y co-ordinate of P.

(b) Find the equation of the tangent at P.

(c) Find the equation of the normal at P.

(iii) Find the values of x for which the curve has a gradient of 13.

5 The sketch shows the graph of $y = x^3 - 9x^2 + 23x - 15$.

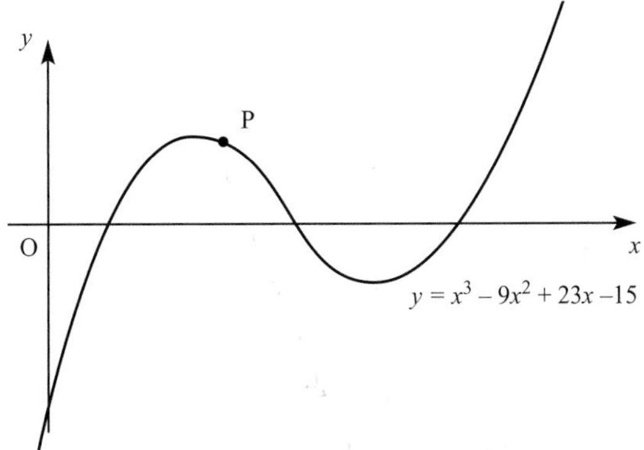

The point P marked on the curve has its x co-ordinate equal to 2. Find

(i) the gradient function $\frac{dy}{dx}$

(ii) the gradient of the curve at P

(iii) the equation of the tangent at P

(iv) the co-ordinates of another point, Q, on the curve at which the tangent is parallel to the tangent at P

(v) the equation of the tangent at Q.

6 The point $(2, -8)$ is on the curve $y = x^3 - px + q$.

(i) Use this information to find a relationship between p and q.

(ii) Find the gradient function $\frac{dy}{dx}$.

The tangent to this curve at the point $(2, -8)$ is parallel to the x axis.

(iii) Use this information to find the value of p.

(iv) Find the co-ordinates of the other point where the tangent is parallel to the x axis.

(v) State the co-ordinates of the point P where the curve crosses the y axis.

(vi) Find the equation of the normal to the curve at the point P.

7 The sketch shows the graph of $y = x^2 - x - 1$.

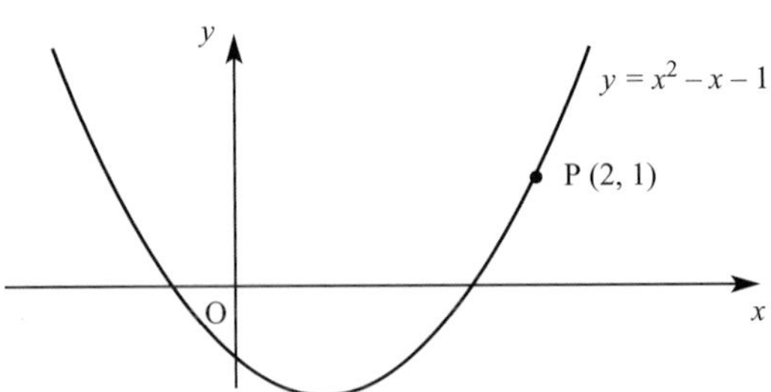

(i) Find the equation of the tangent at the point P(2, 1).

The normal at a point Q on the curve is parallel to the tangent at P.

(ii) State the gradient of the tangent at Q.

(iii) Find the co-ordinates of the point Q.

8 A curve has the equation $y = (x - 3)(7 - x)$.

(i) Find the gradient function $\dfrac{dy}{dx}$.

(ii) Find the equation of the tangent at the point (6, 3).

(iii) Find the equation of the normal at the point (6, 3).

(iv) Which one of these lines passes through the origin?

9 A curve has the equation $y = 1.5x^3 - 3.5x^2 + 2x$.

(i) Show that the curve passes through the points (0, 0) and (1, 0).

(ii) Find the equations of the tangents and normals at each of these points.

(iii) Prove that the four lines in **(ii)** form a rectangle.

Stationary points

ACTIVITY 9.4

(i) Plot the graph of $y = x^4 - 3x^3 - x^2 + 3x$, taking values of x from -1.5 to $+3.5$ in steps of 0.5.
You will need your y axis to go from -10 to $+20$.

(ii) How many turning points are there on the graph?

(iii) What is the gradient at each of these turning points?

(iv) One of the turning points is a maximum and the others are minima. Which are of each type?

(v) Is the maximum the highest point of the graph?

(vi) Do the two minima occur exactly at the points you plotted?

(vii) Estimate the lowest value that y takes.

A *stationary point* on a curve is one where the gradient is zero. This means that the tangents to the curve at these points are horizontal. Figure 9.10 shows a curve with four stationary points A, B, C and D.

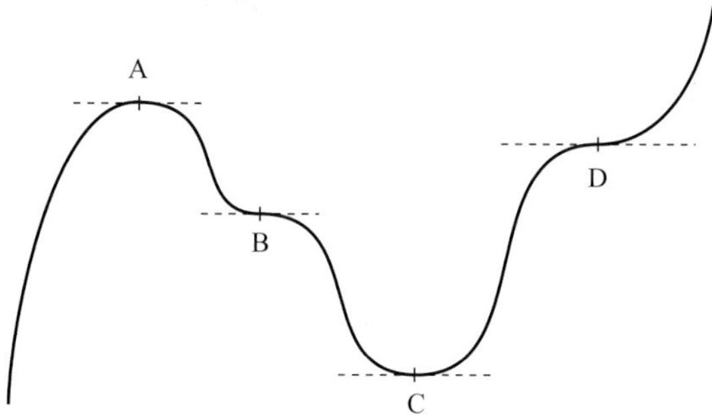

Figure 9.10

The points A and C are *turning points* of the curve since, as the curve passes through these points it changes direction completely: at A the gradient changes from positive to negative and at C from negative to positive. A is called a *maximum* turning point, and C is a *minimum* turning point.

At B the curve does not turn: the gradient is negative both to the left and to the right of this point. B is a *stationary point of inflection.*

What can you say about the gradient to the left and to the right of D?

NOTE

Points where a curve just 'twists', but doesn't have a zero gradient are also called points of inflection, but only *stationary* points of inflection, ones where the gradient is zero, are included in this section. The tangent at a point of inflection both touches and intersects the curve.

ACTIVITY 9.5 Figure 9.11 shows the graph of $y = \cos x$.

Describe the gradient of the curve, using the words 'positive', 'negative', 'zero', 'increasing' and 'decreasing', as x increases from 0° to 360°.

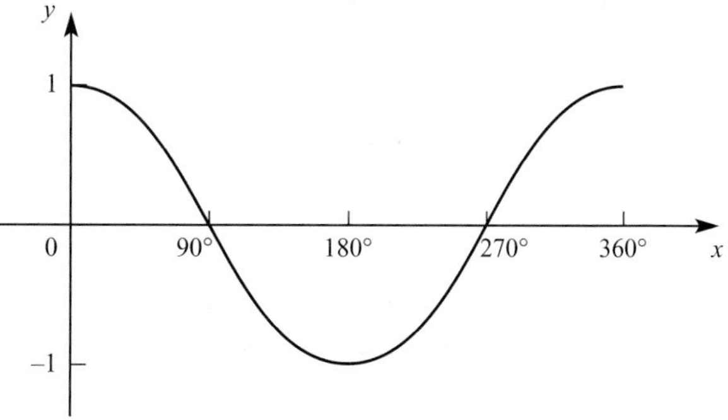

Figure 9.11

Maximum and minimum points

Figure 9.12 shows the graph of $y = 4x - x^2$. It has a *maximum* point at (2, 4).

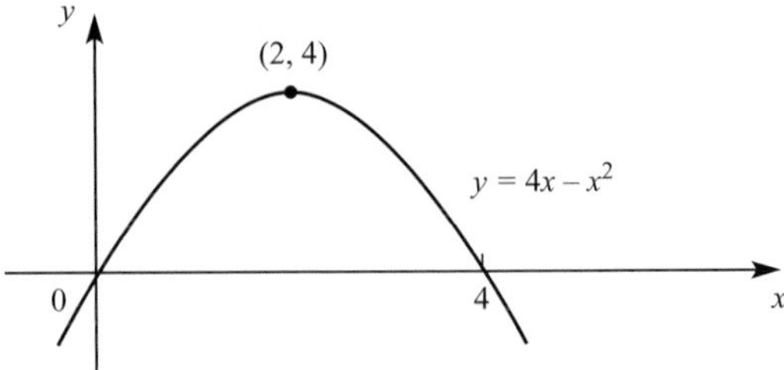

Figure 9.12

You can see that

- at the maximum point the gradient $\frac{dy}{dx}$ is zero
- the gradient is positive to the left of the maximum and negative to the right of it.

This is true for any maximum point (see figure 9.13).

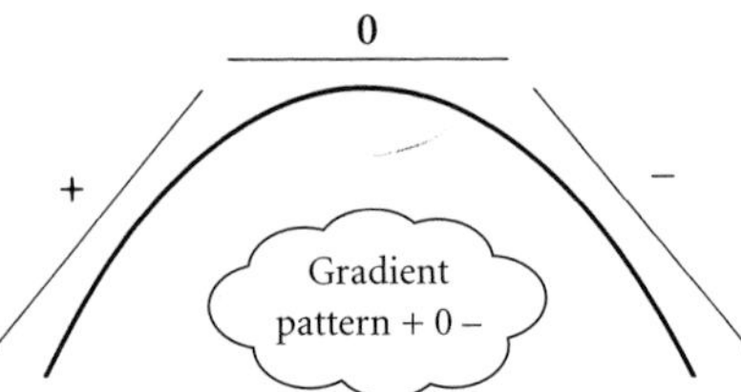

Figure 9.13

In the same way, for any minimum turning point (see figure 9.14)

- the gradient is zero at the minimum
- the gradient goes from negative to zero to positive.

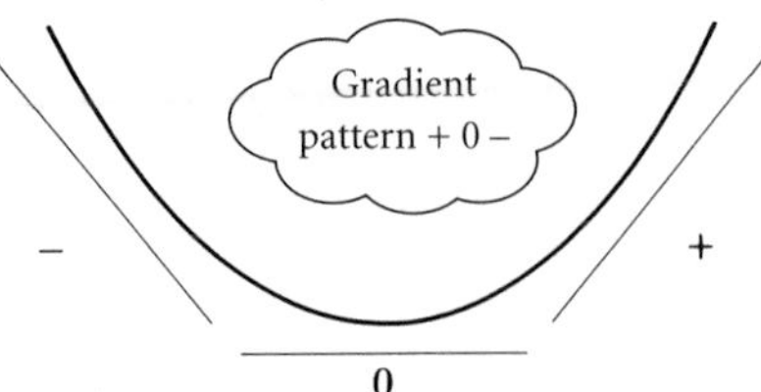

Figure 9.14

Once you have found the the position and type of any stationary points, you can use this information to sketch the curve.

EXAMPLE 9.7

For the curve $y = x^3 - 12x + 3$

(i) find $\frac{dy}{dx}$ and the values of x for which $\frac{dy}{dx} = 0$

(ii) classify the points on the curve with these x values

(iii) find the corresponding y values

(iv) sketch the curve.

SOLUTION

(i) $\frac{dy}{dx} = 3x^2 - 12$

When $\frac{dy}{dx} = 0$

$$3x^2 - 12 = 0$$
$$\Rightarrow \quad 3(x^2 - 4) = 0$$
$$\Rightarrow \quad 3(x + 2)(x - 2) = 0$$
$$\Rightarrow \quad x = -2 \text{ or } x = 2$$

(ii) For $x = -2$

$$x = -3 \quad \Rightarrow \quad \frac{dy}{dx} = 3(-3)^2 - 12 = +15$$
$$x = -1 \quad \Rightarrow \quad \frac{dy}{dx} = 3(-1)^2 - 12 = -9.$$

Gradient pattern + 0 –

$\Rightarrow$ maximum turning point when $x = -2$.

For $x = +2$

$$x = 1 \quad \Rightarrow \quad \frac{dy}{dx} = 3(1)^2 - 12 = -9$$
$$x = 3 \quad \Rightarrow \quad \frac{dy}{dx} = 3(3)^2 - 12 = +15$$

Gradient pattern – 0 +

$\Rightarrow$ minimum turning point when $x = +2$.

(iii) When $x = -2$, $y = (-2)^3 - 12(-2) + 3 = 19$.
When $x = +2$, $y = (2)^3 - 12(2) + 3 = -13$.

There is a maximum at (–2, 19) and a minimum at (2, –13).

(iv) The only other information you need to sketch the curve is the value of y when $x = 0$. This tells you where the curve crosses the y axis.
When $x = 0$, $y = (0)^3 - 12(0) + 3 = 3$.

The graph of $y = x^3 - 12x + 3$ is shown in figure 9.15.

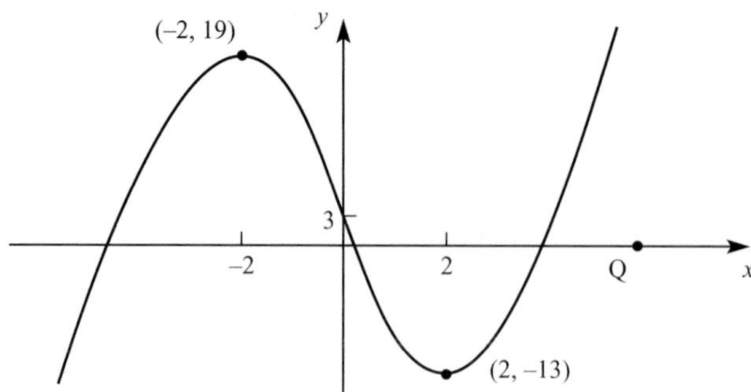

Figure 9.15

? Why can you be confident about continuing the sketch of the curve beyond the x values of the turning points?

? In Example 9.7 you did not find the co-ordinates of the points where the curve crosses the x axis.

(i) Why was this?

(ii) Under what circumstances would you find these points?

EXAMPLE 9.8

Find all the turning points on the curve of $y = t^4 - 2t^3 + t^2 - 2$ and sketch the curve.

SOLUTION

$$\frac{dy}{dt} = 4t^3 - 6t^2 + 2t$$

Turning points occur when $\frac{dy}{dt} = 0$.

$$\Rightarrow \quad 2t(2t^2 - 3t + 1) = 0$$
$$\Rightarrow \quad 2t(2t - 1)(t - 1) = 0$$
$$\Rightarrow \quad t = 0 \text{ or } t = 0.5 \text{ or } t = 1$$

You may find it helpful to summarise your working in a table. You can find the various signs, + or –, by taking a test point in each interval, for example $t = 0.25$ in the interval $0 < t < 0.5$.

	$t < 0$	0	$0 < t < 0.5$	0.5	$0.5 < t < 1$	1	$t > 1$
sign of $\frac{dy}{dx}$	–	0	+	0	–	0	+
turning point		min		max		min	

When $t = 0$: $y = (0)^4 - 2(0)^3 + (0)^2 - 2 = -2$.
When $t = 0.5$: $y = (0.5)^4 - 2(0.5)^3 + (0.5)^2 - 2 = -1.9375$.
When $t = 1$: $y = (1)^4 - 2(1)^3 + (1)^2 - 2 = -2$.

Therefore $(0.5, -1.9375)$ is a maximum turning point and $(0, -2)$ and $(1, -2)$ are minima.

The graph of $y = t^4 - 2t^3 + t^2 - 2$ is shown in figure 9.16.

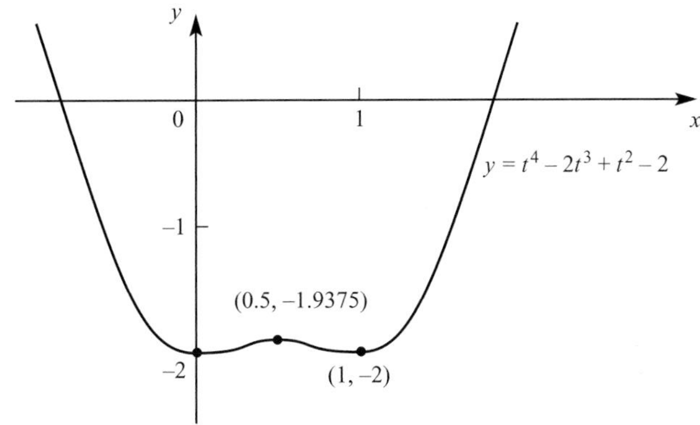

Figure 9.16

EXERCISE 9D

If you have access to a graphic calculator you will find it helpful to use it to check your answers.

1 For each of the curves given below

(a) find $\dfrac{\mathrm{d}y}{\mathrm{d}x}$ and the value(s) of x for which $\dfrac{\mathrm{d}y}{\mathrm{d}x} = 0$

(b) classify the point(s) on the curve with these x values

(c) find the corresponding y value(s)

(d) sketch the curve.

(i) $y = 1 + x - 2x^2$

(ii) $y = 12x + 3x^2 - 2x^3$

(iii) $y = x^3 - 4x^2 + 9$

(iv) $y = x^2(x - 1)^2$

(v) $y = x^4 - 8x^2 + 4$

(vi) $y = x^3 - 48x$

(vii) $y = x^3 + 6x^2 - 36x + 25$

(viii) $y = 2x^3 - 15x^2 + 24x + 8$

2 The graph of $y = px + qx^2$ passes through the point $(3, -15)$ and its gradient at that point is -14.

(i) Find the values of p and q.

(ii) Calculate the maximum value of y and state the value of x at which it occurs.

3 (i) Find the stationary points of the function $\mathrm{f}(x) = x^2(3x^2 - 2x - 3)$ and distinguish between them.

(ii) Sketch the curve $y = \mathrm{f}(x)$.

Points of inflection

Stationary points of inflection also have their own gradient patterns that you can use to classify them.

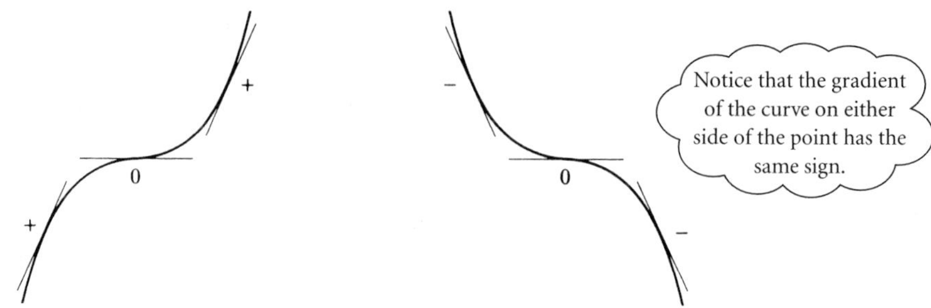

Figure 9.17

EXAMPLE 9.9

(i) Find the stationary values of the function $y = 4x^3 - x^4$ and distinguish between them.

(ii) Sketch the curve.

SOLUTION

(i) $\frac{dy}{dx} = 12x^2 - 4x^3$

Stationary points occur when $12x^2 - 4x^3 = 0$.

$\Rightarrow \quad 4x^2(3 - x) = 0$

$\Rightarrow \quad x = 0$ or $x = 3$

For $x = 0$

$$x = -1 \quad \Rightarrow \quad \frac{dy}{dx} = 12(-1)^2 - 4(-1)^3 = +16$$

$$x = 1 \quad \Rightarrow \quad \frac{dy}{dx} = 12(1)^2 - 4(1)^3 = +8$$

Gradient pattern $+ \; 0 \; + \Rightarrow$ point of inflection when $x = 0$.

For $x = 3$

$$x = 2 \quad \Rightarrow \quad \frac{dy}{dx} = 12(2)^2 - 4(2)^3 = +16$$

$$x = 4 \quad \Rightarrow \quad \frac{dy}{dx} = 12(4)^2 - 4(4)^3 = -64$$

Gradient pattern $+ \; 0 \; - \Rightarrow$ maximum turning point when $x = 3$.

Substituting the x co-ordinates in the equation of the curve gives the *stationary values* of the function.

$x = 0 \Rightarrow y = 0$ and $x = 3 \Rightarrow y = 27$

(ii) The curve crosses the y axis at $(0, 0)$.
The graph of $y = 4x^3 - x^4$ is shown in figure 9.18.

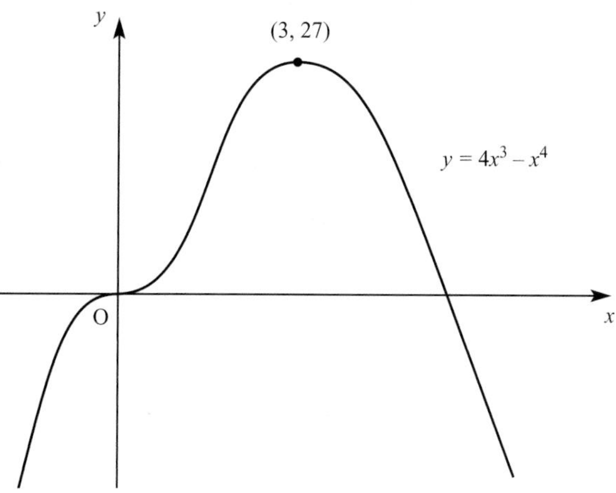

Figure 9.18

EXERCISE 9E

If you have access to a graphic calculator you will find it helpful to use it to check your answers.

1 For each of the curves given below

(a) find $\frac{dy}{dx}$ and the value(s) of x for which $\frac{dy}{dx} = 0$

(b) classify the point(s) that correspond to these x values

(c) find the corresponding y values(s)

(d) sketch the curve.

(i) $y = x^3 + 6x^2 + 12x + 8$ **(ii)** $y = 3x^4 + 4x^3$

(iii) $y = 3 + 4x^3 - x^4$ **(iv)** $y = 3x^5 - 5x^3$

(v) $y = 4x^3 (2 - x)$ **(vi)** $y = x^3 - 3x^2 + 3x + 1$

2 (i) Find the position and nature of any stationary points of the curve $y = x^3 - 3x^2 + 3x + 2$.

(ii) sketch the curve.

3 The function $y = px^4 + qx^3$, where p and q are constants, has a stationary point at $(1, -3)$.

(i) Using the fact that $(1, -3)$ lies on the curve, form an equation involving p and q.

(ii) Differentiate y and, using the fact that $(1, -3)$ is a stationary point, form another equation involving p and q.

(iii) Solve these two equations simultaneously to find the values of p and q.

(iv) Determine the nature of the stationary point at $(1, -3)$.

(v) Locate and classify the other stationary point on the curve.

KEY POINTS

1 $y = kx^n \Rightarrow \dfrac{dy}{dx} = nkx^{n-1}$

$y = c \Rightarrow \dfrac{dy}{dx} = 0$

where n is a positive integer and k and c are constants.

2 $y = f(x) + g(x) \Rightarrow \dfrac{dy}{dx} = f'(x) + g'(x)$

3 For the tangent and normal at (x_1, y_1)

- the gradient of the tangent, m_1 = the value of $\dfrac{dy}{dx}$
- the gradient of the normal, $m_2 = -\dfrac{1}{m_1}$
- the equation of the tangent is $y - y_1 = m_1(x - x_1)$
- the equation of the normal is $y - y_1 = m_2(x - x_1)$.

4 At a stationary point, $\dfrac{dy}{dx} = 0$.

The nature of the stationary point can be determined by looking at the sign of the gradient just either side of it.

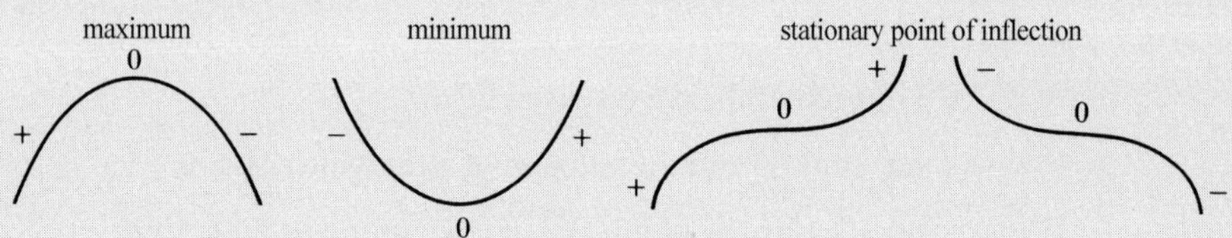

10 Calculus II – integration

Nothing tends so much to the advancement of knowledge as the application of a new instrument.

Sir Humphrey Davy

? Suppose that you know the gradient function, $\frac{dy}{dx}$, of a curve. What other information would you need in order to find the equation of the curve?

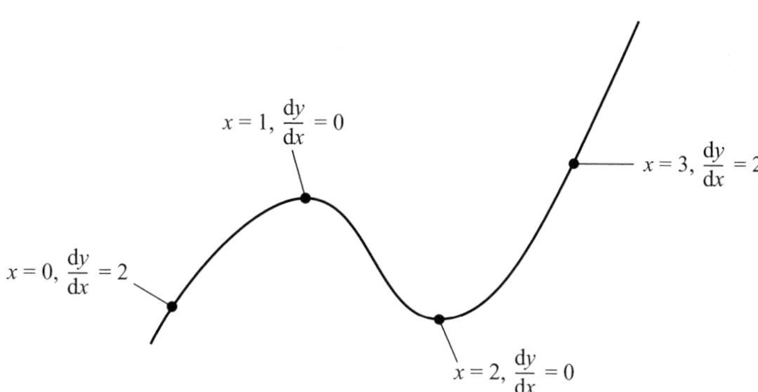

Figure 10.1

Reversing differentiation

ACTIVITY 10.1 **(i)** Differentiate each of the following.

(a) $y = x^3$ **(b)** $y = x^3 + 4$ **(c)** $y = x^3 - 7$

(ii) What do you notice?

The equation $\frac{dy}{dx} = 3x^2$ is an example of a *differential equation*. All the expressions above are solutions of this equation. All that you can say at this point is that if $\frac{dy}{dx} = 3x^2$, then $y = x^3 + c$ where c is described as an *arbitrary constant*. An arbitrary constant can take any value, positive, negative or zero.

? For how many equations is $\dfrac{dy}{dx} = 3x^2$?

$y = x^3 + c$ is called the *general solution* of the differential equation $\dfrac{dy}{dx} = 3x^2$ and the process of solving the equation in this way is called *integration.*

A solution such as this, containing an arbitrary constant, would give a family of curves as in figure 10.2. Each curve corresponds to a particular value of c.

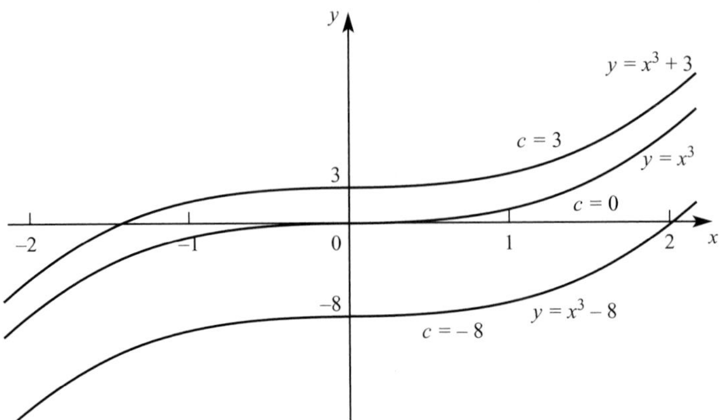

Figure 10.2

Suppose that in the previous activity you were also told that the solution curve passes through the point (1, 4). Substituting these co-ordinates in $y = x^3 + c$ gives

$$4 = 1^3 + c \quad \Rightarrow \quad c = 3$$

This example shows that if you know a point on a curve in the family, you can find the value of c, and so the equation of the curve. This is called the *particular solution.*

The particular solution in this case is $y = x^3 + 3$, which is one of the curves shown in figure 10.2.

The rule for integrating x^n where n is a positive integer

The rule for differentiation is usually given as

$$y = x^n \Rightarrow \frac{dy}{dx} = nx^{n-1}.$$

It can also be given as $$y = x^{n+1} \Rightarrow \frac{dy}{dx} = (n+1)x^n.$$

This is the same as $$y = \frac{1}{n+1}x^{n+1} \Rightarrow \frac{dy}{dx} = x^n.$$

Reversing this gives you the rule for integration.

Integrating x^n with respect to x gives $\dfrac{x^{n+1}}{n+1} + c.$

NOTE

As with differentiation, in this book if you are asked to integrate an expression f(x), take this to mean integrate with respect to x unless otherwise stated.

You may find it helpful to remember the rule as

add 1 to the power and divide by the new power, then add a constant.

Differentiating x gives 1, so integrating 1 gives $x + c$. How does this fit into the pattern above?

EXAMPLE 10.1

Integrate the following.

(i) x^6 **(ii)** $5x^4$ **(iii)** 7

SOLUTION

(i) $\dfrac{x^7}{7} + c$ **(ii)** $5 \times \dfrac{x^5}{5} + c = x^5 + c$ **(iii)** $7x + c$

Don't forget to include the arbitrary constant, c, until you have enough information to find a value for it.

EXAMPLE 10.2

Given that $\dfrac{dy}{dx} = 6x^2 + 2x - 5$

(i) find the general solution of this differential equation

(ii) find the equation of the curve with this gradient function that passes through the point with co-ordinates (1, 7).

SOLUTION

(i) By integration

$$y = 6 \times \frac{x^3}{3} + 2 \times \frac{x^2}{2} - 5x + c$$
$$= 2x^3 - x^2 - 5x + c, \quad \text{where } c \text{ is a constant.}$$

(ii) Since the graph passes through (1, 7),

$$7 = 2(1)^3 + 1^2 - 5 + c$$
$$\Rightarrow \quad c = 9$$
$$\Rightarrow \quad y = 2x^3 + x^2 - 5x + 9.$$

EXAMPLE 10.3

Find f(x) given that $f'(x) = 2x + 4$ and $f(2) = -4$.

SOLUTION

By integration

$$f(x) = \frac{2x^2}{2} + 4x + c$$
$$= x^2 + 4x + c, \quad \text{where } c \text{ is a constant.}$$

$$f(2) = -4 \quad \Rightarrow \quad -4 = (2)^2 + 4(2) + c$$
$$\Rightarrow \quad c = -16$$
$$\Rightarrow \quad f(x) = x^2 + 4x - 16.$$

EXAMPLE 10.4

A curve passes through (3, 5). The gradient of the curve is given by $\frac{dy}{dx} = x^2 - 4$.

(i) Find y in terms of x.

(ii) Find the co-ordinates of any stationary points of the graph of y.

(iii) Sketch the curve.

SOLUTION

(i) $\frac{dy}{dx} = x^2 - 4 \quad \Rightarrow \quad y = \frac{x^3}{3} - 4x + c$

When $x = 3$,

$$5 = 9 - 12 + c$$
$$\Rightarrow \quad c = 8$$

So the equation of the curve is $y = \frac{x^3}{3} - 4x + 8$.

(ii) At all stationary points $\frac{dy}{dx} = 0$.

Remember that you were given $\frac{dy}{dx}$.

$$\Rightarrow \quad x^2 - 4 = 0$$
$$\Rightarrow \quad (x + 2)(x - 2) = 0$$
$$\Rightarrow \quad x = -2 \text{ or } x = 2$$

Substitute into the equation of the curve to find y.

The stationary points are $\left(-2, 13\frac{1}{3}\right)$ and $\left(2, 2\frac{2}{3}\right)$.

(iii) The curve is a cubic with a positive x^3 term and two turning points, so it has this shape.

It crosses the y axis when $x = 0$, $y = 8$.

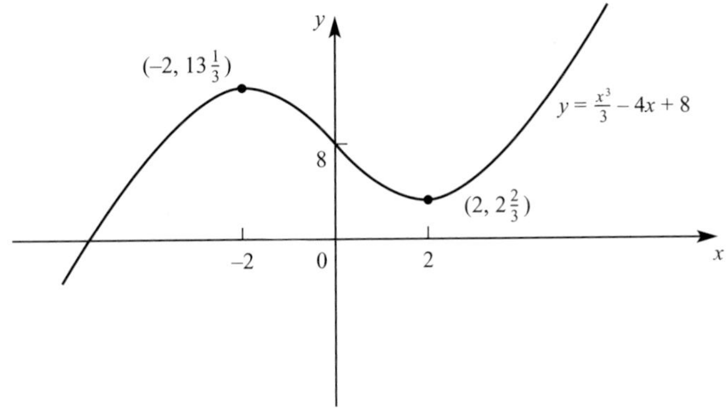

Figure 10.3

The integral symbol

An alternative way of writing $\frac{dy}{dx} = 2x \Rightarrow y = x^2 + c$ is

$$\int 2x \, dx = x^2 + c.$$

Read this as 'the integral of $2x$ with respect to x'. It is called an *indefinite integral* because it includes a constant c whose value is not known.

EXAMPLE 10.5

Find $\int (x^3 - 2x^2) \, dx$.

SOLUTION

$$\int (x^3 - 2x^2) \, dx = \frac{x^4}{4} - \frac{2x^3}{3} + c.$$

? What would you need to do first before you could integrate $(2x + 1)(x - 4)$?

EXERCISE 10A

1 For each of these gradient functions find $y = f(x)$.

(i) $\frac{dy}{dx} = 4x + 2$
(ii) $\frac{dy}{dx} = 6x^2 - 5x - 1$
(iii) $\frac{dy}{dx} = 3 - 5x^3$
(iv) $\frac{dy}{dx} = (x - 2)(3x + 2)$
(v) $f'(x) = 5x + 3$
(vi) $f'(x) = x^4 + 2x^3 - x + 8$
(vii) $f'(x) = (x - 4)(x^2 + 2)$
(viii) $f'(x) = (x - 7)^2$

2 Find the following indefinite integrals.

(i) $\int 5x^3 \, dx$
(ii) $\int (2x - 3) \, dx$
(iii) $\int (3x^3 - 4x + 3) \, dx$
(iv) $\int (3 - x)^2 \, dx$
(v) $\int 4 \, dx$
(vi) $\int (2x + 1)(x - 3) \, dx$
(vii) $\int (x + 1)^2 \, dx$
(viii) $\int (2x - 1)^2 \, dx$

3 For each of the following gradient functions find the equation of the curve $y = \mathrm{f}(x)$ that passes through the given point.

(i) $\dfrac{\mathrm{d}y}{\mathrm{d}x} = 2x - 3;$ $(2, 4)$

(ii) $\dfrac{\mathrm{d}y}{\mathrm{d}x} = 4 + 3x^3;$ $(4, -2)$

(iii) $\dfrac{\mathrm{d}y}{\mathrm{d}x} = 5x - 6;$ $(-2, 4)$

(iv) $\mathrm{f}'(x) = x^2 + 1;$ $(-3, -3)$

(v) $\mathrm{f}'(x) = (x + 1)(x - 2);$ $(6, -2)$

(vi) $\mathrm{f}'(x) = (2x + 1)^2;$ $(1, -1)$

4 You are given that $\dfrac{\mathrm{d}y}{\mathrm{d}x} = 2x + 3$.

(i) Find the general solution of the differential equation.

(ii) Find the equation of the curve with gradient function $\dfrac{\mathrm{d}y}{\mathrm{d}x}$ and which passes through $(2, -1)$.

(iii) Hence show that $(-1, -13)$ also lies on the curve.

5 The curve C passes through the point $(3, 21)$ and its gradient at any point is given by $\dfrac{\mathrm{d}y}{\mathrm{d}x} = 3x^2 - 4x + 1$.

(i) Find the equation of the curve C.

(ii) Show that the point $(-2, -9)$ lies on the curve.

6 (i) Find the general solution of the differential equation $\dfrac{\mathrm{d}y}{\mathrm{d}x} = 4x - 1$.

(ii) Find the particular solution that passes through the point $(-1, 4)$.

(iii) Does this curve pass above, below or through the point $(2, 4)$?

7 The curve $y = \mathrm{f}(x)$ passes through the point $(2, -4)$ and $\mathrm{f}'(x) = 2 - 3x^2$. Find the value of $\mathrm{f}(-1)$.

8 A curve has stationary points at the points where $x = 0$ and $x = 2$.

(i) Explain why $\dfrac{\mathrm{d}y}{\mathrm{d}x} = x^2 - 2x$ is a possible expression for the gradient of the curve and give an alternative expression for $\dfrac{\mathrm{d}y}{\mathrm{d}x}$.

(ii) The curve passes through the point $(3, 2)$.
Taking $\dfrac{\mathrm{d}y}{\mathrm{d}x}$ as $x^2 - 2x$, find the equation of the curve.

Definite integrals

So far the integrals you have met have all been indefinite integrals: they have finished with '+ c'. You may or may not have had some extra information to allow you to find a value for c.

Another form of integral, called a *definite integral*, is shown in Example 10.6.

EXAMPLE 10.6 Find $\int_1^3 3x^2\,dx$

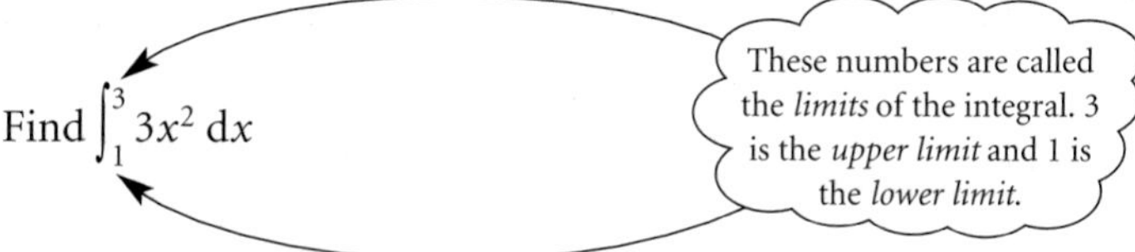

SOLUTION

$$\int 3x^2\,dx = x^3 + c$$

To find the definite integral, you work out the value of the integral when $x = 3$ and subtract the value when $x = 1$.

This gives $\int_1^3 3x^2\,dx = [3^3 + c] - [1^3 + c] = 26.$

NOTE

Notice how the c vanishes when you work out the expression above. When evaluating definite integrals it is common practice to omit the c and write

$$\int_1^3 3x^2\,dx = \left[x^3\right]_1^3 = [3^3] - [1^3] = 26.$$

Notice how the limits move to the right of the first set of square brackets.

The definite integral is defined as

$$\int_a^b f'(x)\,dx = \left[f(x)\right]_a^b = f(b) - f(a).$$

EXAMPLE 10.7 Evaluate $\int_1^4 (x^2 + 3)\,dx$.

SOLUTION

$$\begin{aligned}\int_1^4 (x^2 + 3)\,dx &= \left[\frac{x^3}{3} + 3x\right]_1^4 \\ &= \left(\frac{4^3}{3} + 3 \times 4\right) - \left(\frac{1^3}{3} + 3 \times 1\right) \\ &= 30\end{aligned}$$

❓ The word 'evaluate' was used to start Example 10.7.
What does that word mean?
Why is it appropriate to use it here?

EXAMPLE 10.8 Evaluate $\int_{-1}^{3} (x+1)(x-3)\,dx$.

SOLUTION

$$\int_{-1}^{3} (x+1)(x-3)\,dx = \int_{-1}^{3} (x^2 - 2x - 3)\,dx.$$

$$= \left[\frac{x^3}{3} - x^2 - 3x\right]_{-1}^{3}$$

$$= \left(\frac{3^3}{3} - 3^2 - 3\times 3\right) - \left(\frac{(-1)^3}{3} - (-1)^2 - 3\times(-1)\right)$$

$$= -10\tfrac{2}{3}$$

Notice how you need to expand $(x+1)(x-3)$ before integrating it.

ACTIVITY 10.2 Evaluate

(i) $\int_{1}^{3} x^2\,dx$ and $\int_{3}^{1} x^2\,dx$

(ii) $\int_{-1}^{4} (x+3)\,dx$ and $\int_{4}^{-1} (x+3)\,dx$.

What do you notice?

What is the relationship between $\int_{a}^{b} f(x)\,dx$ and $\int_{b}^{a} f(x)\,dx$?

EXERCISE 10B

1 Evaluate the following definite integrals

(i) $\int_{1}^{2} 3x^2\,dx$

(ii) $\int_{1}^{4} 4x^3\,dx$

(iii) $\int_{-1}^{1} 6x^2\,dx$

(iv) $\int_{1}^{5} 4\,dx$

(v) $\int_{2}^{4} (x^2+1)\,dx$

(vi) $\int_{-2}^{3} (2x+5)\,dx$

(vii) $\int_{2}^{5} (4x^3-2x+1)\,dx$

(viii) $\int_{5}^{6} (x^2-5)\,dx$

(ix) $\int_{1}^{3} (x^2-3x+1)\,dx$

(x) $\int_{-1}^{2} (x^2+3)\,dx$

(xi) $\int_{-4}^{-1} (16-x^2)\,dx$

(xii) $\int_{1}^{3} (x+1)(3-x)\,dx$

(xiii) $\int_{2}^{4} (3x(x+2))\,dx$

(xiv) $\int_{-1}^{1} (x+1)(x-1)\,dx$

(xv) $\int_{-1}^{2} (x+4x^2)\,dx$

(xvi) $\int_{-1}^{1} x(x-1)(x+1)\,dx$

(xvii) $\int_{-1}^{3} (x^3+2)\,dx$

(xviii) $\int_{-3}^{1} (9-x^2)\,dx$

ACTIVITY 10.3 Figure 10.4 shows the line $y = 2x + 1$.
The shaded region is bounded by $y = 2x + 1$, the x axis and the lines $x = 2$ and $x = 4$.

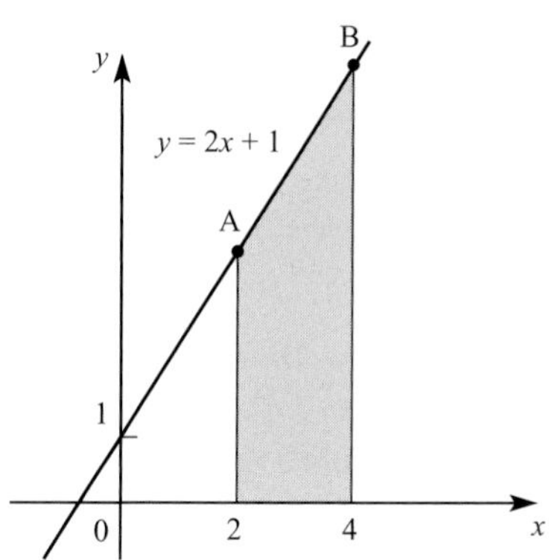

Figure 10.4

(i) Find the co-ordinates of the points A and B in figure 10.4.
(ii) Use the formula for the area of a trapezium to find the area of the shaded region.
(iii) Evaluate $\int_2^4 (2x + 1)\,dx$ and confirm that your answer is the same as in **(ii)**.

ACTIVITY 10.4 *If possible, part* **(ii)** *of this activity should be done using a spreadsheet.*

Figure 10.5(a) shows the graph of $y = x^2$ with the area between the curve, the x axis and the lines $x = 0$ and $x = 3$ shaded.

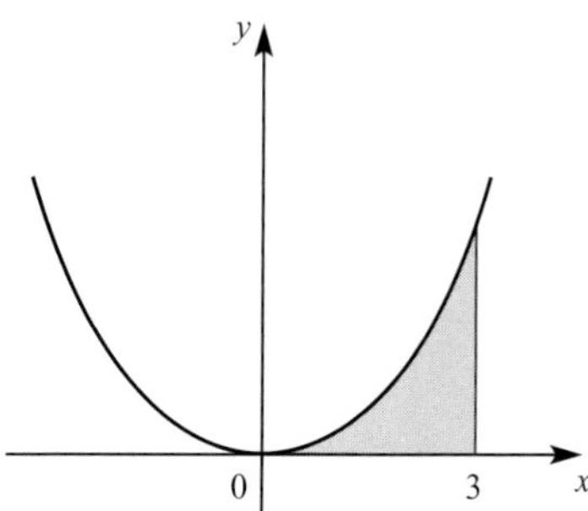

Figure 10.5 (a)

(i) Calculate the area of the rectangles shaded in figures 10.5(b) and 10.5(c) and, for each one, say if you would expect this area to be larger or smaller than the area shaded in figure 10.5(a).

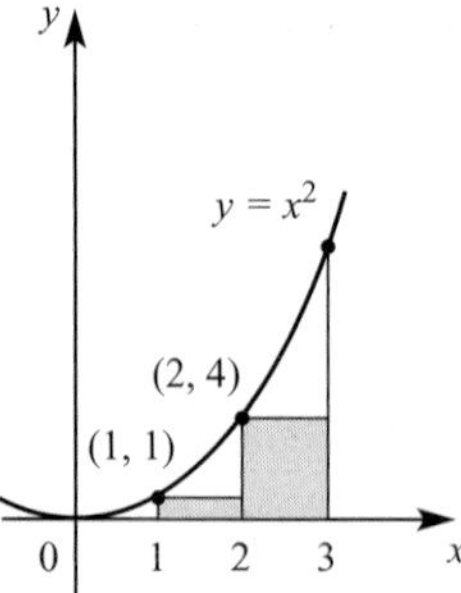

Figure 10.5(b)

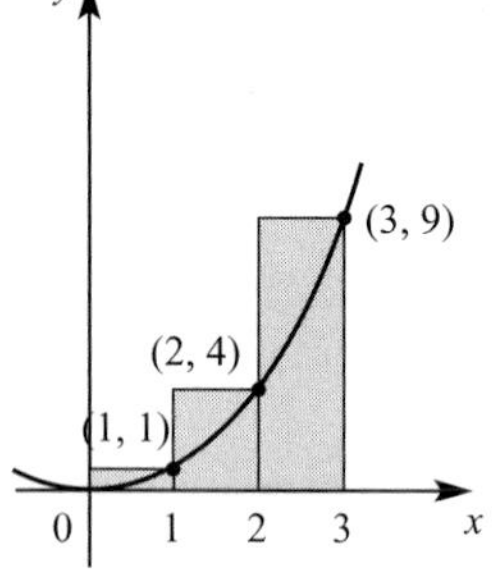

Figure 10.5(c)

(ii) Now calculate the area of the rectangles when the width of the rectangles is reduced to

(a) 0.5

(b) 0.1.

(iii) Evaluate $\int_0^3 x^2 \, dx$.

(iv) What do you notice?

These activities illustrate a very important result: the area under a graph is given by a definite integral.

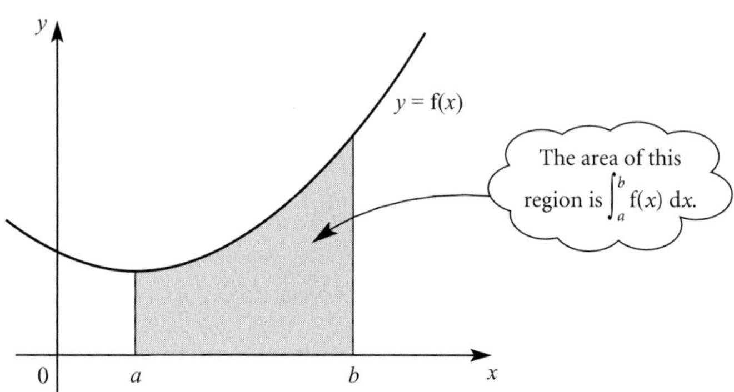

Figure 10.6

NOTE

The proof of this formula in general is not required at this level. It is an important result known as the fundamental theorem of calculus and the proof can be found in more advanced texts on pure mathematics.

EXAMPLE 10.9

Figure 10.7 shows a sketch of the curve $y = 4 - x^2$.
Find the area of the shaded region.

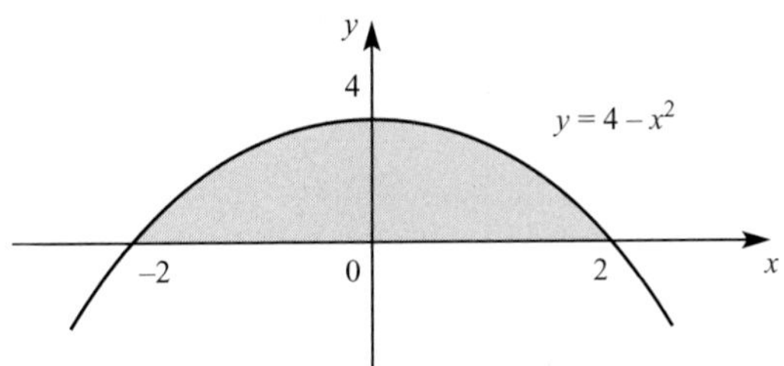

Figure 10.7

SOLUTION

$$\text{Area} = \int_{-2}^{2} (4 - x^2)\,dx = \left[4x - \frac{x^3}{3}\right]_{-2}^{2}$$

$$= \left[4 \times 2 - \frac{2^3}{3}\right] - \left[4 \times (-2) - \frac{(-2)^3}{3}\right]$$

$$= 10\tfrac{2}{3} \text{ units}^2.$$

EXERCISE 10C

1 Find the area of each of the shaded regions.

(i)

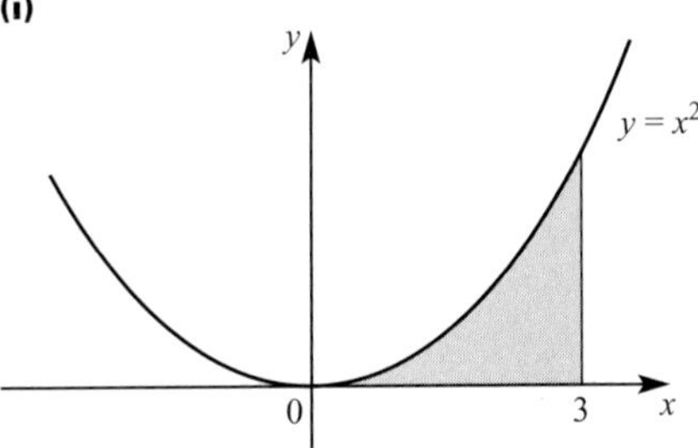

(ii)

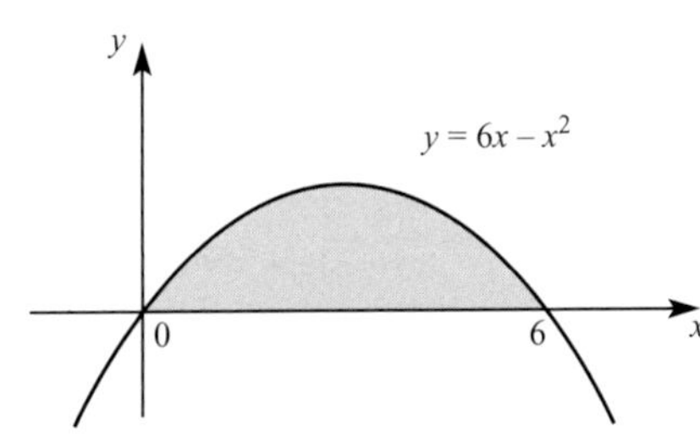

(iii)

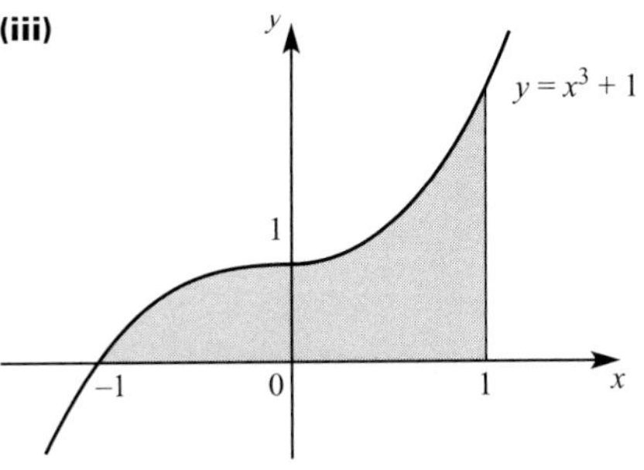

(iv)

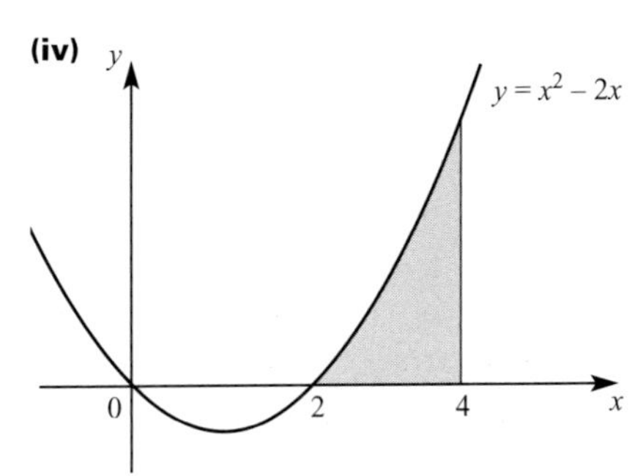

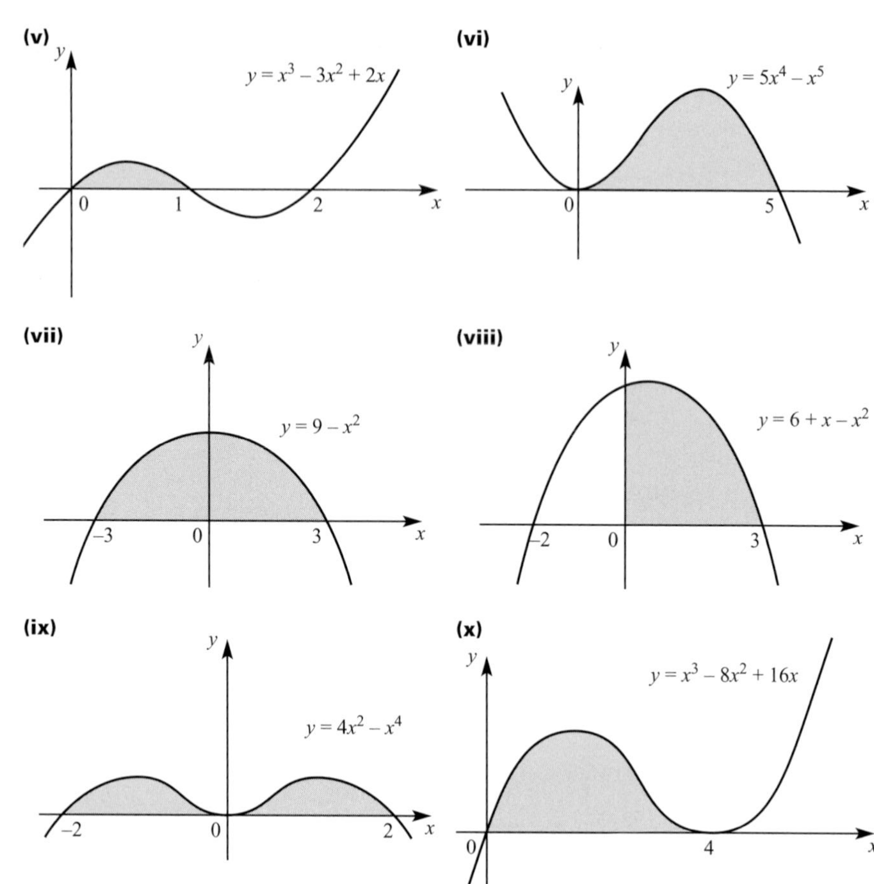

ACTIVITY 10.5 Figure 10.8 shows the line $y = x$ together with two areas marked A and B.

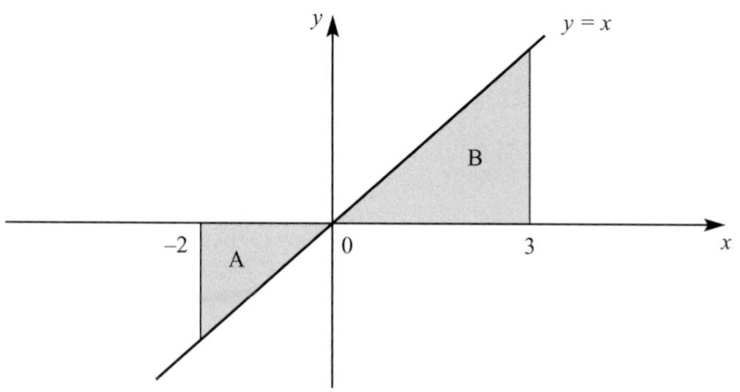

Figure 10.8

(i) Calculate the areas A and B using the formula for the area of a triangle.

(ii) Evaluate $\int_{-2}^{0} x \, dx$ and $\int_{0}^{3} x \, dx$. What do you notice?

(iii) Evaluate $\int_{-2}^{3} x \, dx$. What do you notice?

EXAMPLE 10.10 Figure 10.9 shows a sketch of the curve $y = x(x-2)(x+2)$.

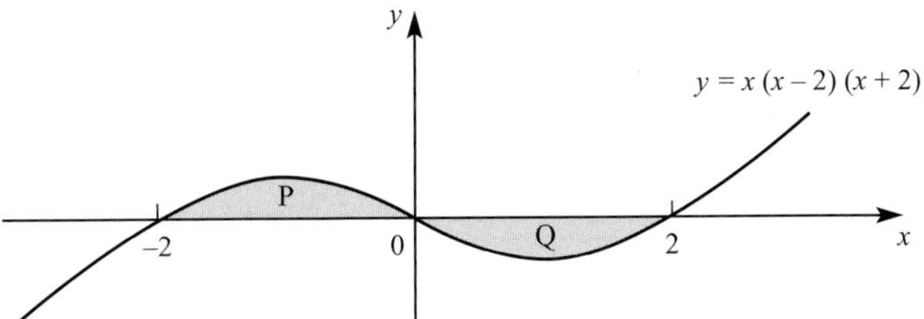

Figure 10.9

(i) Use integration to find the areas of each of the shaded regions P and Q.

(ii) Evaluate $\int_{-2}^{2} x(x-2)(x+2)\,dx$.

(iii) What do you notice?

SOLUTION

(i) $$\int_{-2}^{0} x(x-2)(x+2)\,dx = \int_{-2}^{0} (x^3 - 4x)\,dx$$
$$= \left[\frac{x^4}{4} - 2x^2\right]_{-2}^{0}$$
$$= 0 - \left[\frac{(-2)^4}{4} - 2\times(-2)^2\right]$$
$$= 4$$

So P has an area of 4 units^2.

$$\int_{0}^{2} x(x-2)(x+2)\,dx = \int_{0}^{2} (x^3 - 4x)\,dx$$
$$= \left[\frac{x^4}{4} - 2x^2\right]_{0}^{2}$$
$$= \left[\frac{2^4}{4} - 2\times 2^2\right]$$
$$= -4$$

The negative sign indicates that the area is *below* the x axis.

So Q also has an area of 4 units^2.

? How could you have deduced the area of Q without any further calculation?

(ii) $\int_{-2}^{2} x(x-2)(x+2)\,dx = \int_{-2}^{2} (x^3 - 4x)\,dx$

$$= \left[\frac{x^4}{4} - 2x^2\right]_{-2}^{2}$$

$$= \left[\frac{2^4}{4} - 2\times 2^2\right] - \left[\frac{(-2)^4}{4} - 2\times(-2)^2\right]$$

$$= 0$$

(iii) The areas of P and Q have 'cancelled out'.

In Example 10.10 would you say that the area between the curve and the x axis is 0 units^2 or 8 units^2?

Always draw a sketch graph when you are going to calculate areas. This will avoid any cancelling out between areas above and below the x axis.

EXERCISE 10D

1 The sketch shows the curve $y = x^3 - x$.
Calculate the area of the shaded region.

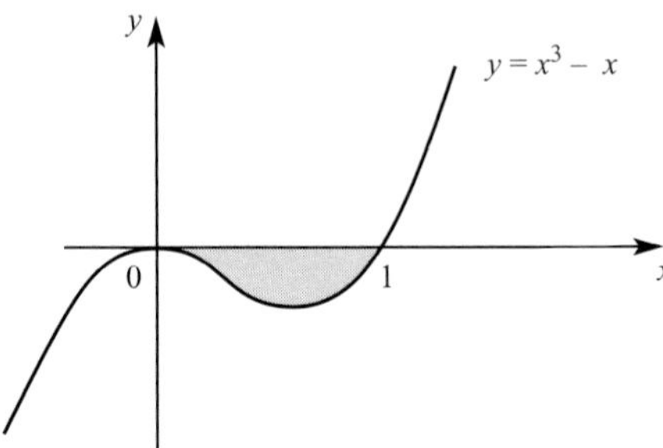

2 The sketch shows the curve $y = x^3 - 4x^2 + 3x$.

(i) Calculate the area of the shaded regions P and Q.

(ii) State the area enclosed between the curve and the x axis.

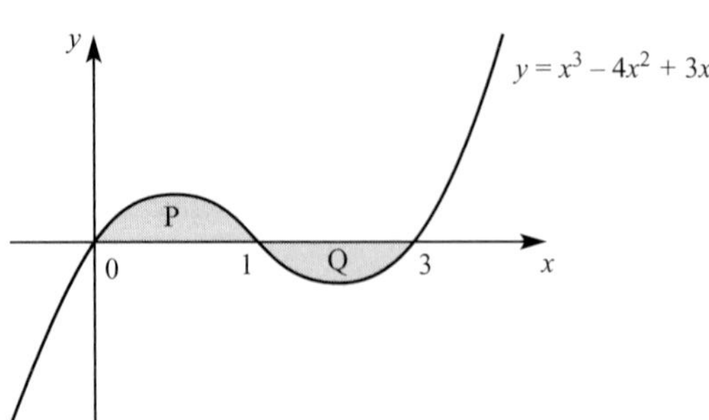

3 The sketch shows the curve $y = x^4 - 2x$.

(i) Find the co-ordinates of the point A.

(ii) Calculate the area of the shaded region.

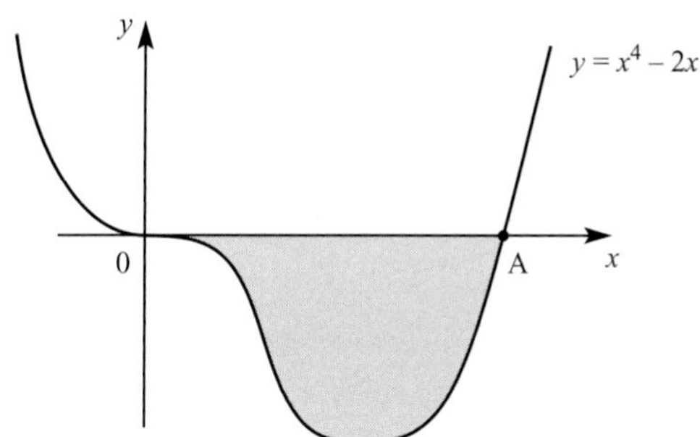

4 The sketch shows the curve $y = x^3 + x^2 - 6x$.
Find the area between the curve and the x axis.

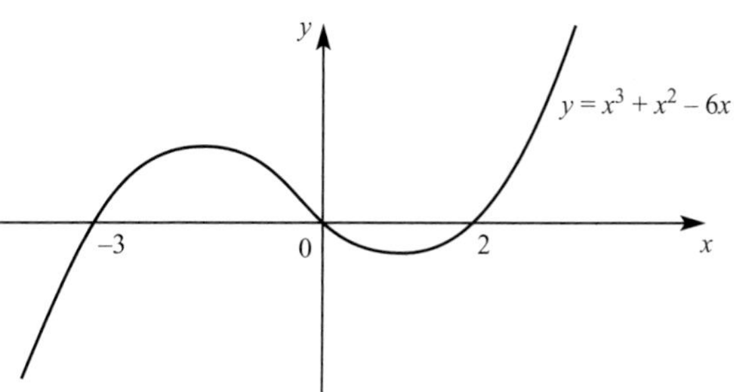

5 (i) Sketch the curve $y = x^2$ for $-3 \leqslant x \leqslant 3$.

(ii) Shade the area bounded by the curve, the lines $x = -1$ and $x = 2$ and the x axis.

(iii) Find, by integration, the area of the region you have shaded.

6 (i) Sketch the curve $y = x^2 - 2x$ for $-1 \leqslant x \leqslant 3$.

(ii) For what values of x does the curve lie below the x axis?

(iii) Find the area between the curve and the x axis.

7 (i) Sketch the curve $y = x^3$ for $-3 \leqslant x \leqslant 3$.

(ii) Shade the area between the curve, the x axis and the line $x = 2$.

(iii) Find, by integration, the area of the region you have shaded.

(iv) Without any further calculation, state, with reasons, the value of $\int_{-2}^{2} x^3 \, dx$.

8 (i) Shade, on a suitable sketch, the region with an area given by

$$\int_{-1}^{2} (x^2 + 1) \, dx.$$

(ii) Evaluate this integral.

9 **(i)** Evaluate $\int_1^4 (2x + 1)\,dx$.

(ii) Interpret this integral on a sketch graph.

10 **(i)** Sketch the curve $y = (x - 1)(x - 3)$ and find the co-ordinates of the points where the curve crosses the x axis.

(ii) Calculate the area between the curve and the x axis.

(iii) Without any further calculation, explain why

$$\int_0^1 (x - 1)(x - 3)\,dx = \int_3^4 (x - 1)(x - 3)\,dx.$$

11 The cross-section of a river bed has the equation $y = \frac{x^4}{40\,000}$, where x and y are measured in metres as indicated in the diagram.
The river is 40 m wide at surface level.

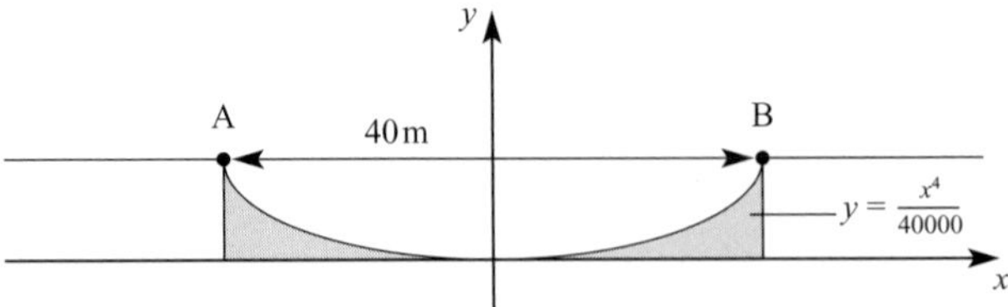

(i) Find the values of x at A and B.

(ii) Find the shaded area.

(iii) Hence find the area of the cross-section of the river bed.

(iv) The water flows at a speed of $1.6\,\text{ms}^{-1}$.
How many cubic metres of water pass a particular point each hour?

12 The diagram shows a flower bed set into a rectangular garden. The rest of the garden is lawn. x and y are measured in metres.
The equation of the curved edge of the flower bed is $y = \frac{1}{4}x^2 - 2x + 22$.

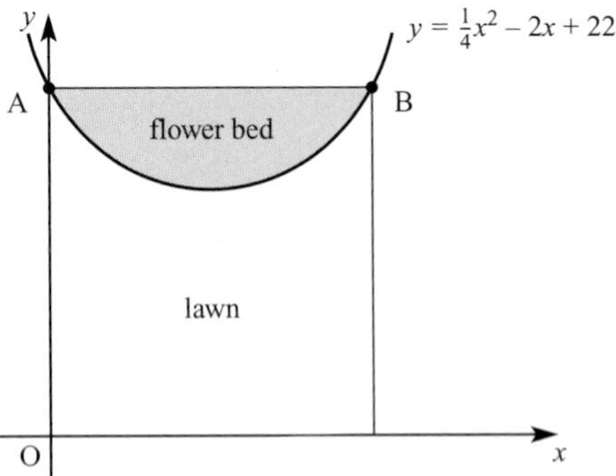

(i) Find the y co-ordinate of the point A.

(ii) Hence find the x co-ordinate of the point B.

(iii) Find the area of the lawn.

(iv) Hence find the area of the flower bed.

(v) The flower bed is to be covered with top soil to a depth of 30 cm.
How many cubic metres of top soil are needed?

13 The child's slide shown in the diagram is bounded by parts of the curves $x = 0$, $y = 0.25$, $y = 0.125x^2 - 0.5x + 0.75$, $y = 2.25$ and $2x + y = 15.25$ as indicated.

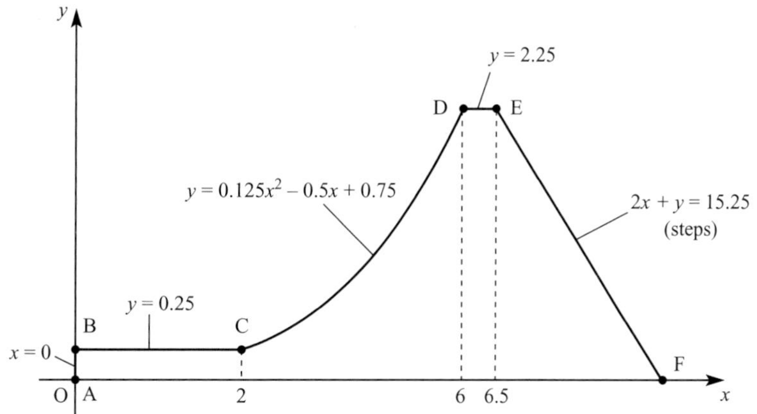

Children have been running underneath the slide and this is considered dangerous, so the Council has decided to board in the space ABCDEF and cover it with a mural.

(i) Find the co-ordinates of the point F.

(ii) Find the area of the board required to enclose one side of the slide.

The area between two curves

EXAMPLE 10.11 Find the area enclosed between the line $y = 5 - x$ and the curve $y = x^2 - 3x + 5$.

SOLUTION

You need to draw a sketch graph, but first you have to find where the curves intersect.

At the points of intersection

$$x^2 - 3x + 5 = 5 - x$$
$$\Rightarrow \quad x^2 - 2x = 0$$
$$\Rightarrow \quad x(x - 2) = 0$$
$$\Rightarrow \quad x = 0 \text{ or } x = 2.$$

The curves intersect at (0, 5) and (2, 3).

$y = 5 - x$ is a line of gradient –1 passing through (0, 5).

$y = x^2 - 3x + 5$ is a ∪-shaped quadratic also passing through (0, 5).

The sketch is shown in figure 10.10.

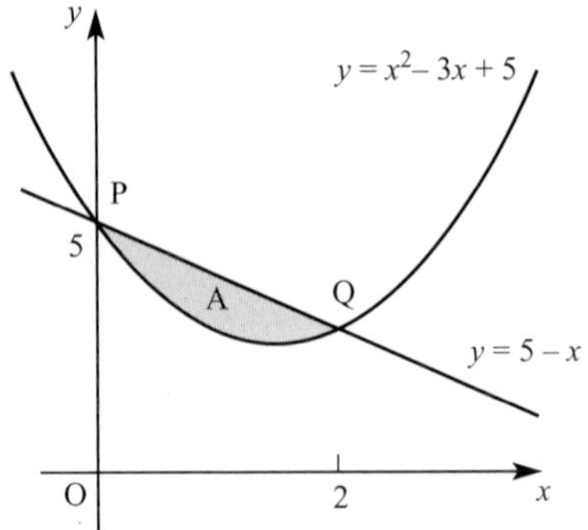

Figure 10.10

The shaded area can now be found in two ways.

Method 1

The area A can be treated as the difference between two areas B and C as shown in figure 10.11.

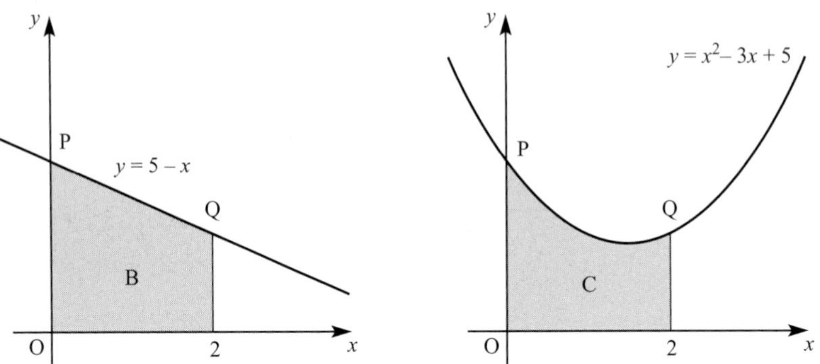

Figure 10.11

$$\begin{aligned} A &= B - C \\ &= \int_0^2 (5 - x)\,dx - \int_0^2 (x^2 - 3x + 5)\,dx \\ &= \left[5x - \frac{x^2}{2}\right]_0^2 - \left[\frac{x^3}{3} - \frac{3x^2}{2} + 5x\right]_0^2 \\ &= [(10 - 2) - 0] - \left[\left(\frac{8}{3} - 6 + 10\right) - 0\right] \\ &= 1\tfrac{1}{3}\text{ units}^2. \end{aligned}$$

Alternatively, use the formula for the area of a trapezium for B.

Method 2

Notice how method 1 started by calculating

$$\int_0^2 (\text{top curve})\ \mathrm{d}x - \int_0^2 (\text{bottom curve})\ \mathrm{d}x.$$

These two integrals have the same limits, 0 and 2, so can be combined as

$$\int_0^2 (\text{top curve} - \text{bottom curve})\ \mathrm{d}x$$

$$= \int_0^2 ((5 - x) - (x^2 - 3x + 5))\ \mathrm{d}x$$

$$= \int_0^2 (2x - x^2)\ \mathrm{d}x$$

$$= \left[x^2 - \frac{x^3}{3}\right]_0^2$$

$$= \left[\left(4 - \tfrac{8}{3}\right) - 0\right] = 1\tfrac{1}{3}\ \text{units}^2.$$

? How does combining the integrals make your calculations easier?

The big advantage of the second method is apparent when the area you require lies partly above and partly below the x axis, as in Example 10.12.

EXAMPLE 10.12

Figure 10.12 shows the curve $y = x^3 - 4x$ and the line $y = 8x + 16$ which is a tangent to the curve at the point A(–2, 0).

The tangent meets the curve again at B(4, 48).

Find the area enclosed between the line and the curve.

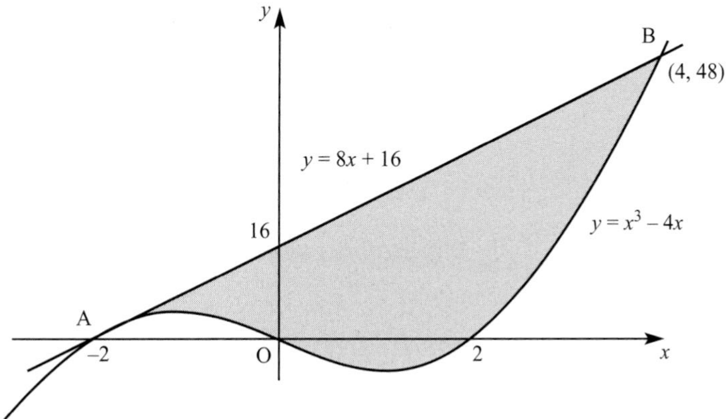

Figure 10.12

SOLUTION

$$\text{Area} = \int_{-2}^{4} (\text{top curve} - \text{bottom curve})\,dx$$

$$= \int_{-2}^{4} ((8x + 16) - (x^3 - 4x))\,dx$$

$$= \int_{-2}^{4} (12x + 16 - x^3)\,dx$$

$$= \left[6x^2 + 16x - \frac{x^4}{4}\right]_{-2}^{4}$$

$$= [(96 + 64 - 64) - (24 - 32 - 4)]$$

$$= 108 \text{ units}^2.$$

? Why was the second method so useful in this example?

EXERCISE 10E

1 For each of parts **(i)** to **(x)**

(a) sketch a graph of the given curves

(b) find the x co-ordinates of the points of intersection

(c) find the area enclosed by the two curves.

(i)	$y = 4 - x$;	$y = (x - 1)(x - 4)$
(ii)	$y = x + 1$;	$y = x^2 - 3x - 4$
(iii)	$y = 1 - x$;	$y = (x - 1)^2$
(iv)	$y = 1 - x^2$;	$y = x^2 - 1$
(v)	$y = x + 2$;	$y = x^2 + x - 2$
(vi)	$x + y = 9$;	$y = x^2 - 2x + 3$
(vii)	$y = x(x - 5)$;	$y = x(10 - x)$
(viii)	$y = 16 - x^2$;	$y = x^2 - 5x + 13$
(ix)	$y = x^2 - 16$;	$y = 4x - x^2$
(x)	$y = x + 1$;	$y = 5x - x^2 + 6$

2 A decorative mirror is bounded by the curves

$$y = \frac{x^2}{8} \text{ and } y = 80 - \frac{x^2}{8}$$

and the lines

$$x = 16 \text{ and } x = -16.$$

(i) Sketch the mirror.

(ii) Find the area of the mirror glass.

3 A sculpture is to be made up from a number of copper sheets, one of which is shown in the diagram.

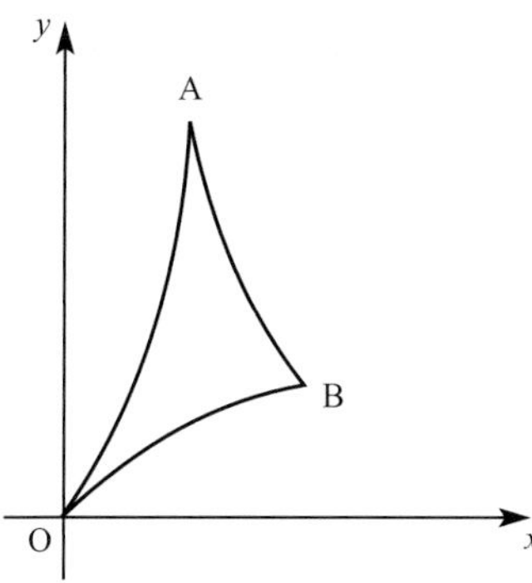

The equations of the sides are

$$y = x(x + 3),\ \ y = x - \tfrac{1}{4}x^2 \text{ and } y = x^2 - 6x + 9.$$

All dimensions are in metres.

(i) Find which side has which equation.

(ii) Find the co-ordinates of A and B.

(iii) Calculate the area of the shape.

4 Some table mats are being designed in the shape of a flower, as shown in the diagram.
Each mat is made up of six equal sectors and all dimensions are in centimetres.

Line OC has equation $y = \sqrt{3}x$ and curve ABC has equation $y = 16.2 - \dfrac{x^2}{12}$.
Line OB is a line of symmetry.

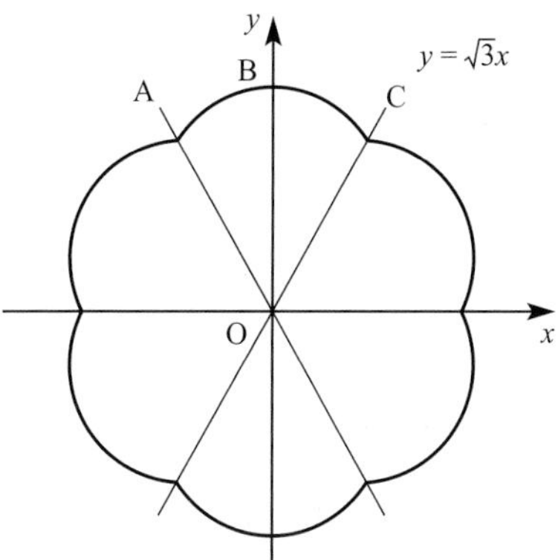

(i) Find the co-ordinates of the point C.

(ii) Find the area of OBC.

(iii) Hence find the total area of the table mat.

KEY POINTS

1 $\dfrac{dy}{dx} = x^n \quad \Rightarrow \quad y = \dfrac{x^{n+1}}{n+1} + c$

2 $\displaystyle\int_a^b x^n \, dx = \left[\frac{x^{n+1}}{n+1}\right]_a^b = \frac{b^{n+1} - a^{n+1}}{n+1}$

3 Area A $= \displaystyle\int_a^b y \, dx = \int_a^b f(x) \, dx$

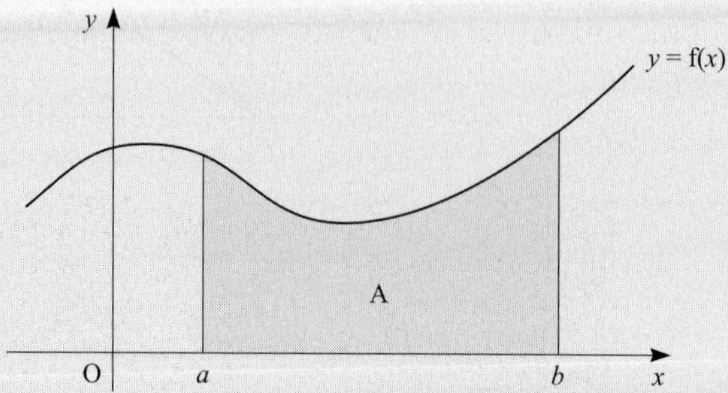

4 Areas below the x axis give rise to negative values for the integral.

5 Area B $= \displaystyle\int_a^b (\text{top curve} - \text{bottom curve}) \, dx = \int_a^b (f(x) - g(x)) \, dx$

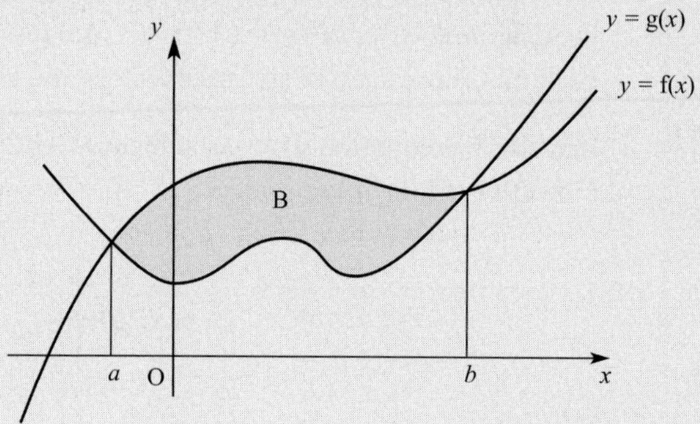

11 Calculus III – applications to kinematics

If I have seen further than others, it is by standing upon the shoulders of giants.

Isaac Newton (1642–1727)

Think about your most recent car journey. Describe the first few minutes of it using words involving direction, position, time, speed, and saying whether the car is speeding up or slowing down.

Historical note

Sir Isaac Newton was one of the greatest mathematicians of all times and much of his early work on calculus and mechanics forms the basis of advanced level mathematics. The 'giants' referred to above include Descartes, Kepler and Galileo. From Descartes, Newton inherited analytical geometry, which he found difficult at first, from Kepler, three fundamental laws of planetary motion, discovered empirically after twenty-two years of inhuman calculation, while from Galileo he acquired the first two of the three laws of motion which were to be the cornerstone of his own dynamics. Newton studied at Trinity College, Cambridge, and, at the age of twenty-six, became Professor of Mathematics there. In 1701–2 he represented Cambridge University in Parliament, and in 1703 he was elected President of the Royal Society. He is buried in Westminster Abbey.

In Chapter 9 you met the expression $\frac{dy}{dx}$. It is the rate of change of y with respect to x. It gives the gradient of the x–y graph, where x is plotted on the horizontal axis and y on the vertical axis.

Now look at the graph in figure 11.1. It represents the distance, s metres, travelled by a cyclist along a country road in time, t seconds. Time is measured along the horizontal axis and distance from the starting point is measured on the vertical axis. When he reaches E the cyclist takes a short break and then returns home along the same road.

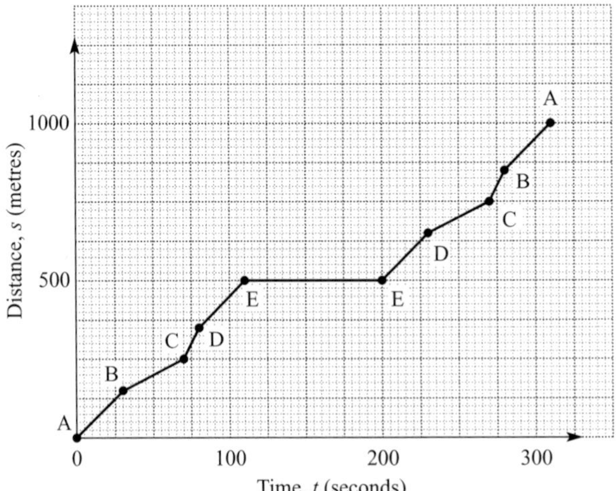

Figure 11.1

Speed is given by the gradient of the distance–time graph. How do you write this using calculus?

What can you say about the speed of the cyclist on his journey to E along the sections of the road represented by AB, BC, CD and DE?

What do you think is the nature of the road for each of these parts of the graph?

ACTIVITY 11.1

(i) Draw a graph showing the distance from the starting point in the positive direction (called the *displacement*) instead of the distance travelled.

(ii) *Velocity* is the gradient of the displacement–time graph.
Figure 11.2 shows the first part of a velocity–time graph for this journey. Complete this graph.

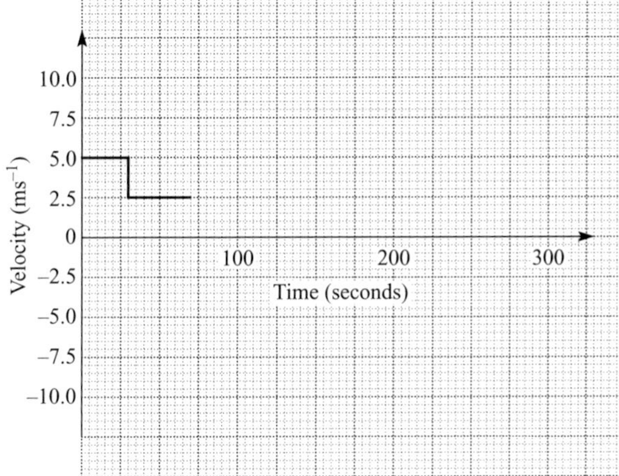

Figure 11.2

(i) What quantity is given by the area under a velocity–time graph?

(ii) The gradient of a velocity–time graph is

$$\frac{\text{change in velocity}}{\text{time taken}}.$$

What is the name of this quantity?

Motion in a straight line

In the work that follows you will use displacement, which measures position, rather than distance travelled.

NOTE

Kinematics is the branch of mathematics that studies the motion of objects, without making reference to any forces that are needed to produce that motion. Throughout this section you will be treating each object as a *particle*, i.e. an object with a mass but with *no dimensions*.

Before doing anything else you need to make two important decisions:

(i) where you will take your origin

(ii) which direction you will take as positive.

Some options are shown in figure 11.3.

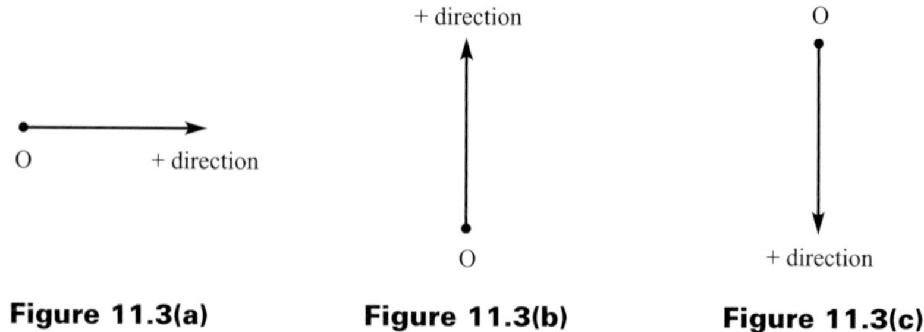

Figure 11.3(a) **Figure 11.3(b)** **Figure 11.3(c)**

Think about the motion of a tennis ball that is thrown up vertically and allowed to fall to the ground, as in the diagram in figure 11.4. Assume that the ball leaves your hand at a height of 1 m above the ground and rises a further 2 m to the highest point. At this point the ball is *instantaneously at rest.*

? What do you think is meant by the phrase 'instantaneously at rest?

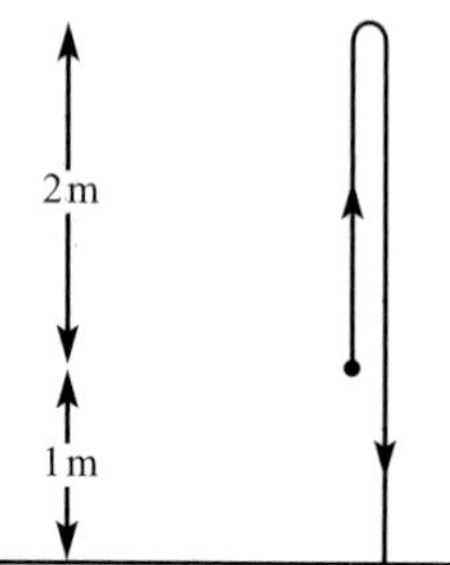

Figure 11.4

? **(i)** There are two obvious positions that could be used as the origin. What are they?

(ii) Which direction would you take as positive?

Figure 11.5 shows the displacement–time graph of the ball's flight. For this graph, displacement is measured from ground level, with upwards as the positive direction.

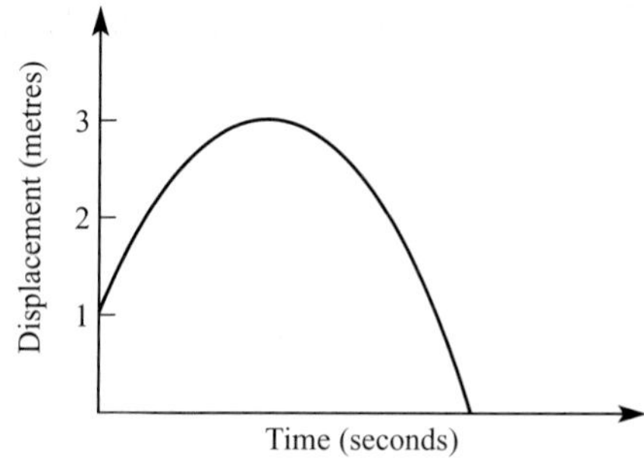

Figure 11.5

What would the graph look like if you took the starting point as the origin?

You need to be careful about the terms velocity and speed. Speed has magnitude (size) but not direction. *Velocity* has direction as well as a magnitude.

Taking upwards as the positive direction, a speed of 3 ms^{-1} upwards is a velocity +3 ms^{-1}.
A speed of 3 ms^{-1} downwards is a velocity of –3 ms^{-1}.

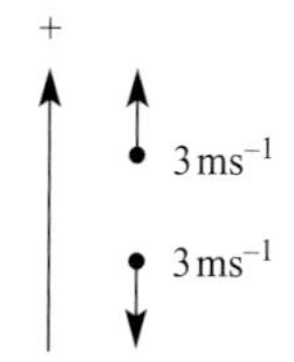

Figure 11.6

Look at this table. It gives the terms that you will be using, together with their definitions, units and the letters that are commonly used for those quantities. Standard units of metres and seconds are used here.

Quantity	Definition	S.I. unit	Symbol	Notation
Time	Measured from a fixed origin	second	s	t
Distance	Distance travelled in a given time	metre	m	x (or y)
Speed	Rate of change of distance	metre per second	ms^{-1}	$v = \frac{dx}{dt}$
Displacement	Distance from a fixed origin	metre	m	s (or h)
Velocity	Rate of change of displacement	metre per second	ms^{-1}	$v = \frac{ds}{dt}$
Acceleration	Rate of change of velocity	metre per second per second	ms^{-2}	$a = \frac{dv}{dt}$

You have seen that velocity can be either positive or negative. This is also true for acceleration. Suggest two different ways in which you might interpret a negative acceleration.

The constant acceleration formulae

Sometimes motion involves constant acceleration; at other times the acceleration is variable. This section deals with the special case of constant acceleration. The general case of variable acceleration is covered in the next section.

Figure 11.7

In figure 11.7 the children are trying to go down the slide as quickly as possible, so they have polished the slide and are pushing themselves off from the top. Oliver has calculated that they have an acceleration of $0.5\,\mathrm{ms^{-2}}$ as they slide, and he is able to give himself an initial velocity of $1\,\mathrm{ms^{-1}}$.

Figure 11.8 shows the velocity–time graph for the first three seconds of Oliver's motion.

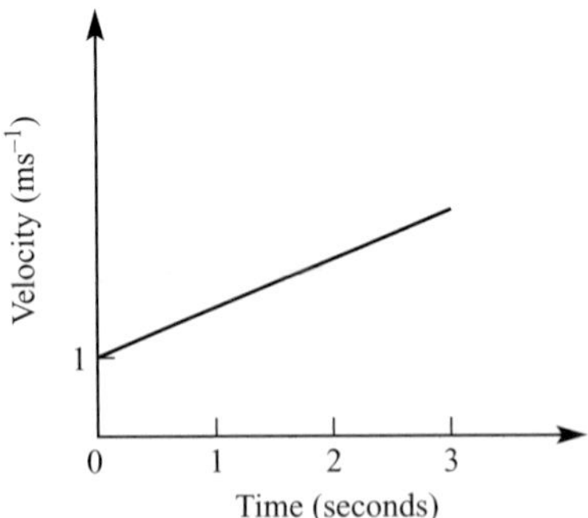

Figure 11.8

(i) Why is the graph in figure 11.8 a straight line?

(ii) What is Oliver's velocity after 3 seconds?

Figure 11.9 shows the general case.

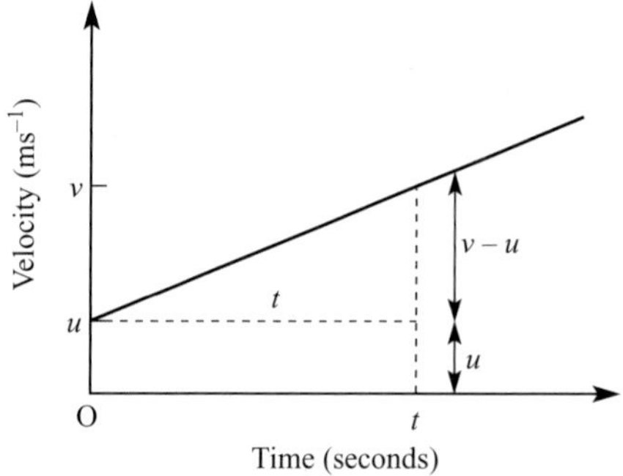

Figure 11.9

When the initial velocity is u ms^{-1} and the velocity at a time t seconds later is v ms^{-1}, the constant acceleration a ms^{-2} is given by

$$\frac{v-u}{t} = a$$

$$\Rightarrow \quad v = u + at.$$ ①

The area under the graph represents the distance, s metres, travelled, so

$$s = \frac{u+v}{2} \times t.$$ ②

Equations ① and ② can be manipulated to give other relationships.

Using equation ① to substitute for v in equation ② gives

$$s = \frac{2u + at}{2} \times t.$$

$$\Rightarrow \quad s = ut + \tfrac{1}{2}at^2.$$ ③

From ① and ②

$$as = \frac{v-u}{t} \times \frac{u+v}{2} \times t$$

$$\Rightarrow \quad as = \frac{v^2 - u^2}{2}$$

$$\Rightarrow \quad v^2 = u^2 + 2as.$$ ④

These four equations are sometimes called the *suvat* equations and can be used whenever an object is moving with a *constant acceleration.*

 Only use the *suvat* equations when the acceleration is constant.

Remember to specify the positive and negative directions clearly.

EXAMPLE 11.1

A particle is moving in a straight line from O to A with a constant acceleration of $2\,\text{ms}^{-2}$.

It takes 20 seconds to reach A and its velocity there is $45\,\text{ms}^{-1}$.

(i) Find its initial velocity.

(ii) Find the distance OA.

SOLUTION

Start by listing the known quantities.

(i) $a = 2\,\text{ms}^{-2},\ t = 20\,\text{s},\ v = 45\,\text{ms}^{-1}$

To find the initial velocity, u, use the equation $v = u + at$.

$\Rightarrow \quad 45 = u + 2 \times 20$

$\Rightarrow \quad u = 5\,\text{ms}^{-1}$

(ii) Now use $s = ut + \frac{1}{2}at^2$ to find the distance, s m.

$$s = (5 \times 20) + \tfrac{1}{2} \times 2 \times 20^2$$
$$= 500\,\text{m}$$

EXAMPLE 11.2

Maisie is cycling at a steady speed of $6\,\text{ms}^{-1}$ when she comes to a hill which causes her to slow down at a rate of $0.5\,\text{ms}^{-2}$.

(i) How far does she travel up the hill before her speed is reduced to zero?

(ii) How long does it take before her speed is reduced to $0.5\,\text{ms}^{-1}$?

SOLUTION

Slowing down is represented by a negative acceleration.

(i) $u = 6\,\text{ms}^{-1},\ a = -0.5\,\text{ms}^{-2},\ v = 0\,\text{ms}^{-1}$

To find the distance, sm, use $v^2 = u^2 + 2as$.

$\Rightarrow \quad 0 = 6^2 + 2 \times (-0.5) \times s$

$\Rightarrow \quad s = 36$

$\Rightarrow$ She travels 36 m up the hill.

(ii) Now $u = 6\,\text{ms}^{-1},\ a = -0.5\,\text{ms}^{-2},\ v = 0.5\,\text{ms}^{-1}$

To find the time, ts, use $v = u + at$.

$\Rightarrow \quad 0.5 = 6 + (-0.5)t$

$\Rightarrow \quad 0.5t = 5.5$

$\Rightarrow \quad t = 11$

$\Rightarrow$ After 11 seconds her speed is reduced to $0.5\,\text{ms}^{-1}$.

Acceleration due to gravity

This is a special case of constant acceleration. Ignoring air resistance, all bodies falling freely under gravity fall with the same constant acceleration, g. The value of g is about $9.8\,\text{ms}^{-2}$ on Earth, although it does vary slightly from place to place. It is often more convenient to use $g = 10\,\text{ms}^{-2}$ as in Example 11.3.

EXAMPLE 11.3

A stone is dropped from a bridge into the river below.
The bridge is 30 m above the river and the acceleration due to gravity can be taken as $10\,\text{ms}^{-2}$.

(i) Find the time it takes the stone to reach the water.

(ii) Find the speed of impact.

? Which word in the question tells you the value of the initial velocity?

SOLUTION

(i) Taking downwards as the positive direction and listing the known quantities gives $u = 0\,\text{ms}^{-1}$, $a = 10\,\text{ms}^{-1}$, $s = 30\,\text{m}$.

To find the time use the equation $s = ut + \frac{1}{2}at^2$.

$\Rightarrow \quad 30 = (0 \times t) + (\frac{1}{2} \times 10 \times t^2)$

$\Rightarrow \quad 30 = 5t^2$

$\Rightarrow$ The stone takes 2.45 s (3 s.f.)

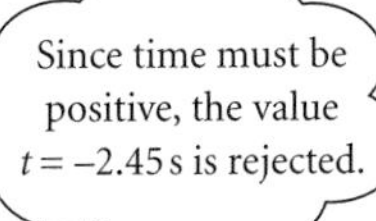

? What happens if you take upwards as the positive direction?

(ii) To find the speed of impact use the equation $v^2 = u^2 + 2as$.

$\Rightarrow \quad v^2 = 0^2 + (2 \times 10 \times 30)$

$\Rightarrow \quad v^2 = 600$

$\Rightarrow$ The speed of impact is $24.5\,\text{ms}^{-1}$ (3 s.f.)

? Which other equation could be used for **(ii)**?

EXERCISE 11A

1 Decide which *suvat* equation to use in each of these situations.

(i) Given u, t, a; find s. **(ii)** Given u, a, s; find v.
(iii) Given u, a, t; find v. **(iv)** Given u, v, t; find s.
(v) Given u, a, v; find t. **(vi)** Given a, s, v; find u.
(vii) Given u, s, t; find a. **(viii)** Given v, a, t; find u.
(ix) Given s, u, v; find t. **(x)** Given u, v, s; find a.

2 **(i)** Find s when $u = 2$, $t = 5$ and $a = 4$.
(ii) Find s when $u = 0$, $v = 8$ and $t = 3$.
(iii) Find v when $u = 3$, $a = 5$ and $t = 2$.
(iv) Find a when $u = 7$, $v = 3$ and $s = 4$.
(v) Find u when $s = 24$, $t = 3$ and $a = 4$.

3 A particle moves in a straight line with a uniform accelerations of $4\,\text{ms}^{-2}$.
It starts from rest when $t = 0$ seconds.
Find its velocity after 5 seconds.

4 A particle moves in a straight line with a constant acceleration.
Initially its velocity is $2\,\text{ms}^{-1}$ and after 5 seconds its velocity is $19\,\text{ms}^{-1}$.
Find the acceleration.

5 A particle with an initial velocity of $10\,\text{ms}^{-1}$ is moving in a straight line with a constant retardation of $2\,\text{ms}^{-2}$.
How long will it be before it is instantaneously at rest?

6 A car is moving at a speed of $15\,\text{ms}^{-1}$ when the driver sees a child run into the road 25 m ahead.
What deceleration will the driver need if he is to stop before hitting the child?

7 A book falls off a table that is 0.9 m high. Taking g as $9.8\,\text{ms}^{-2}$, find
(i) the speed at which the book is travelling when it lands on the floor
(ii) the time that it takes to reach the floor.

8 A particle is moving along a straight line with a constant acceleration.
It starts from A with a velocity of $4\,\text{ms}^{-1}$ and passes points B and C after 3 seconds and 5 seconds respectively.
The distance AC is 45 m.
(i) Represent this information on a velocity–time graph.
(ii) Find the velocity of the particle when it passes C.
(iii) Find the acceleration of the particle.
(iv) Find the velocity of the particle when it passes B.

9 A ball is thrown vertically upwards with a speed of $10\,\text{ms}^{-1}$ and caught when it returns to its starting point.
Taking g as $10\,\text{ms}^{-2}$, find
(i) the time taken to the highest point
(ii) the greatest height of the ball above the point of projection
(iii) the total time that it is in the air.

10 An object moves along a straight line.
It starts at the origin with a velocity of $6\,\mathrm{ms}^{-1}$ and has a constant acceleration of $-4\,\mathrm{ms}^{-2}$.

(i) After what time is the object instantaneously at rest?
(ii) How far is it from the origin at this time?
(iii) When does the object next pass through the origin?
(iv) When is the object 20 m from the origin?

11 A car is travelling along a straight road.
It accelerates from rest at $4\,\mathrm{ms}^{-2}$ until it reaches a speed of $20\,\mathrm{ms}^{-1}$.
It maintains this speed for 2 minutes and then decelerates uniformly to rest at $5\,\mathrm{ms}^{-2}$.

(i) Represent this information on a velocity–time graph.
(ii) Find the total time for the journey.
(iii) Find the total distance travelled.

Motion with variable acceleration: the general case

In the special case of constant acceleration you can use the *suvat* equations. However, where the motion involves variable acceleration you must use calculus. The two key relationships are

$$v = \frac{\mathrm{d}s}{\mathrm{d}t} \quad \text{and} \quad a = \frac{\mathrm{d}v}{\mathrm{d}t}.$$

Their use is illustrated in the following examples.

EXAMPLE 11.4

For the first 4 seconds, the displacement, in metres, of a sports car from its initial position is given by

$$s = 12t^2 - t^3.$$

Find

(i) an expression for the velocity in terms of t
(ii) the initial velocity
(iii) the velocity after 4 seconds
(iv) an expression for the acceleration in terms of t
(v) the accelerations after 4 seconds.

The national speed limit in Great Britain is 70 mph.

(vi) At the end of 4 seconds, would the driver be breaking the law?

SOLUTION

(i) $$v = \frac{\mathrm{d}s}{\mathrm{d}t} = 24t - 3t^2$$

(ii) When $t = 0$, $v = 0\,\mathrm{ms}^{-1}$.

(iii) When $t = 4$,

$$v = 24 \times 4 - 3 \times 4^2$$
$$= 48\,\text{ms}^{-1}.$$

(iv) $a = \dfrac{\mathrm{d}v}{\mathrm{d}t}$

$= 24 - 6t$

Notice that a is not constant so it would have been wrong to use the *suvat* equations.

(v) When $t = 4$,

$$a = 24 - 6 \times 4$$
$$= 0\,\text{ms}^{-2}.$$

(vi) $48\,\text{ms}^{-1} = \dfrac{48 \times 60 \times 60}{1000}$

$= 172.8\,\text{km}\,\text{h}^{-1}$

$= \frac{5}{8} \times 172.8$

$= 108$ mph (approx.)

$\Rightarrow$ The driver would be breaking the law.

EXAMPLE 11.5

A particle moves in a straight line such that at time t its displacement, s, from a fixed point O on that line is given by $s = 5 + 2t^3 - 3t^2$.

Find

(i) expressions for the velocity and acceleration in terms of t

(ii) the times when it is at rest and sketch the velocity–time graph

(iii) how far it is from O when it is at rest

(iv) the initial acceleration of the particle.

SOLUTION

(i) $v = \dfrac{\mathrm{d}s}{\mathrm{d}t} = 6t^2 - 6t$

$a = \dfrac{\mathrm{d}v}{\mathrm{d}t} = 12t - 6$

? How can you tell that a is not constant?

(ii) The particle is at rest when $v = 0$.

$\Rightarrow$ $6t^2 - 6t = 0$

$\Rightarrow$ $6t(t - 1) = 0$

$\Rightarrow$ $t = 0$ or $t = 1$

$\Rightarrow$ The particle is at rest initially and after 1 second.

The graph of v against t is a U-shaped curve, crossing the t axis at $t = 0$ and $t = 1$.

Figure 11.10 shows a sketch of the velocity–time graph.

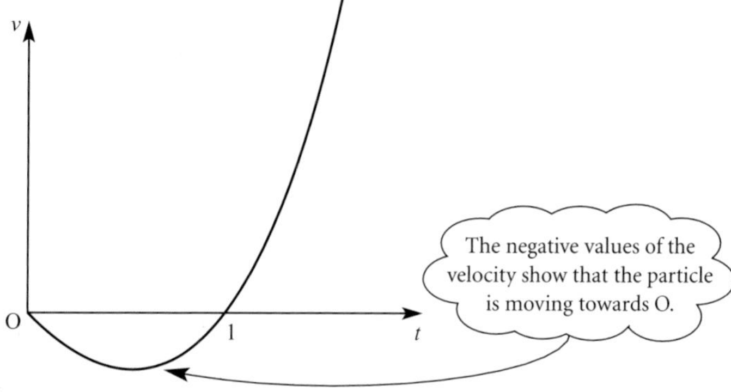

Figure 11.10

(iii) When $t = 0$, $s = 5$.
When $t = 1$,

$$s = 5 + 2 - 3$$
$$= 4.$$

$\Rightarrow$ Initially the particle is at rest 5 m from O and, after 1 second, it is instantaneously at rest 4 m from O.

(iv) When $t = 0$, $a = -6\,\text{ms}^{-2}$.

? How would you interpret the negative acceleration in Example 11.5?

EXERCISE 11B

1 In each of the following cases

(a) find expressions for the velocity and acceleration at time t

(b) use these expressions to find the initial position, velocity and acceleration

(c) find the time and position when the velocity is zero.

(i) $s = 5t^2 - t + 3$

(ii) $s = 3t - t^3$

(iii) $s = t^4 - 4t - 6$

(iv) $s = 4t^3 - 3t + 5$

(v) $s = 5 - 2t^2 + t$

2 A body is projected in a straight line from a point O.
After t seconds its displacement, s metres, from O is given by $s = 3t^2 - t^3$.

(i) Find expressions for the velocity and acceleration at time t

(ii) Find the times when the body is instantaneously at rest.

(iii) Find the distance travelled between these times.

(iv) Find the velocity when $t = 4$ and interpret your result.

(v) Find the initial acceleration.

3 A ball is thrown upwards and its height, h metres, above ground after t seconds is given by $h = 1 + 4t - 5t^2$.

(i) From what height was the ball projected?

(ii) Find an expression for the velocity of the ball at time t.

(iii) When is the ball instantaneously at rest?

(iv) What is the greatest height reached by the ball?

(v) After what length of time does the ball hit the ground?

(vi) Sketch the graph of h against t.

(vii) At what speed is the ball travelling when it hits the ground?

4 In the early stages of its motion the height of a rocket, h metres, is given by $h = \frac{1}{6}t^4$, where t seconds is the time after blast-off.

(i) Find expressions for the velocity and acceleration of the rocket at time t.

(ii) After how long is the acceleration of the rocket $72\,\text{ms}^{-2}$?

(iii) Find the height and velocity of the rocket at this time.

5 The velocity of a moving object at time t seconds is given by $v\,\text{ms}^{-1}$, where $v = 15t - 2t^2 - 25$.

(i) Find the times when the object is instantaneously at rest.

(ii) Find the acceleration at these times.

(iii) Find the velocity when the acceleration is zero.

(iv) Sketch the graph of v against t.

Finding displacement from velocity and velocity from acceleration

In the previous section, you used the result $v = \frac{ds}{dt}$. This meant that when s was given as an expression in t, you differentiated s to get v. Reversing this, when v is given as an expression in t, integrating v will give you an expression for s.

$$s = \int v\,dt$$

Similarly, the result $a = \frac{dv}{dt}$ reverses to give

$$v = \int a\,dt.$$

EXAMPLE 11.6

A particle P moves in a straight line so that at time t seconds its acceleration is $(6t + 2)\,\text{ms}^{-2}$.

P passes through a point O at time $t = 0$ with a velocity of $3\,\text{ms}^{-1}$.

Find

(i) the velocity of P in terms of t

(ii) the distance of P from O when $t = 2$.

Is the acceleration constant?

SOLUTION

(i) $v = \int a \, dt$

$\Rightarrow \quad v = \int (6t + 2) \, dt$

$\Rightarrow \quad v = 3t^2 + 2t + c$

$v = 3$ when $t = 0$

$\Rightarrow \quad c = 3$

$\Rightarrow \quad v = 3t^2 + 2t + 3$

What does the value of c represent?

(ii) $s = \int v \, dt$

$\Rightarrow \quad s = \int (3t^2 + 2t + 3) \, dt$

$\Rightarrow \quad s = t^3 + t^2 + 3t + k$

$s = 0$ when $t = 0$

$\Rightarrow \quad k = 0$

$\Rightarrow \quad s = t^3 + t^2 + 3t$

When $t = 2$,

$$s = 8 + 4 + 6$$
$$= 18.$$

The particle is 18 m from O.

What does the value of k represent?

EXAMPLE 11.7

The acceleration of a particle, in ms^{-2}, at time t seconds is given by $a = 6 - 2t$. When $t = 0$, the particle is at rest at a point 4 m from the origin O.

(i) Find expressions for the velocity and displacement in terms of t.

(ii) Find when the particle is next at rest, and its displacement from O at that time.

? Is the acceleration constant?

SOLUTION

(i) $v = \int a\,\mathrm{d}t$

$\Rightarrow \quad v = \int (6 - 2t)\,\mathrm{d}t$

$\Rightarrow \quad v = 6t - t^2 + c$

$v = 0$ when $t = 0$

$\Rightarrow \quad c = 0$

$\Rightarrow \quad v = 6t - t^2$

$s = \int v\,\mathrm{d}t$

$\Rightarrow \quad s = \int (6t - t^2)\,\mathrm{d}t$

$\Rightarrow \quad s = 3t^2 - \frac{t^3}{3} + k$

$s = 4$ when $t = 0$

$\Rightarrow \quad k = 4$

$\Rightarrow \quad s = 3t^2 - \frac{t^3}{3} + 4$

(ii) $v = 6t - t^2 \quad \Rightarrow$ the particle is at rest when $6t - t^2 = 0$.

$t(6 - t) = 0 \quad \Rightarrow \quad t = 0$ or $t = 6$

The particle is next at rest after 6 seconds.

When $t = 6$,

$$s = 3 \times 6^2 - \frac{6^3}{3} + 4$$
$$= 40.$$

The particle is 40 m from O.

EXAMPLE 11.8

A particle is projected along a straight line.
At time t its velocity is given by $v = 2t + 3$.
The units are metres and seconds.

(i) Sketch the graph of v against t.
(ii) Find the distance the particle moves in the third second.

SOLUTION

(i) $v = 2t + 3$ is a straight line with gradient 2, passing through (0, 3).

Figure 11.11 shows a sketch of the graph of v against t.

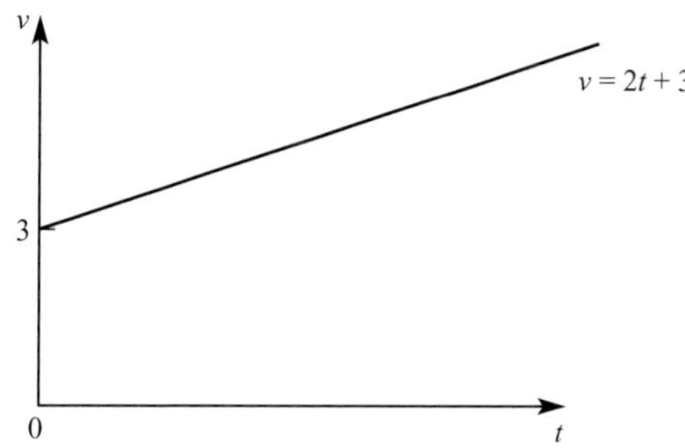

Figure 11.11

(ii) *Method 1*

The graph shows that the velocity is always positive, so velocity and speed are the same.

The distance travelled is equal to the area under the graph.

The third second starts when $t = 2$ and finishes when $t = 3$.

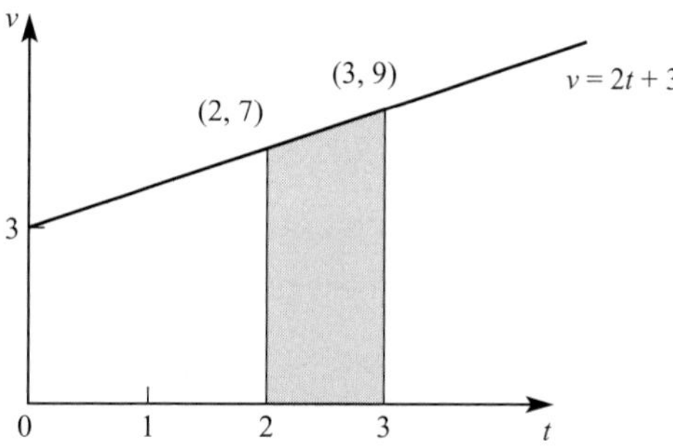

Figure 11.12

Using the formula for the area of a trapezium,

$$\begin{aligned}\text{distance} &= \tfrac{1}{2}(7 + 9) \times 1 \\ &= 8\,\text{m}.\end{aligned}$$

Method 2

The area under a graph can also be found using integration.

$$\begin{aligned}\text{Distance} &= \int_a^b v\,\mathrm{d}t \\ &= \int_2^3 (2t + 3)\,\mathrm{d}t \\ &= \Big[t^2 + 3t\Big]_2^3 \\ &= [9 + 9] - [4 + 6] \\ &= 8\,\text{m}.\end{aligned}$$

? Which method did you prefer to use in Example 11.8?

? **(i)** Which method would you need to use if v was given by $v = 3t^2 + 2$?
(ii) Is acceleration constant in this case?
How can you tell?
(iii) Could you have used the *suvat* equations?

? Can you use calculus when acceleration is constant?

EXERCISE 11C

1 In each of the following cases find expressions for the velocity, v, and displacement, s, at time t.

(i) $a = 2 - 6t$; when $t = 0$, $v = 1$ and $s = 0$.
(ii) $a = 4t$; when $t = 0$, $v = 4$ and $s = 3$.
(iii) $a = 12t^2 - 4$; when $t = 0$, $v = 2$ and $s = 1$.
(iv) $a = 2$; when $t = 0$, $v = 2$ and $s = 4$.
(v) $a = 4 + t$; when $t = 0$, $v = 1$ and $s = 3$.

2 A particle P sets off from the origin, O, with a velocity of $9\,\text{ms}^{-1}$ and moves along the x axis.
At time t seconds its acceleration is given by

$$a = (6t - 12)\,\text{ms}^{-2}.$$

(i) Find expressions for the velocity and displacement at time t.
(ii) Find the time when the particle returns to its starting point.

3 A particle P starts from rest at a fixed origin O when $t = 0$.
The acceleration $a\,\text{ms}^{-2}$ at time t seconds is given by

$$a = 6t - 6.$$

(i) Find the velocity of the particle after 1 second.
(ii) Find the next time when the particle is instantaneously at rest, and the distance travelled to this point.

4 During braking, the speed, $v\,\text{ms}^{-1}$, of a car is given by

$$v = 30 - 5t$$

where t seconds is the time since the brakes were applied.
(i) Sketch a graph of v, on the vertical axis, against t.
(ii) How long does the car take to stop?
(iii) How far does it travel while braking?

5 A particle P moves in a straight line, starting from rest at the point O. At time t seconds after leaving O, the acceleration, $a\,\text{ms}^{-2}$, of P is given by

$$a = 4 + 12t.$$

(i) Find an expression for the velocity of the particle at time t.

(ii) Calculate the distance travelled by P in the third second.

6 The velocity $v\,\text{ms}^{-1}$, of a particle P at time t seconds is given by

$$v = t^3 - 4t^2 + 4t + 2.$$

P moves in a straight line.

(i) Find an expression for the acceleration, $a\,\text{ms}^{-2}$, in terms of t.

(ii) Find the times at which the acceleration is zero, and say what is happening between these times.

(iii) Find the distance travelled in the first three seconds.

KEY POINTS

Standard units of metres and seconds are used here.

Quantity	Definition	S.I. unit	Symbol	Notation
Time	Measured from a fixed origin	second	s	t
Distance	Distance travelled in a given time	metre	m	x (or y)
Speed	Rate of change of distance	metre per second	$\mathrm{ms^{-1}}$	$v = \frac{dx}{dt}$
Displacement	Distance from a fixed origin	metre	m	s (or h)
Velocity	Rate of change of displacement	metre per second	$\mathrm{ms^{-1}}$	$v = \frac{ds}{dt}$
Acceleration	Rate of change of velocity	metre per second per second	$\mathrm{ms^{-2}}$	$a = \frac{dv}{dt}$

1 When the acceleration is constant and the initial velocity is u:

- $v = u + at$
- $s = \frac{u + v}{2} \times t$
- $s = ut + \frac{1}{2} at^2$
- $v^2 = u^2 + 2as$

2 Motion under gravity, with no air resistance, is subject to an acceleration of g.

The value of g is about $9.8\,\mathrm{ms^{-2}}$ on Earth.

3 For general motion

- $v = \frac{ds}{dt}$ (Velocity is the gradient of a displacement–time graph.)
- $a = \frac{dv}{dt}$
- $s = \int v\,dt$ (Displacement is the area under a velocity–time graph.)
- $v = \int a\,dt$.

Answers

Chapter 1

❓ (Page 3)

A small ice cream costs 80p, a large ice cream costs £1.20.

❓ (Page 3)

You just get $0 = 0$.

❓ (Page 4)

Express as a product of factors. Factorise *fully* means that you cannot split any of the factors into yet more factors.

Exercise 1A (Page 5)

1 **(i)** $10a - b - 2c$
(ii) $6x - 3y - 4z$
(iii) $19x + 5y$
(iv) $p + 14q$
(v) $5x$
(vi) $2a^2 + 12a - 12$
(vii) $3q^2 - 3p^2$
(viii) $10fg + 10fh - 5gh$

2 **(i)** $2(4 - 5x^2)$
(ii) $2b(3a + 4c)$
(iii) $2a(a + 2b)$
(iv) $pq(q^2 - p^2)$
(v) $3xy(x + 2y^3)$
(vi) $2pq(3p^2 - 2pq + q^2)$
(vii) $3lm^2(5 - 3l^2m + 4lm^2)$
(viii) $12a^4b^4(7a - 8b)$

3 **(i)** $4(5x - 4y)$
(ii) $6(x + 1)$
(iii) $z(x - y)$
(iv) $2q(p - r)$
(v) $k(l + n)$
(vi) $-4(a + 2)$
(vii) $3(x^2 + 2y^2)$
(viii) $2(a + 4)$

4 **(i)** $10a^3b^4$
(ii) $12p^3q^4r$
(iii) lm^2n^2p
(iv) $36r^4s^3$
(v) $64ab^2c^2d^2e$
(vi) $60x^3y^3z^3$
(vii) $84a^5b^9$
(viii) $42p^3q^8r^5$

5 **(i)** $2a$
(ii) pq
(iii) $\frac{4b}{a}$
(iv) $\frac{bd}{ac}$
(v) $\frac{2xy^2z}{3}$
(vi) $\frac{5}{2a^2b^2}$
(vii) $\frac{7p^2r^3}{6q^4s^2}$

6 **(i)** $\frac{11a}{12}$
(ii) $\frac{13x}{20}$
(iii) $\frac{7p}{12}$
(iv) $\frac{s}{3}$
(v) $\frac{5b}{12}$
(vi) $\frac{7a}{3b}$
(vii) $\frac{5q - 3p}{2pq}$
(viii) $-\frac{5x}{6y}$

❓ (Page 6)

An equation is how you use algebra to show that two expressions or numbers are equal, for example $5x + 2 = 3x + 8$. An equation must contain an equals sign.
When you solve an equation you find values for any variables in the equation.

❓ (Page 7)

It is the value of the variable that is required.

Exercise 1B (Page 8)

1 **(i)** $x = 7$
(ii) $a = -2$
(iii) $x = 2$
(iv) $y = 2$
(v) $c = 5$
(vi) $p = 10$
(vii) $x = -5$
(viii) $x = -6$
(ix) $y = 7$
(x) $k = 42$
(xi) $t = 60$
(xii) $p = -55$
(xiii) $p = 0$

2 **(i)** $2l + 2(l - 80) = 600$
(ii) $l = 190$; Area $= 20\,900\,\text{m}^2$

3 **(i)** $2(j + 4) + j = 17$
(ii) Louise and Molly, 7 years; Jonathan, 3 years

4 **(i)** $5c - a$
(ii) $5c - 15 = 40$; $c = 11$

5 **(i)** John $(3m + x)$ years; Michael $(m + x)$ years where Michael is m years old now.

(ii) $3m + x = 2(m + x); x = m$

6 **(i)** $8a$

(ii) $6a + 6$

(iii) $a = 3$

7 **(i)** $m - 2, m - 1, m, m + 1, m + 2$

(ii) $m - 2 + (m - 1) + m + (m + 1) + (m + 2) = 105; m = 21$

(iii) 19, 20, 21, 22, 23

8 **(i)** $2(x + 2) = 5(x - 3); x = 6\frac{1}{3}$

(ii) $16\frac{2}{3}\,\text{cm}^2$

❓ (Page 9)

The subject appears only once in a formula, on its own on the left-hand side.

❓ (Page 10)

Omit it since length must be positive.

Exercise 1C (Page 10)

1 **(i)** $u = v - at$

(ii) $a = \frac{v - u}{t}$; an equation of motion

2 $b = \frac{2A}{h}$; area of a triangle

3 $l = \frac{P - 2b}{2}$; perimeter of a rectangle

4 $r = \sqrt{\frac{A}{\pi}}$; area of a circle

5 $c = \frac{2A - bh}{h}$; area of a trapezium

6 $h = \frac{A - \pi r^2}{2\pi r}$; surface area of a cylinder with a base but no top

7 $l = \frac{\lambda e}{T}$; tension of a spring or string

8 **(i)** $u = \frac{2s - at^2}{2t}$

(ii) $a = \frac{2(s - ut)}{t^2}$; an equation of motion

9 $x = \frac{\sqrt{\omega^2 a^2 - v^2}}{\omega}$; speed of a particle on an oscillating spring

10 $g = \frac{4\pi^2 l}{T^2}$; period of a pendulum

11 $f = \frac{uv}{u + v}$; focal length of a lens

12 $g = \frac{2E - mv^2}{2mh}$; total energy of a body (potential energy plus kinetic energy)

❓ (Page 11)

When the brackets are removed there are terms in x^2 and x and a number, but no other terms $\left(\text{e.g. no } x^3, \text{ no } \sqrt{x}, \text{ no } \frac{1}{x}\right)$.

❓ (Page 13)

Yes, except that the rows and columns could be interchanged.

❓ (Page 13)

Yes, but the brackets will be in the reverse order. (Work it through to check for yourself.)

Exercise 1D (Page 15)

1 **(i)** $x^2 + 9x + 20$

(ii) $x^2 + 4x + 3$

(iii) $a^2 + 9a - 5$

(iv) $6p^2 + 5p - 6$

(v) $x^2 + 6x + 9$

(vi) $4x^2 - 9$

(vii) $14m - 3m^2 - 8$

(viii) $12 + 4t - 5t^2$

(ix) $16 - 24x + 9x^2$

(x) $m^2 - 6mn + 9n^2$

2 **(i)** $(x + 2)(x + 3)$

(ii) $(y - 1)(y - 4)$

(iii) $(m - 4)^2$

(iv) $(m - 3)(m - 5)$

(v) $(x + 5)(x - 2)$

(vi) $(a + 12)(a + 8)$

(vii) $(x - 3)(x + 2)$

(viii) $(y - 12)(y - 4)$

(ix) $(k + 6)(k + 4)$

(x) $(k - 12)(k + 2)$

3 **(i)** $(x + 2)(x - 2)$

(ii) $(a + 5)(a - 5)$

(iii) $(3 + p)(3 - p)$

(iv) $(x + y)(x - y)$

(v) $(t + 8)(t - 8)$

(vi) $(2x + 1)(2x - 1)$

(vii) $(2x + 3)(2x - 3)$

(viii) $(2x + y)(2x - y)$

(ix) $(4x + 5)(4x - 5)$

(x) $(3a + 2b)(3a - 2b)$

4 **(i)** $(2x + 1)(x + 2)$

(ii) $(2a - 3)(a + 7)$

(iii) $(5p - 1)(3p + 1)$

(iv) $(3x - 1)(x + 3)$

(v) $(5a + 1)(a - 2)$

(vi) $(2p - 1)(p + 3)$

(vii) $(4x - 1)(2x + 3)$

(viii) $(2a - 9)(a + 3)$

(ix) $(3x - 5)^2$

(x) $(2x + 5)(2x - 3)$

❓ (Page 18)

Start by taking the coefficient of x^2 out as a factor, e.g. $2x^2 - 4x + 3 = 2\left(x^2 - 2x + \frac{3}{2}\right)$

Activity 1.1 (Page 20)

(i)

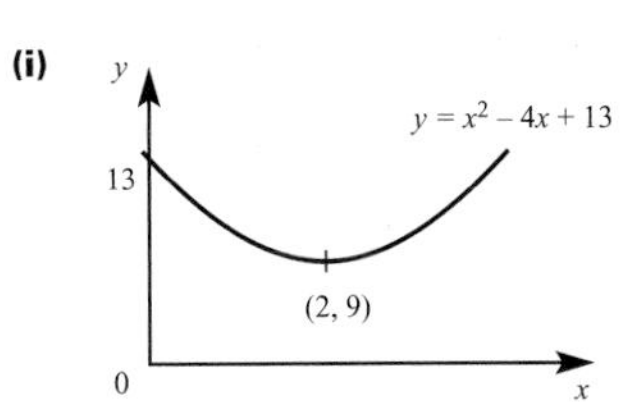

(ii) The number under the square root sign is negative.

Exercise 1E (Page 21)

1 **(i)** $x = 2$ or $x = 6$

(ii) $m = 2$ (repeated)

(iii) $p = 5$ or $p = -3$

(iv) $a = -2$ or $a = -9$

(v) $x = -2$ or $x = -\frac{1}{2}$

(vi) $x = 1$ or $x = -1\frac{3}{4}$

(vii) $t = \frac{1}{5}$ or $t = -\frac{1}{3}$

(viii) $r = -\frac{1}{8}$ or $r = -\frac{2}{3}$

(ix) $x = \frac{1}{3}$ or $x = -3$

(x) $p = \frac{2}{3}$ or $p = 4$

2 **(i)** $x = 4.32$ or $x = -2.32$

(ii) $x = 1.37$ or $x = -4.37$

(iii) $x = 2.37$ or $x = -3.37$

(iv) $x = 1.77$ or $x = -2.27$

(v) $x = 1.68$ or $x = -2.68$

(vi) $x = 2.70$ or $x = -3.70$

(vii) $x = 4.24$ or $x = -0.24$

(viii) $x = 3.22$ or $x = 0.78$

3 **(i)** $x = -0.23$ or $x = -1.43$

(ii) $x = -0.41$ or $x = -1.84$

(iii) $x = 0.34$ or $x = -5.84$

(iv) $x = 1.64$ or $x = 0.61$

(v) $x = 1.89$ or $x = 0.11$

(vi) $x = -1.23$ or $x = -2.43$

4 **(i)** $x = 3.19$ or $x = 0.31$

(ii) No solution

(iii) $x = 2.87$ or $x = -4.87$

(iv) $x = 1.37$ or $x = -1.70$

(v) No solution

(vi) $x = 0.62$ or $x = -1.62$

5 3 cm, 4 cm, 5 cm

6 $x = 1.5$

7 $x = 9$

8 **(i)** $t = 1$ s and $t = 2$ s

(ii) 3 seconds

9 **(ii)** 17 cm

10 **(i)** **(a)** $(x + 6)$ cm

(b) $(x - 10)$ cm

(c) $(x - 16)$ cm

(iii) Length = 34 cm, width = 28 cm

? (Page 23)

Infinitely many. x can take any value and, in this example, the corresponding value of y is $4 - x$.

? (Page 26)

In this example the correct solution would be found, but in some cases, e.g. if the curve had equation $y^2 = 4x$, additional values that are not part of the solution can be obtained. Always substitute into the equation of the line. For example

$$y = x - 2$$
$$y^2 = 4x - 8$$

has $x = 2$, $y = 0$ and $x = 6$, $y = 4$ as its solution.

Substituting into the equation of the curve would also give the pair of values $x = 6$, $y = -4$.

? (Page 27)

You subtract if the coefficients of the variable to be eliminated have the same sign. You add if they have opposite signs.

Exercise 1F (Page 29)

1 **(i)** $x = 5, y = 2$

(ii) $x = 4, y = -1$

(iii) $x = 2\frac{1}{4}, y = 6\frac{1}{2}$

(iv) $x = -2, y = -3$

(v) $x = 1\frac{1}{2}, y = 4$

(vi) $x = -\frac{1}{2}, y = -6\frac{1}{2}$

2 **(i)** $x = 2, y = 3$

(ii) $x = 4, y = 3$

(iii) $x = 6, y = 2$

(iv) $x = -\frac{3}{7}, y = 3\frac{2}{7}$

(v) $x = 2, y = 5$

(vi) $x = -1, y = -2$

3 **(i)** $x = 1, y = 4$ or $x = 4, y = 1$

(ii) $x = 2, y = 3$ or $x = -\frac{2}{3}, y = \frac{1}{3}$

(iii) $x = 4, y = -2$ or $x = -1, y = -7$

(iv) $x = 1, y = 5$ or $x = 11, y = 25$

(v) $x = 4, y = 2$ or $x = -4, y = -2$

(vi) $x = 1, y = -2$ or $x = -2\frac{3}{7}, y = -\frac{2}{7}$

4 **(i)** $3c + 4l = 72$, $5c + 2l = 64$; a chew costs 8p and a lollipop costs 12p.

(ii) $x + 5m = 500$, $x + 7m = 660$; $m = 80$, $x = 100$; £3.40

(iii) $3c + 2n = 145$, $2c + 5n = 225$; $n = 35$, $c = 25$; £1.65

(iv) $2a + c = 3750$, $a + 3c = 375$; $c = 750$, $a = 1500$; £67.50

5 A(3, 4), B(4, 3)

Chapter 2

? (Page 31)

An equation contains an = sign and any solution will consist of one or more particular values of any variables involved.

An inequality contains any of the signs $<$, $\leqslant$, $>$, $\geqslant$ and any solution will consist of a range of values of its variable(s).

? (Page 31)

$8 \times 10^7 \leqslant x \leqslant 3.8 \times 10^8$

? (Page 32)

An inequality may be rearranged using addition, subtraction, multiplication by a positive number and division by a positive number in the same way as an equation. Multiplication or division by a negative number reverses the inequality.

? (Page 32)

$2 < 3$, but $-2 > -3$; $5 > 1$, but $-5 < -1$.

Exercise 2A (Page 33)

1 **(i)** $x < 5$

(ii) $x \geqslant 2$

(iii) $y \leqslant 4$

(iv) $y < 4$

(v) $x \geqslant -3$

(vi) $b \geqslant -3$

(vii) $x > 3$

(viii) $x < 12$

(ix) $x \geqslant -5$

(x) $x \leqslant -4$

(xi) $2 \leqslant x \leqslant 4$

(xii) $2 \leqslant x \leqslant 5$

(xiii) $-3 < x < 1$

(xiv) $1 < x < 2$

Exercise 2B (Page 35)

1 **(i)** $x < 1$ or $x > 5$

(ii) $-4 \leqslant a \leqslant 1$

(iii) $-1\frac{1}{2} < y < 1$

(iv) $-2 \leqslant y \leqslant 2$

(v) $x < 2$ or $x > 2$

(vi) $1 \leqslant p \leqslant 2$

(vii) $a < -3$ or $a > 2$

(viii) $-4 \leqslant a \leqslant 2$

(ix) $y < -1$ or $y > \frac{1}{3}$

(x) $y \leqslant -1$ or $y \geqslant 5$

? (Page 35)

A fraction in arithmetic is one number divided by another number. The definition of a fraction in algebra is the same but with 'number' replaced by 'expression'.

? (Page 36)

You can cancel a fraction in arithmetic when the numerator and denominator have a common factor. It is the same for fractions in algebra. A factor in arithmetic is a number that divides exactly into the given number, i.e. there is no remainder. The definition of a factor in algebra is the same but with 'number' replaced by 'expression'.

? (Page 36)

x is not a factor of the numerator $(2x + 2)$ or the denominator $(3x + 3)$. The correct answer involves factorising both the numerator and the denominator:

$$\frac{2(x+1)}{3(x+1)}$$

Cancelling $(x + 1)$ gives $\frac{2}{3}$.

? (Page 36)

Neither a nor a^2 is a factor of the numerator and denominator. The correct answer involves factorising to get

$$\frac{(a-3)(a+2)}{(a-3)(a+5)} = \frac{a+2}{a+5}$$

? (Page 37)

Individual terms have been cancelled rather than factors. The correct answer is

$$\frac{(2n+3)(2n-3)}{(n+1)} \times \frac{(n+1)(n-1)}{(2n+3)}$$

$$= (2n-3)(n-1)$$

? (Page 37)

A denominator that is the same for both fractions. For example in

$$\frac{1}{2} + \frac{1}{x} = \frac{x}{2x} + \frac{2}{2x}$$

$$= \frac{x+2}{2x}$$

the common denominator is $2x$.

? (Page 37)

(a) 12

(b) $(x-1)(x+1)(x-3)$

Exercise 2C (Page 38)

1 **(i)** $\frac{1}{2}$

(ii) $\dfrac{4}{x+8}$

(iii) $\dfrac{3}{x-y}$

(iv) $\dfrac{2x}{3y}$

(v) $\dfrac{1}{3-p}$

(vi) $\dfrac{2b^2}{5a^2}$

(vii) $\dfrac{x-1}{2}$

(viii) $\dfrac{x}{x-y}$

(ix) $\dfrac{1}{a-3}$

(x) $\dfrac{3}{2}$

(xi) $\dfrac{3x-1}{3}$

(xii) $\dfrac{x}{2y}$

2 **(i)** $\dfrac{b}{2}$

(ii) x

(iii) $\dfrac{x}{8(x-1)}$

(iv) $2(a+1)$

(v) $\dfrac{(x-2)}{x(x+2)}$

(vi) $\dfrac{(2x-1)(x+2)}{(2x+1)(x-1)}$

(vii) $4(p+3)$

(viii) $\dfrac{3(x-1)(x^2-3)}{(x-3)^2}$

3 **(i)** $\dfrac{7a}{20}$

(ii) $-\dfrac{7}{3a}$

(iii) $\dfrac{(m-3n)}{(m+n)(m-n)}$

(iv) $\dfrac{5(p+2)}{(p-2)(2p+1)}$

(v) $\dfrac{(5a+1)}{a(a+1)(a-1)}$

(vi) 2

(vii) $\dfrac{1}{(p+1)(p-1)}$

(viii) $\dfrac{2(a^2+b^2)}{(a+b)(a-b)}$

? (Page 39)

When a fraction is simplified, the answer is an expression, but when an equation is solved the answer is the value of the variable.

? (Page 39)

If you multiplied both the numerator and the denominator the multiplier would cancel and the fraction would be unchanged.

? (Page 39)

It allows you to cancel out all the fractions.

? (Page 40)

Not all of the left-hand side has been multiplied by 30.

? (Page 40)

It enables you to see what is best to multiply through by; in this case $2(a+1)(a-1)$, rather than $2(a+1)(a^2-1)$.

Exercise 2D (Page 41)

1 **(i)** $x = \frac{5}{6}$

(ii) $a = \frac{5}{8}$

(iii) $x = \pm\frac{1}{\sqrt{3}}$

(iv) $x = 1\frac{5}{13}$

(v) $x = 2$ or $x = -6\frac{1}{3}$

(vi) $a = -2$

(vii) $p = \frac{1}{3}$ or $p = 3$

(viii) $x = 0$ or $x = 3$

2 30 cm

3 **(i)** $\dfrac{20}{x}$

(ii) $\dfrac{20}{(x-5)}$

(iii) $\dfrac{20}{(x-5)} - \dfrac{20}{x} = \dfrac{1}{3}$; $x = 20$

(iv) 1 hour

4 **(i)** $\dfrac{2400}{x} + 1$

(ii) $\dfrac{2400}{(x+2)} + 1$

(iii) $x = 48$

(iv) 51 lamp-posts

5 **(i)** $\dfrac{184}{n}$

(ii) $\dfrac{225}{(n+1)}$

(iii) $\dfrac{225}{(n+1)} - \dfrac{184}{n} = 2$; $n = 8$

(iv) 25

? (Page 43)

A number of the form $\frac{a}{b}$ where a and b are integers. So a rational number can be a fraction or an integer (when $b = 1$); it can be positive or negative.

Exercise 2E (Page 44)

1 **(i)** $4\sqrt{2}$

(ii) $5\sqrt{5}$

(iii) $5\sqrt{3}$

(iv) $\sqrt{2}$

(v) $3\sqrt{3}$

(vi) $7\sqrt{2}-3$

(vii) $10\sqrt{2}$

(viii) $36+3\sqrt{3}$

(ix) $16\sqrt{5}$

(x) $\sqrt{3}$

2 **(i)** $3-2\sqrt{2}$

(ii) $3+2\sqrt{5}$

(iii) $3\sqrt{7}-9$

(iv) 2

(v) $11-\sqrt{2}$

(vi) $5-3\sqrt{7}$

(vii) $24-13\sqrt{3}$

(viii) $8-2\sqrt{15}$

(ix) $13\sqrt{2}-17$

(x) $17+12\sqrt{2}$

3 (i) $\frac{\sqrt{3}}{3}$

(ii) $\sqrt{5}$

(iii) $\frac{4\sqrt{6}}{3}$

(iv) $\frac{\sqrt{6}}{3}$

(v) 1

(vi) $\frac{\sqrt{21}}{7}$

(vii) $3\sqrt{7}$

(viii) $\frac{\sqrt{5}}{3}$

(ix) $\frac{\sqrt{15}}{5}$

(x) $\frac{\sqrt{2}}{2}$

Chapter 3

? (Page 45)

$y = x^3 - 4x$ since this factorises to give $y = x(x + 2)(x - 2)$.

? (Page 46)

$x^5 + 3x^3 + 2x - 1$ and $z^3 + z^2 + z + 1$. In the other two expressions, not all the powers of x are positive whole numbers. One contains $\sqrt{x}$ and the other $\frac{3}{x}$.

? (Page 47)

All answers are given to 1 decimal place.

$y = x^3 - 5x^2 + 6x$: max (0.8, 2.1), min (2.5, –0.6)

$y = 4 + x - x^4$: infl (0, 4), max (0.6, 4.5)

$y = x^5 - 5x^3 + 4x$: max (–1.6, 3.6), min (–0.5, –1.5) max (0.5, 1.5) min (1.6, –3.6)

$y = x^3 + 3x$ infl (0, 0)

? (Page 47)

A reflection of the original curve in the x axis.

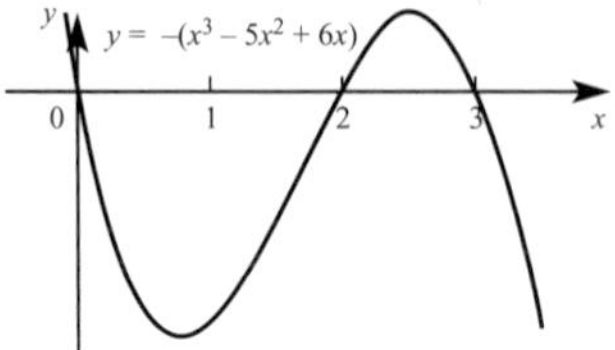

Exercise 3A (Page 50)

1 (i) 3

(ii) 5

(iii) 6

(iv) 4

2 (i) $5x^3 + 2x^2 + 4x + 3$

(ii) $5x^4 + x^3 - 3x^2 + 8x - 4$

3 (i) $3x^3 + x^2 - 2$

(ii) $-3x^4 + x^3 + 6x^2 + 6x - 4$

4 (i) $x^5 - x^4 + 2x^3 - 3x^2 + x - 2$

(ii) $x^6 + 2x^5 - 3x^4 - 4x^3 + 5x^2 + 6x - 3$

(iii) $2x^5 - 4x^4 - x^3 + 11x^2 - 13x + 5$

(iv) $x^6 - 1$

(v) $x^3 + 4x^2 + x - 6$

(vi) $2x^3 + 5x^2 - 14x - 8$

(vii) $2x^3 - 17x - 18$

(viii) $2x^2 - 2x$

(ix) $x^2 + 3x + 2$

(x) $x^2 - 2x + 1$

(xi) $x^2 - 3x - 4$

(xii) $3x^2 + x - 2$

? (Page 51)

The answer is zero in both cases.

? (Page 52)

The x^3 term on the left-hand side is x^3 and on the right-hand side is kx^3.

? (Page 52)

± 1 and ± 3 are the only factors of –3.

? (Page 53)

$f(1) = -4$

No, since you would only try factors of the constant term –1.

? (Page 54)

The number left over when a whole number is divided by another whole number.

? (Page 54)

Yes.

The remainder will always be smaller than the number that you are dividing by. If the remainder is larger you can divide the number into it at least one more time. For example: rather than writing $49 \div 8 = 5$ remainder 9, you would write $48 \div 8 = 6$ remainder 1.

Activity 3.1 (Page 54)

(i) $(x^3 + 3x^2 - 2x + 1) \div (x + 1) = (x^2 + 2x - 4)$ remainder 5

(ii) $f(-1) = 5$

(iii) $f(-1)$ = the remainder from **(i)**

(iv) $(x^3 + 3x^2 - 2x + 1) \div (x - 2) = (x^2 - 5x + 8)$ remainder 17

Yes, the remainder is the same as $f(2)$.

? (Page 55)

No, since when a is a number, $f(a)$ will also be just a number.

Exercise 3B (Page 56)

1 (i) Factor
(ii) Not a factor
(iii) Not a factor
(iv) Not a factor
(v) Factor
(vi) Factor

2 (i) $(x-1)(x+1)(x-3)$
(ii) $(x+1)(x+2)(x-3)$
(iii) $x(x+1)(x-2)$
(iv) $(x+1)(2x+1)(x+2)$
(v) $(x+1)(2x-1)(x-2)$
(vi) $(x+1)(x-1)(x+5)$
(vii) $(x-2)(x+2)^2$
(viii) $(x-1)(x+3)^2$
(ix) $(x-1)(2x+3)(x+2)$
(x) $(x-4)(3x-2)(x+2)$

3 (i) $(x-2)(x^2+x+1)$
(ii) $(x-1)(x^2+1)$
(iii) $(x-2)(x^2-3x+4)$
(iv) $(x-2)(x^2+2x+4)$
(v) $(x+2)(x^2+x+1)$
(vi) $(x-3)(2x^2+x+1)$
(vii) $(x+2)(x^2+4x+5)$
(viii) $(x+1)(3x^2+2x+1)$
(ix) $(x+5)(2x^2-x+5)$
(x) $(x+9)(x^2+x+1)$

4 (i) 18
(ii) 29
(iii) -2
(iv) -4
(v) $-1\frac{3}{4}$
(vi) 86
(vii) $2a^3$
(viii) $-2a^3$

5 $a=2$

6 $k=1$

7 46

8 $a=-3$

9 $16+8p+2q+36=0$
$81-27p-3q+36=0$
$p=13, q=-78$

10 (i) $k=-7$
(ii) $x=1, x=-3$

11 (i) $a=-1, b=-7$
(ii) $x=1$ $x=1\frac{1}{2}, x=-2$

12 $a=3$

13 $a=100$

14 (i) $\frac{8}{x^2}$
(iv) $x=4, x=1.46$

Chapter 4

? (Page 59)

10

Using sequences NNNEE, NNENE, etc. where there are three Ns for the three blocks north and two Es for the two blocks east.

20

For n steps north and e steps east, the number of routes is the number of different arrangements of e Es and n Ns.

Activity 4.1 (Page 60)

(i) 1
(ii) $1+x$
(iii) $1+2x+x^2$
(iv) $1+3x+3x^2+x^3$
(v) $1+4x+6x^2+4x^3+x^4$

? (Page 60)

10 is the coefficient of x^2 in the expansion of $(1+x)^5$.

20 is the coefficient of x^3 in the expansion of $(1+x)^6$.

? (Page 61)

1 7 21 35 35 21 7 1

Numbers are symmetrical about a vertical line.

Sequences include:

All the outside numbers are 1.

The second number in the row is the power of $(1+x)$.

Each number is the sum of the two immediately above it.

? (Page 62)

The x terms arise when you take x out of one bracket and 1 out of the other (2 ways).

The x^2 terms arise when you take x out of any two brackets and 1 out of the other (3 ways).

The x^r terms in the expansion of $(1+x)^n$ arise when you take x from r brackets and 1 from the other $(n-r)$ brackets.

P to Q: coefficient of x^2 in $(1+x)^5$. Taking x out of two of the five brackets is like going east for two of the five steps.

P to R: coefficient of x^3 in $(1+x)^6$. Taking x out of three of the six brackets is like going east for three of the six steps.

? (Page 62)

(i) ${}^{10}C_4$ is the number of ways you can take x from four brackets and 1 from six brackets in the expansion of $(1+x)^{10}$.

(ii) ${}^{10}C_0$ is the number of ways you can take x from no brackets, i.e. 1 from all ten brackets in the expansion of $(1 + x)^{10}$. There is one way of doing this.

(iii) ${}^{10}C_{10}$ is the number of ways you can take x from all ten brackets in the expansion of $(1 + x)^{10}$. There is one way of doing this.

Activity 4.2 (Page 63)

(i) ${}^4C_0 = 1$, ${}^4C_1 = 4$, ${}^4C_2 = 6$, ${}^4C_3 = 4$, ${}^4C_4 = 1$

(ii) **(a)** 4, 6, 4

(b) ${}^4C_1 = \frac{4!}{1! \times 3!}$, ${}^4C_2 = \frac{4!}{2! \times 2!}$, ${}^4C_3 = \frac{4!}{3! \times 1!}$

(c) ${}^4C_0 = \frac{4!}{0! \times 4!} = \frac{1}{0!}$, ${}^4C_4 = \frac{4!}{4! \times 0!} = \frac{1}{0!}$

(d) $0! = 1$

? (Page 63)

${}^{10}C_4 = 210$

Exercise 4A (Page 63)

1 **(i)** $1 - 4x + 6x^2 - 4x^3 + x^4$

(ii) $1 + 9x + 27x^2 + 27x^3$

(iii) $1 - 10x + 40x^2 - 80x^3 + 80x^4 - 32x^5$

(iv) $1 + 8x + 24x^2 + 32x^3 + 16x^4$

(v) $1 - 15x + 90x^2 - 270x^3 + 405x^4 - 243x^5$

(vi) $1 + 6x^2 + 15x^4 + 20x^6 + 15x^8 + 6x^{10} + x^{12}$

2 **(i)** $8x^3 - 12x^2 + 6x - 1$

(ii) $x^5 + 15x^4 + 90x^3 + 270x^2 + 405x + 243$

(iii) $16x^4 - 96x^3 + 216x^2 - 216x + 81$

(iv) $27x^3 + 27x^2y + 9xy^2 + y^3$

(v) $x^3 - 6x^2y + 12xy^2 - 8y^3$

(vi) $27x^3 + 108x^2y + 144xy^2 + 64y^3$

3 **(i)** 126

(ii) 560

(iii) –721 710

(iv) 448

(v) 36

4 $8x + 8x^3$

5 **(i)** $1 + 6x + 12x^2 + 8x^3$

(ii) $1 + 5x + 6x^2 - 4x^3 - 8x^4$

6 **(i)** $a^3 + 3a^2bx + 3ab^2x^2 + b^3x^3$

(ii) $a = 2, b = +3$

7 1.338 225 577 6

8 **(i)** $1 + 3x + 3x^2 + x^3$

(ii) $1 + 3y + 6y^2 + 7y^3 + 6y^4 + 3y^5 + y^6$

? (Page 64)

$${}^5C_2 \times \left(\frac{1}{6}\right)^2 \times \left(\frac{5}{6}\right)^3 = 0.160\,75$$

? (Page 65)

$${}^5C_4 \times \left(\frac{1}{6}\right)^4 \times \left(\frac{5}{6}\right) = 0.003\,22$$

? (Page 65)

The word *and* relates to the operation *multiply*, provided that the events are independent of each other. The word *or* relates to the operation *add*, provided the events are mutually exclusive, i.e. there is no overlap of the events.

? (Page 66)

One possible example is X = the number of heads when a fair coin is tossed 20 times.

? (Page 66)

$P(X = 0) + P(X = 1) + P(X \geqslant 2) = 1$

Exercise 4B (Page 68)

1 0.0046

2 0.269

3 **(i)** 0.0154

(ii) 0.6554

4 **(i)** 3

(ii) 0.2394

(iii) 0.4082

(iv) 0.3524

5 **(i)** 0.4305

(ii) 0.1869

6 **(i)** 0.1088

(ii) 0.1281

7 **(i)** 0.0115

(ii) 0.2182

(iii) 0.0020

8 **(i)** **(a)** 0.0785

(b) 0.0355

(ii) 11

9 **(i)** **(a)** 0.2787

(b) 0.8936

(c) 0.2322

(ii) 6

10 **(i)** 0.2668

(ii) 0.1503

(iii) 0.0060

11 **(i)** 0.9415

(ii) 0.9985

(iii) 0.7857

(iv) 0.0239

12 **(i)** 0.8280

(ii) 0.0988

(iii) 0.0731

Chapter 5

? (Page 73)

(i) 2

(ii) 3

? (Page 74)

Two points on the line *or* one point and the gradient of the line.

Activity 5.1 (Page 75)

Line A: 3

Line B: 0

Line C: $-\frac{2}{5}$

Line D: ∞

? (Page 75)

No, since

$$\frac{y_1 - y_2}{x_1 - x_2} = \frac{-(y_2 - y_1)}{-(x_2 - x_1)} = \frac{y_2 - y_1}{x_2 - x_1}$$

? (Page 76)

When the increase in x is the same for both lines, then the increase in y is also the same for both lines.

Activity 5.2 (Page 77)

(i) As figure 5.7

(ii) $\angle ABP = \angle BCD$ and $\angle BCD + \angle CBD = 90°$

$\Rightarrow \angle ABP = \angle CBD = 90°$ i.e. $\angle ABC = 90°$

(iii) Depends on the individual sketch.

(iv) Follows from (iii).

? (Page 79)

$\sqrt{4a^2 + 16b^2} = \sqrt{4(a^2 + 4b^2)} = 2\sqrt{a^2 + 4b^2}$

Exercise 5A (Page 80)

1 (i) **(a)** 2

(b) $-\frac{1}{2}$

(c) $\sqrt{80} = 4\sqrt{5}$

(d) $(6, 7)$

(ii) **(a)** -3

(b) $\frac{1}{3}$

(c) $\sqrt{90} = 3\sqrt{10}$

(d) $(1\frac{1}{2}, 8\frac{1}{2})$

(iii) **(a)** $-\frac{11}{5}$

(b) $\frac{5}{11}$

(c) $\sqrt{146}$

(d) $(7\frac{1}{2}, -2\frac{1}{2})$

(iv) **(a)** 3

(b) $-\frac{1}{3}$

(c) $\sqrt{490} = 7\sqrt{10}$

(d) $(-2\frac{1}{2}, -3\frac{1}{2})$

(v) **(a)** $7\frac{1}{2}$

(b) $-\frac{2}{15}$

(c) $\sqrt{229}$

(d) $(7, 7\frac{1}{2})$

(vi) **(a)** $2\frac{3}{5}$

(b) $-\frac{5}{13}$

(c) $\sqrt{194}$

(d) $(\frac{1}{2}, 2\frac{1}{2})$

(vii) **(a)** $-\frac{1}{5}$

(b) 5

(c) $\sqrt{26}$

(d) $(-\frac{1}{2}, -6\frac{1}{2})$

(viii) **(a)** $-3\frac{2}{3}$

(b) $\frac{3}{11}$

(c) $\sqrt{130}$

(d) $(5\frac{1}{2}, 1\frac{1}{2})$

2 (i) Gradient AB $= -1$; gradient AC $= 1$; product $= -1$

(ii) $AB = \sqrt{32}$; $AC = \sqrt{8}$; $BC = \sqrt{40}$; $BC^2 = AB^2 + AB^2$

3 Gradient AB $= -\frac{1}{2}$; gradient AC $= 2$; $AB = AC = \sqrt{20}$

4 (i) 19.73 units

(ii) 9 units^2.

5 (i) $PQ = \sqrt{173}$; $QR = \sqrt{173}$; $RS = \sqrt{173}$; $PS = \sqrt{173}$

(ii) $(3\frac{1}{2}, \frac{1}{2})$

(iii) Gradient PQ $= -\frac{2}{13}$; gradient QR $= -\frac{13}{2}$, so PQ is not perpendicular to QR; rhombus

6 (i)

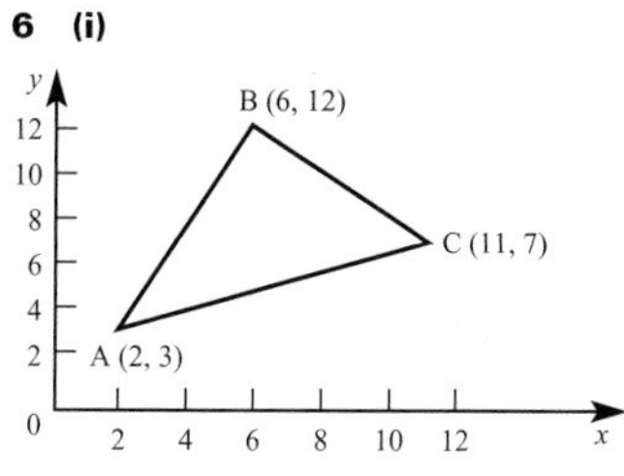

(ii) $AB = AC = \sqrt{97}$; $BC = \sqrt{50}$

(iii) $(8\frac{1}{2}, 9\frac{1}{2})$

(iv) 32.5 units^2

7 (i) $(-\frac{1}{2}, 2)$

(ii) $(0, -1)$

8 (i) Gradient AB $= -2$; gradient BC $= \frac{1}{2}$

(ii) $(7, 4)$

9 (i) $q = 2$

(ii) 1 : 2

10 (i)

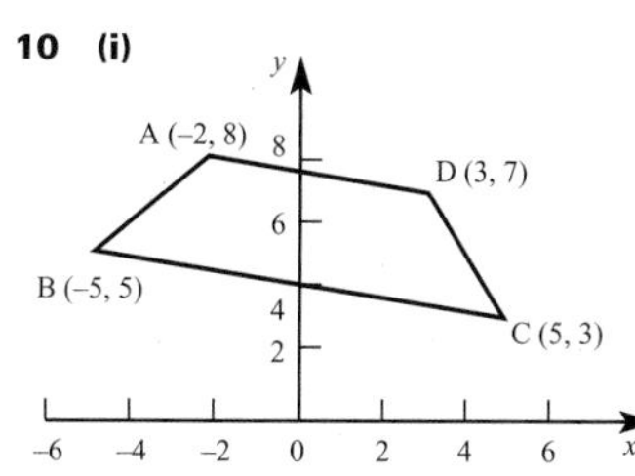

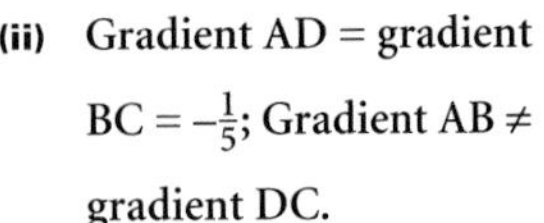

(ii) Gradient AD = gradient BC = $-\frac{1}{5}$; Gradient AB ≠ gradient DC.

(iii) (8, 6)

❓ (Page 82)

Straight means that the gradient of the line is the same along its entire length.

❓ (Page 84)

(i) $\frac{x}{4} + \frac{y}{3} = 1$

(ii) $a = 4, b = 3$

(iii) a is the intercept on the x axis and b is the intercept on the y axis.

Exercise 5B (Page 85)

1 (i)

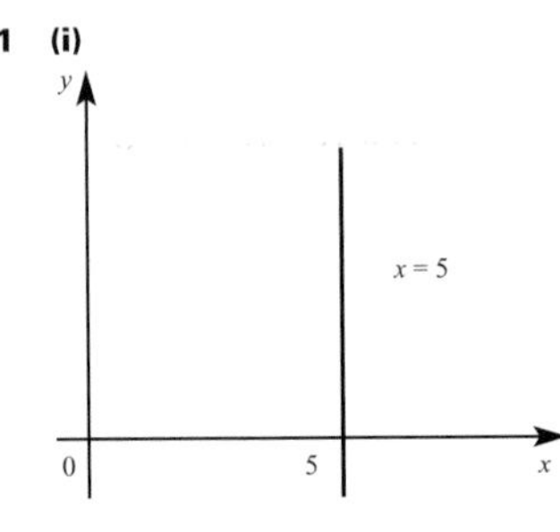

(ii)

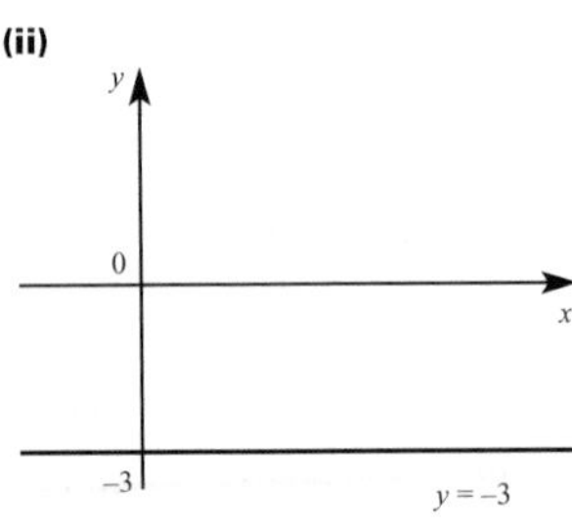

(iii)

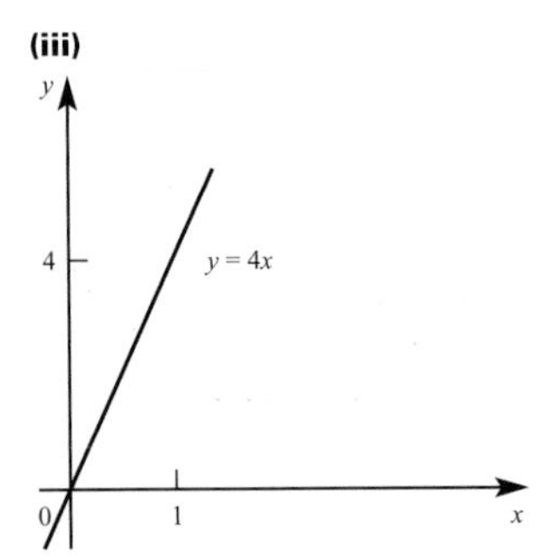

(iv)

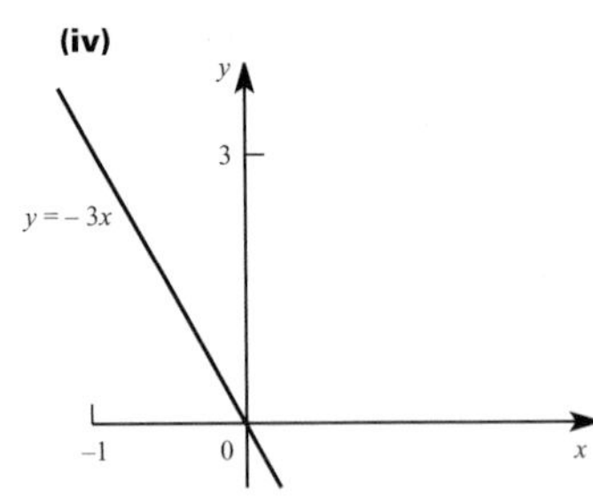

(v)

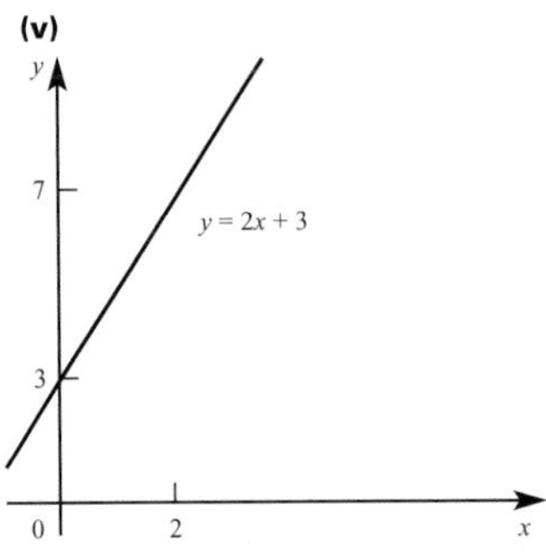

(vi)

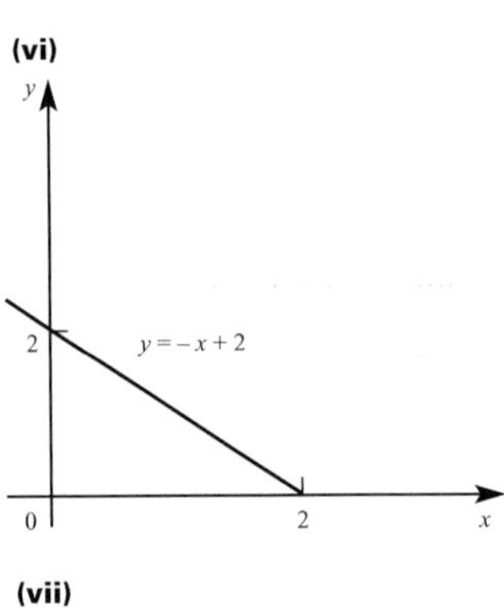

(vii)

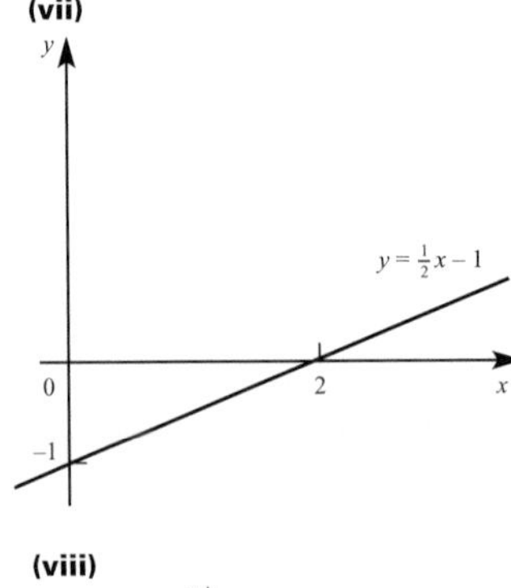

(viii)

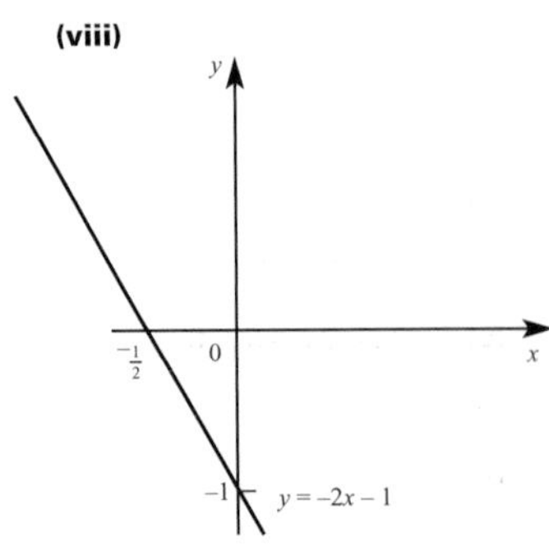

(ix)

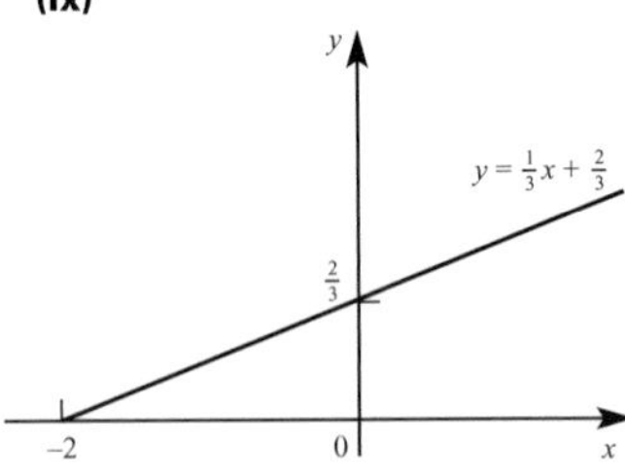

(x)

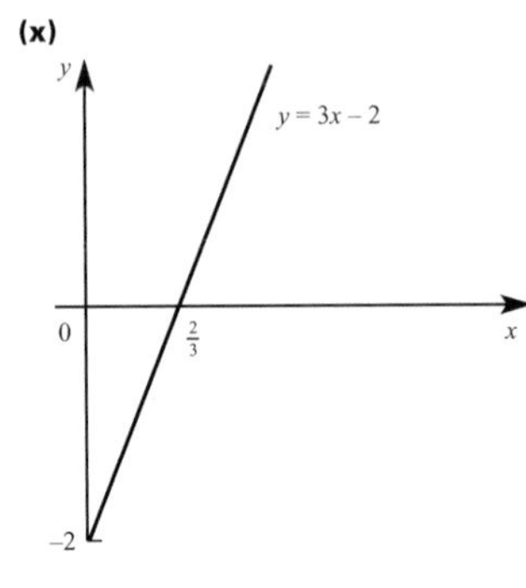

(xi)

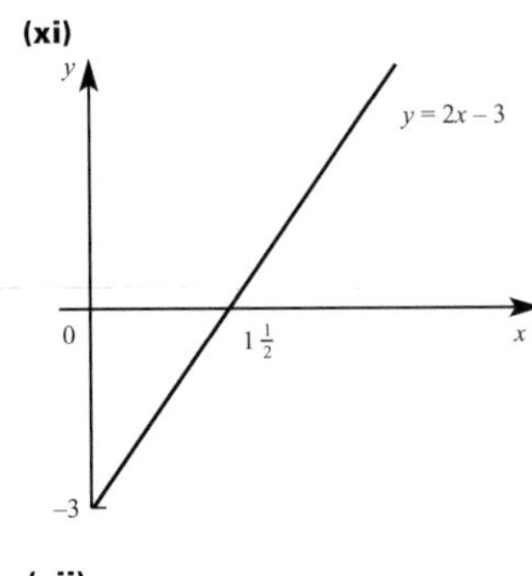

(xii)

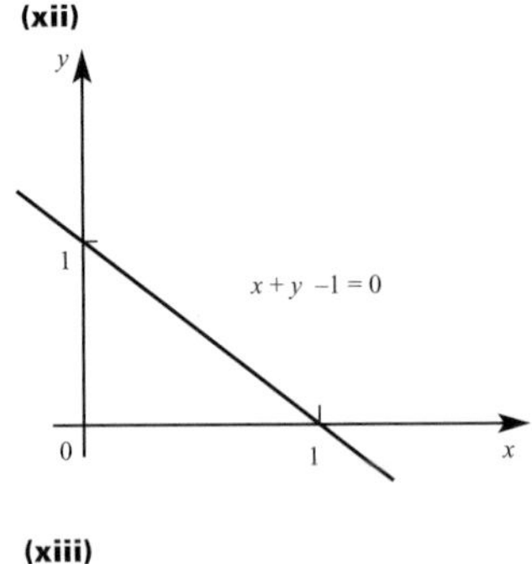

(xiii)

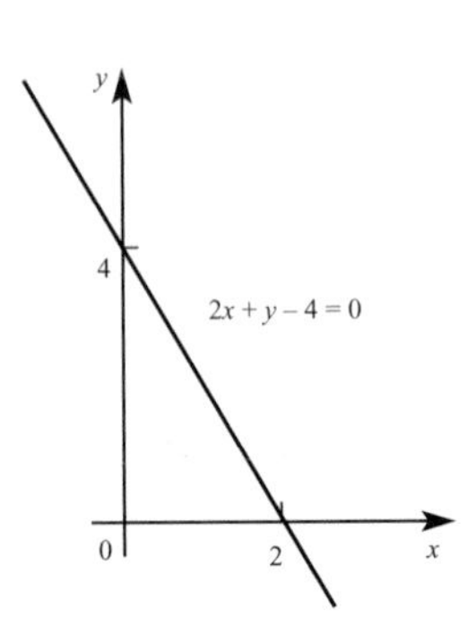

(xiv)

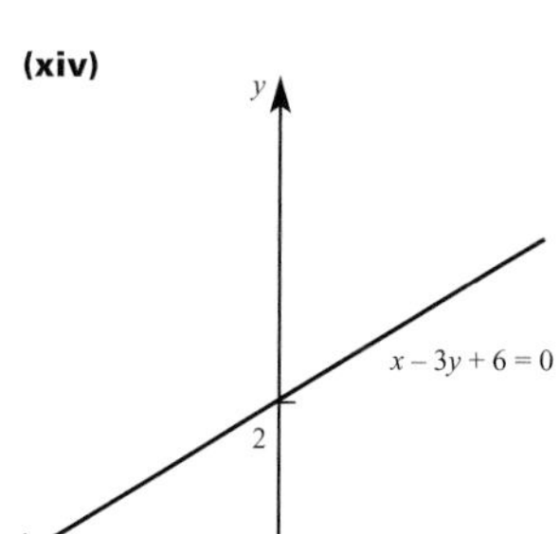
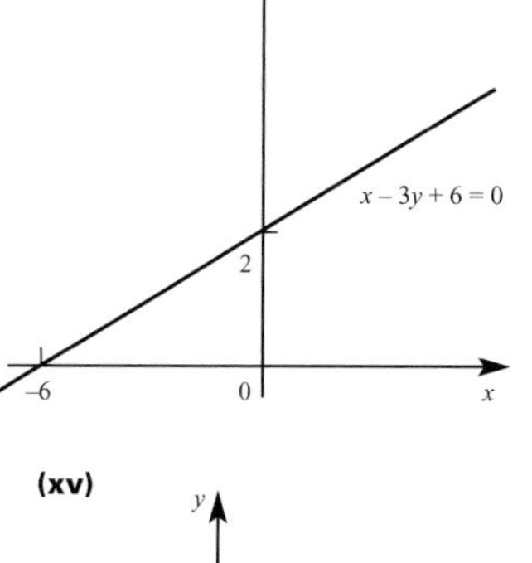

(xv)

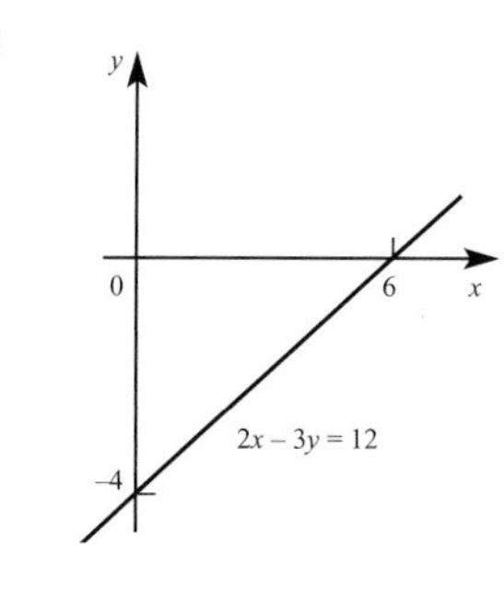

(xvi)

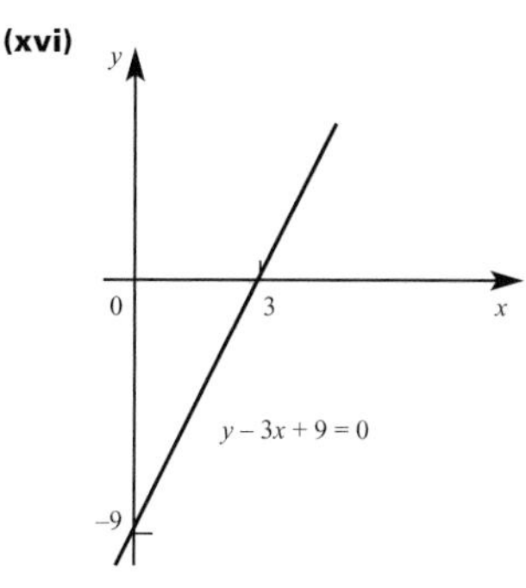

(xvii)

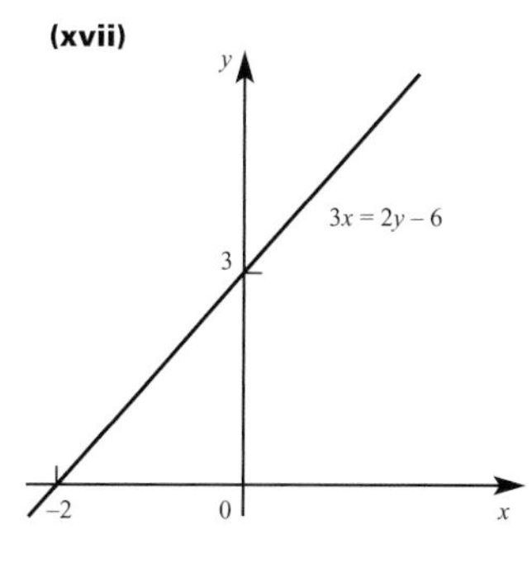

(xviii)

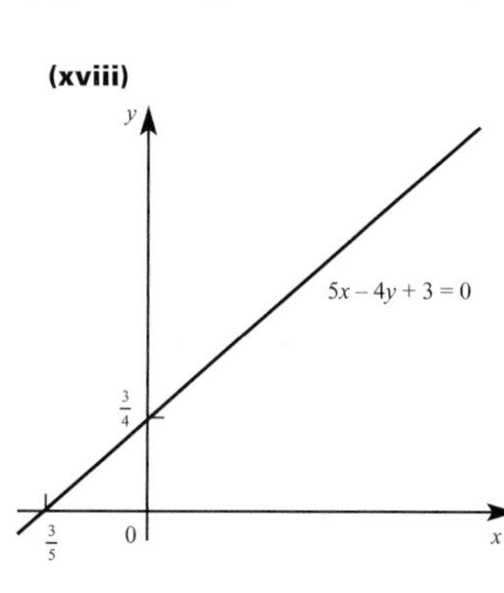

2 **(i)** 0, ∞; perpendicular

(ii) 2, –2; neither

(iii) $-\frac{1}{2}$, 2; perpendicular

(iv) 1, 1; parallel

(v) –4, –3; neither

(vi) –1, 1; perpendicular

(vii) $\frac{1}{2}$, $\frac{1}{2}$; parallel

(viii) $-\frac{1}{3}$, 3; perpendicular

(ix) $\frac{1}{2}$, –2; perpendicular

(x) $-\frac{2}{3}$, $-\frac{2}{3}$; parallel

(xi) $-\frac{1}{3}$, –3; neither

(xii) $\frac{2}{5}$, $-\frac{5}{2}$; perpendicular

Exercise 5C (Page 91)

1 **(a)** $x = -3$

(b) $y = 5$

(c) $y = 2x$

(d) $2x + y = 4$

(e) $2x + 3y = 12$

(f) $x = 5$

(g) $y = -3$

(h) $x + 2y = 0$

(i) $y = x + 4$

(j) $y = 2x - 6$

2 **(i)** $y = 3x - 7$

(ii) $y = 2x$

(iii) $y = 3x - 13$

(iv) $4x - y - 16 = 0$

(v) $3x + 2y + 1 = 0$

(vi) $x + 2y - 12 = 0$

3 **(i)** $x + 2y = 0$

(ii) $x + 3y - 12 = 0$

(iii) $y = x - 4$

(iv) $x + 2y + 1 = 0$

(v) $2x - 3y - 6 = 0$

(vi) $x - 2y - 2 = 0$

4 **(i)** $y = x - 2$

(ii) $5x + 3y - 12 = 0$

(iii) $y = x - 5$

(iv) $3x + 5y - 12 = 0$

(v) $x + 7y + 32 = 0$

(vi) $y = 2x$

5 **(i)** 4

(ii) (4, 3)

(iii) $x + 4y - 16 = 0$

6 **(i)**

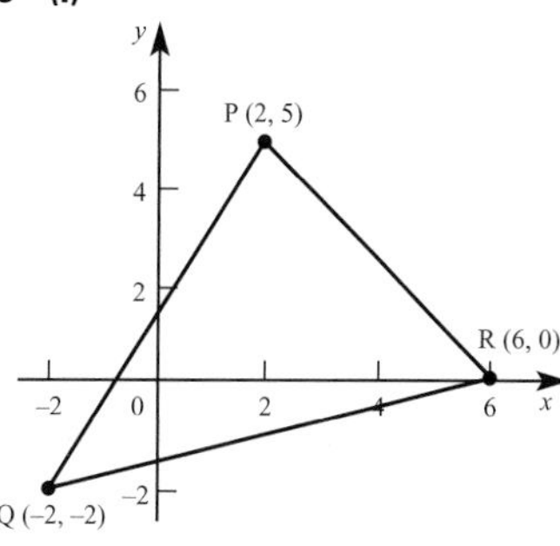

(ii) $L\left(0, 1\frac{1}{2}\right)$, $M(2, -1)$, $N\left(4, 2\frac{1}{2}\right)$

(iii) LR: $x + 4y - 6 = 0$

MP: $x = 2$

NQ: $3x - 4y - 2 = 0$

7 **(i)**

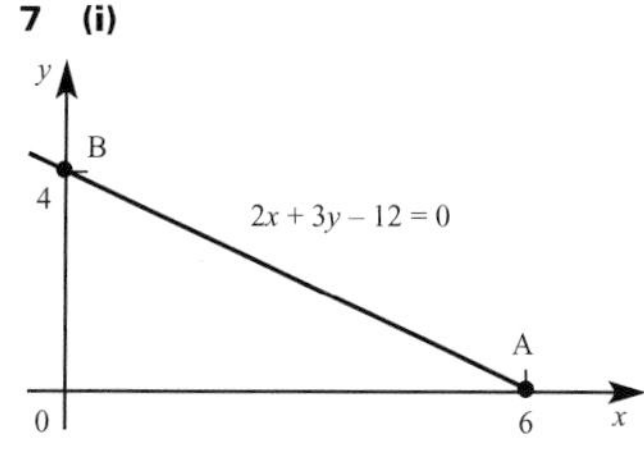

(ii) A(6, 0), B(0, 4)

(iii) 12 units^2

(iv) $3x - 2y = 0$

(v) AB = $\sqrt{52}$ units; shortest distance = 3.33 units (2 d.p.)

8 **(i)**

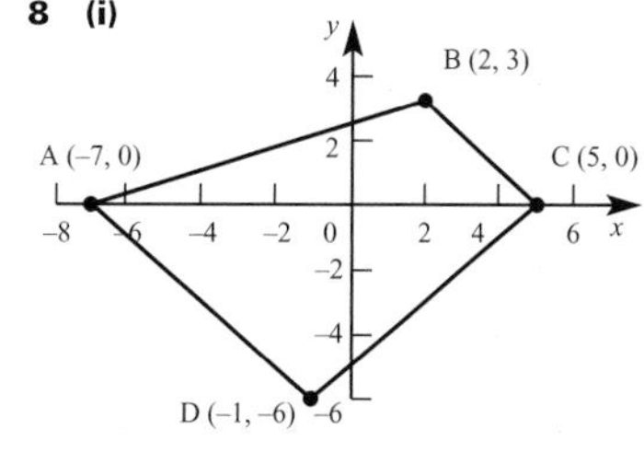

(ii) AB: $\frac{1}{3}$; BC: –1; CD: 1; DA: –1

(iii) AB: $x - 3y + 7 = 0$

BC: $x + y - 5 = 0$

CD: $x - y - 5 = 0$

DA: $x + y + 7 = 0$

(iv) AB: $\sqrt{90}$ units

BC: $3\sqrt{2}$ units

CD: $6\sqrt{2}$ units

DA: $6\sqrt{2}$ units

(v) 54 units^2

9 **(i)**

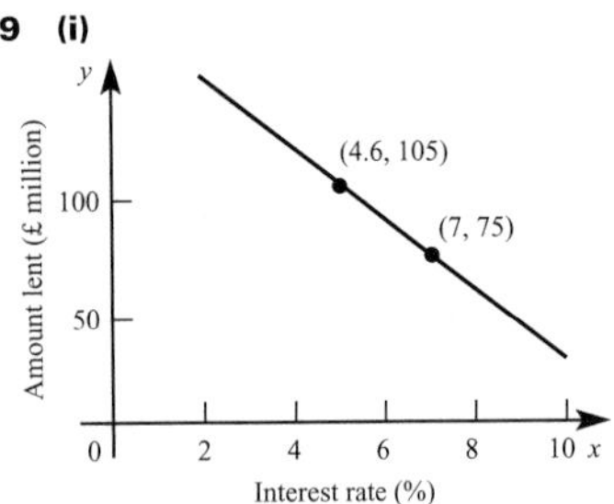

(ii) $25x + 2y - 325 = 0$

(iii) £87.5 million

(iv) 2.6%

10 **(i)**

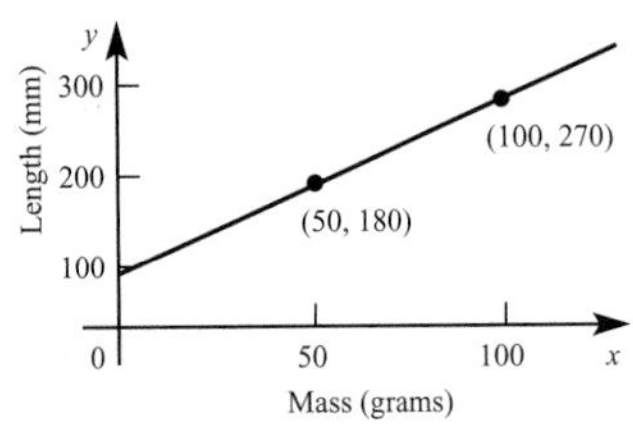

(ii) $9x - 5y + 450 = 0$

(iii) 90 mm

(iv) $83\frac{1}{3}$ g

(v) It is likely to snap.

? (Page 94)

You need to choose a scale that makes it easy to plot the points and read off the co-ordinates of the point of intersection. It is particularly difficult to get an accurate solution when it is not represented by a point on the grid.

? (Page 95)

You can always join two points with a straight line. Using three points alerts you if one of your calculated points is wrong.

? (Page 95)

They won't intersect if they are parallel.

Exercise 5D (Page 95)

1 **(i)** $x = 1, y = 0$

(ii) $x = -1, y = 4$

(iii) $x = 3, y = 2$

(iv) $x = \frac{1}{2}, y = -2$

2 **(i)**

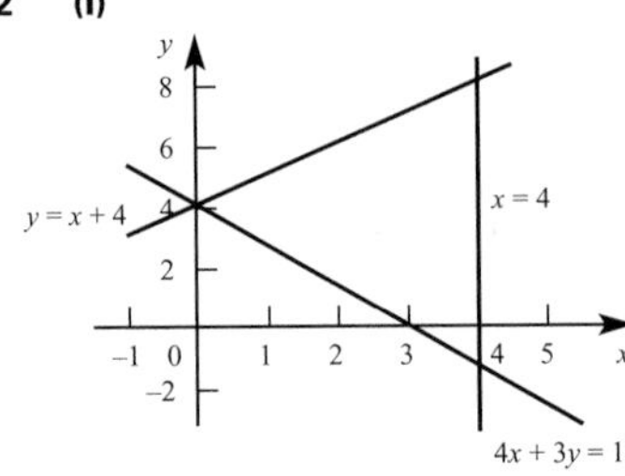

(ii) (4, 8) at the intersection of $x = 4$ and $y = x + 4$

$(4, -1\frac{1}{3})$ at the intersection of $x = 4$ and $4x + 3y = 12$

(0, 4) at the intersection of $y = x + 4$ and $4x + 3y = 12$

(iii) $18\frac{2}{3}$ units^2

3 **(i)**

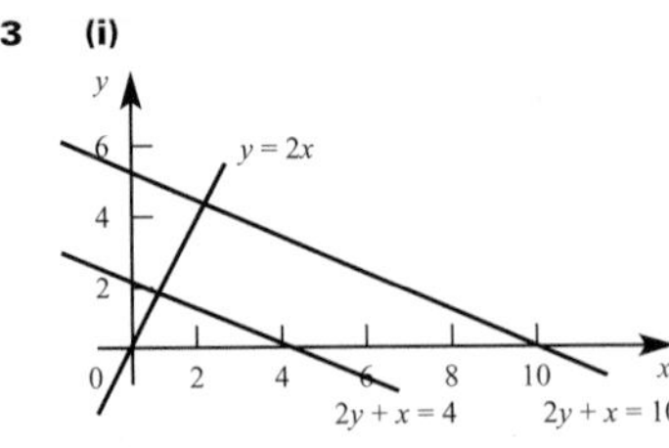

The lines look parallel. They have the same gradient.

(ii) This line looks perpendicular to the first two lines.

The first two lines have a gradient of $-\frac{1}{2}$ and the third line has a gradient of +2. The product of the gradients is –1 so they are perpendicular.

(iii) $(\frac{4}{5}, 1\frac{3}{5})$ at the intersection of $y = 2x$ and $2y + x = 4$

(2, 4) at the intersection of $y = 2x$ and $2y + x = 10$

? (Page 96)

This line uses Pythagoras' theorem to find the length of OP.

Exercise 5E (Page 99)

1 **(i)** $(x-1)^2 + (y-2)^2 = 9$

(ii) $(x-4)^2 + (y+3)^2 = 16$

(iii) $(x-1)^2 + y^2 = 25$

(iv) $(x+2)^2 + (y+2)^2 = 4$

(v) $(x+4)^2 + (y-3)^2 = 1$

2 **(i)** **(a)** (0, 0)

(b) 5

(c)

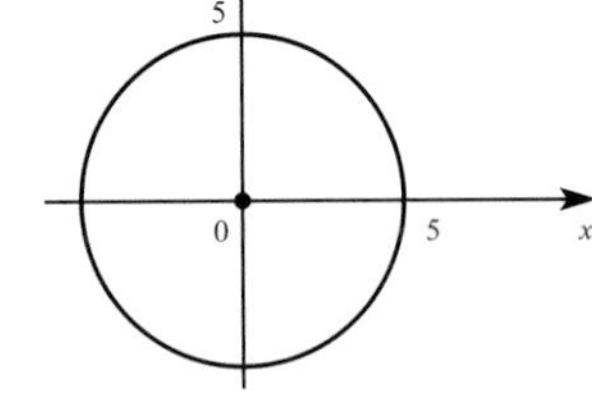

(ii) **(a)** (3, 0)

(b) 3

(c)

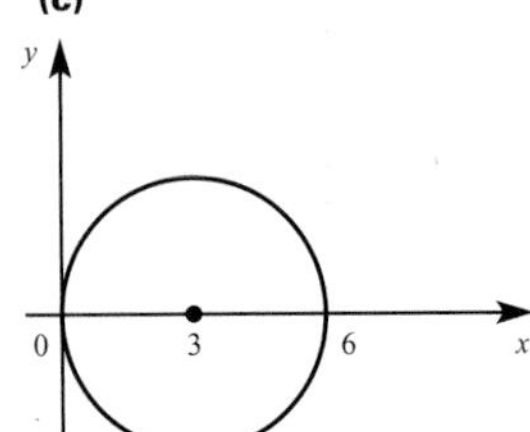

(iii) **(a)** $(-4, 3)$

(b) 5

(c)

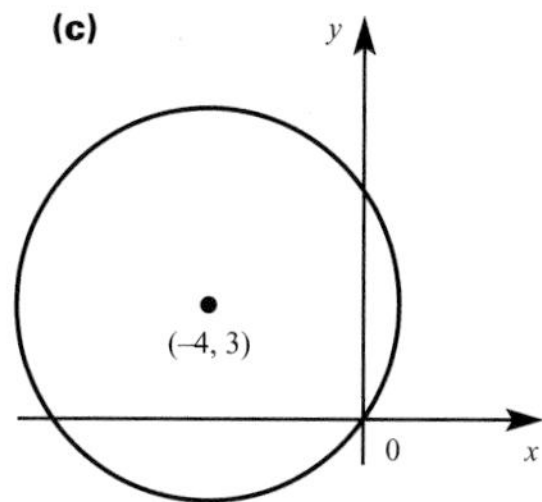

(iv) **(a)** $(-1, -6)$

(b) 6

(c)

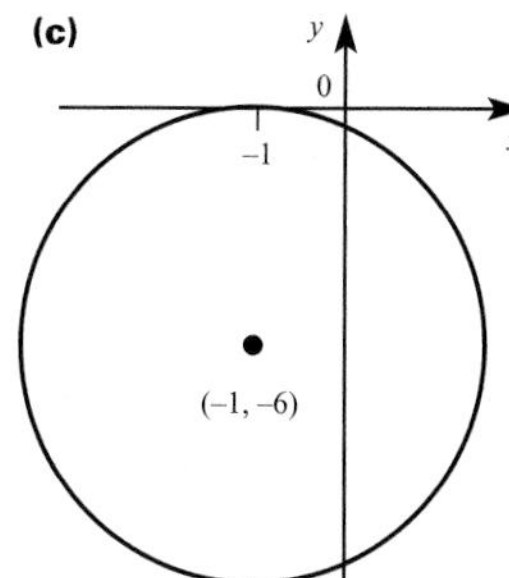

(v) **(a)** $(4, 4)$

(b) 4

(c)

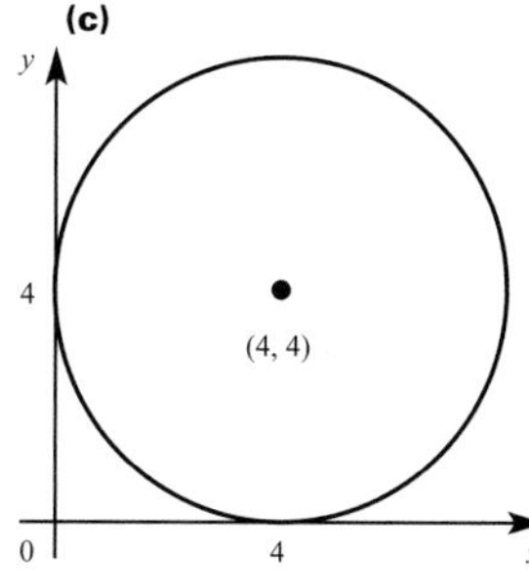

3 $(x-2)^2 + (y+3)^2 = 5$

4 **(i)** $(3, 1)$

(ii) $\sqrt{26}$

(iii) $(x-3)^2 + (y-1)^2 = 26$

5 Centre $(2, 4)$, radius 4

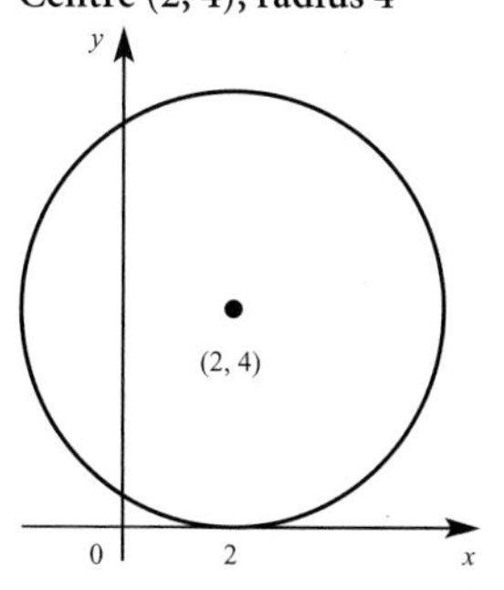

Chapter 6

❓ (Page 101)

$-1\frac{1}{2} \leqslant x \leqslant 1\frac{1}{2}, 8 \leqslant y \leqslant 11$

❓ (Page 102)

If you shade the area you don't want it leaves the area you require clear and uncluttered.

❓ (Page 106)

One of the inequalities was $y > 0$, so $(5, 0)$ cannot be in the feasible region.

Exercise 6A (Page 106)

1 **(i)**

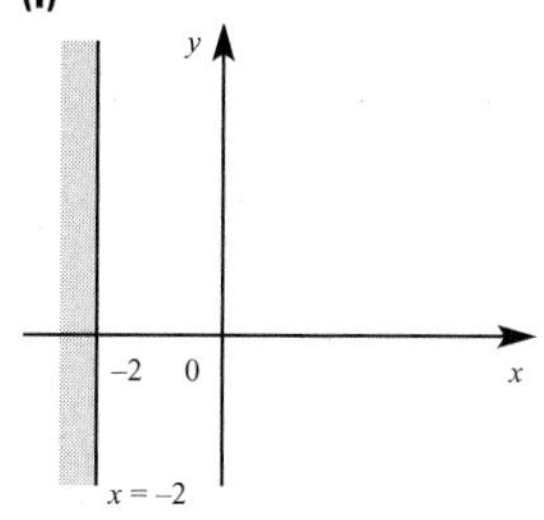

(ii)

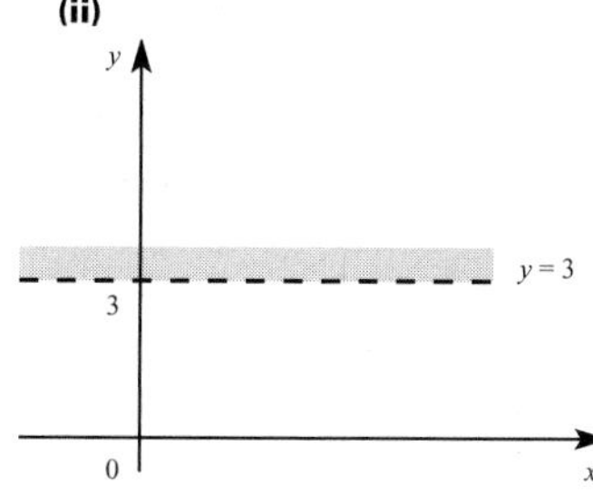

(iii)

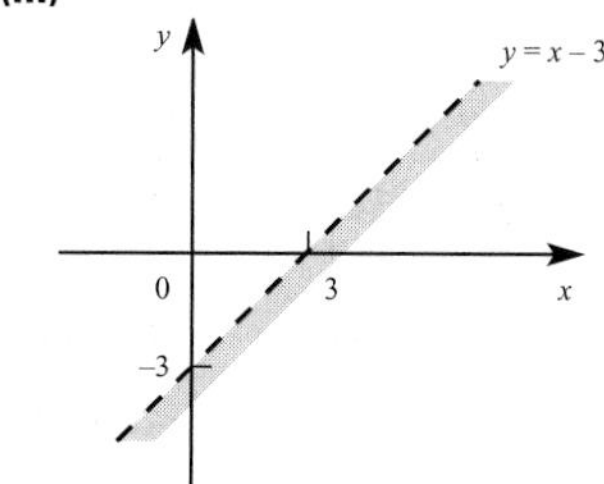

(iv)

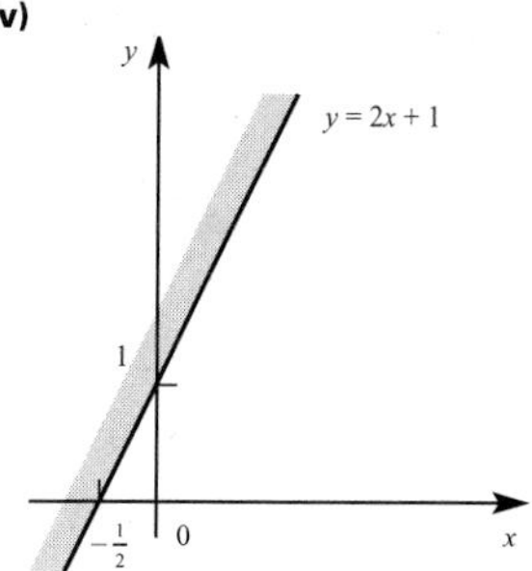

(v)

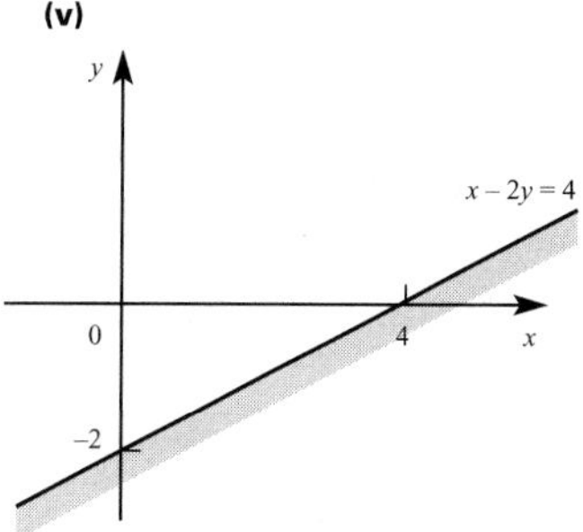

(vi)

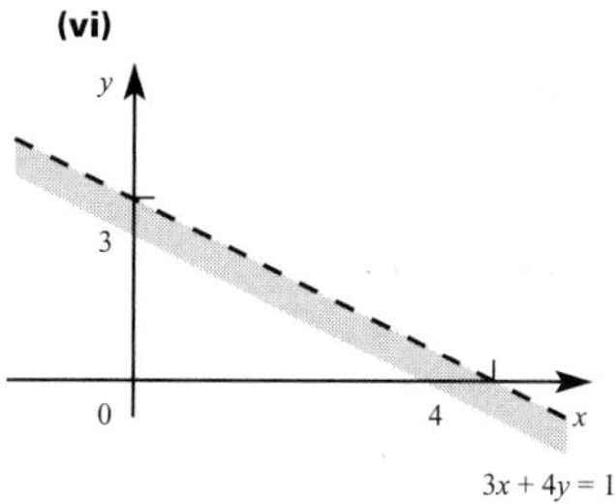

(vii)

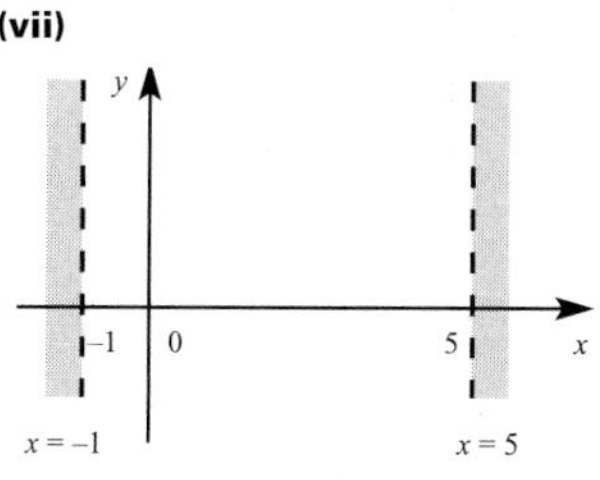

(viii)

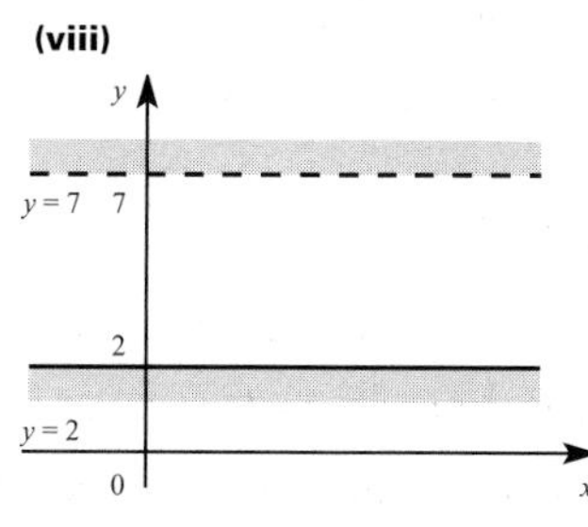

Chapter 6

(ix)

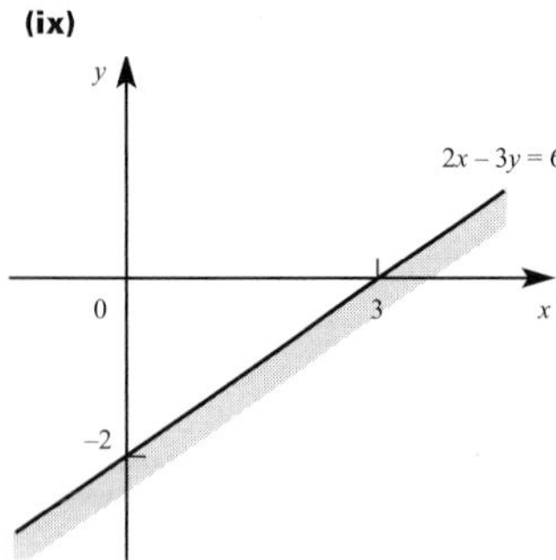

(x)

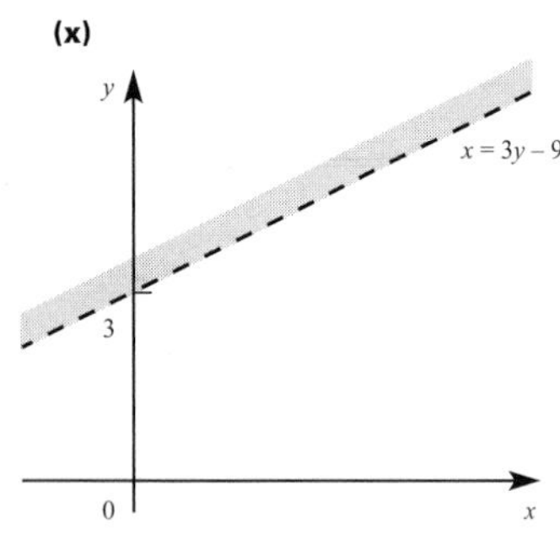

2 (i)

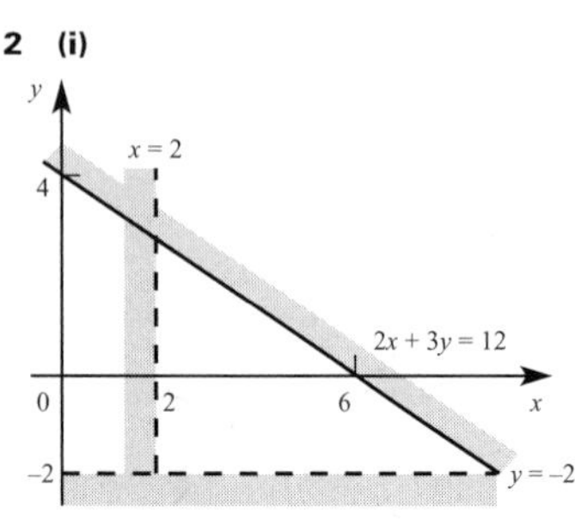

(ii) (a) Yes

(b) No

(c) No

3 $y \leqslant x$, $2x + 3y < 12$, $y > -1$

4

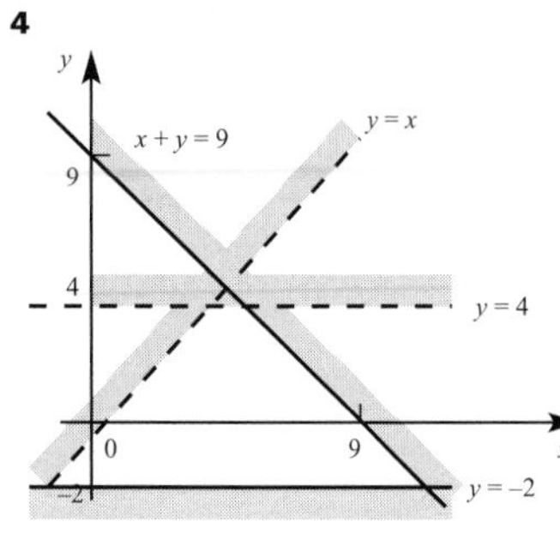

? (Page 107)

Possible answers:

(i) Tiger Woods drives his ball *at least* 300 m.
$x \geqslant 300$ where x m is the length of the drive.

(ii) *At most* 25 000 people can fit into the new sports stadium.
$x \leqslant 25\,000$ where x represents the number of people.

(iii) A mango costs *more than* an apple.
$m > a$ where m pence is the cost of a mango and a pence is the cost of an apple.

(iv) The premature baby weighed *less than* 1 kg.
$m < 1$ where m kg is the mass of the baby.

(v) Simon was *no more than* 10 m from the scene of the accident.
$d \leqslant 10$ where Simon was d metres from the scene of the accident.

(vi) To ride on the Big Dipper you must be *no less than* 120 cm tall.
$h \geqslant 120$ where h is the height in centimetres.

(vii) The record for the 100 m sprint stands at under 10 seconds.
$t < 10$ where t is the time in seconds.

(viii) The winner of the high jump competition was over 2 m tall.
$h > 2$ where h is the height of the winner in metres.

Exercise 6B (Page 108)

1 (i) (a) She would like to buy at least six apples.

(b) She would like to buy at least eight bananas.

(c) She must buy at least 18 pieces of fruit.

(d) She has £4 to spend, so $24a + 20b \leqslant 400$.

(ii)

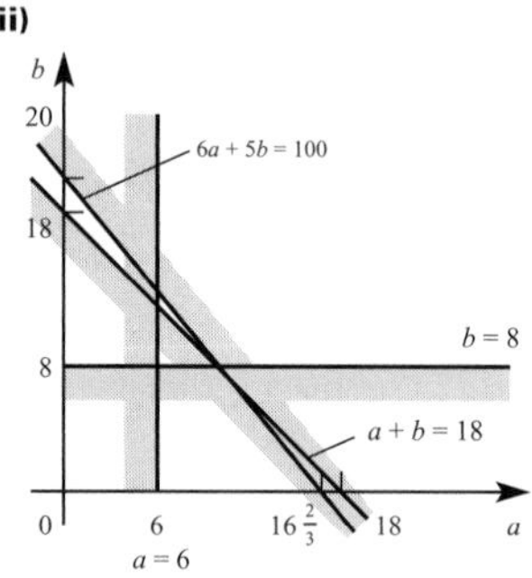

(iii) Crosses marked at (6, 12), (7,11), (8, 10), (9, 9) and (10, 8).

2 (i) (a) $s + p \leqslant 8$

(b) $p \geqslant 2$

(c) $s \geqslant 3$

(d) $s > p$

(ii)

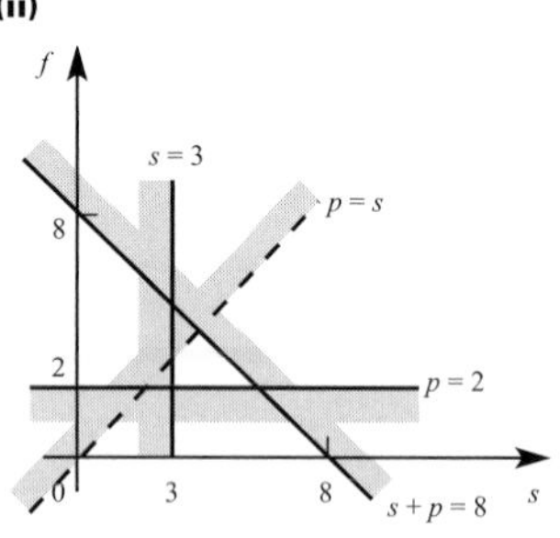

(iii) Crosses marked at (3, 2), (4, 2), (5, 2), (6, 2), (4, 3) and (5, 3).

3 **(i)** **(a)** $f \geqslant 5$

(b) $l \geqslant 4$

(c) $20f + 15l \leqslant 250 \Rightarrow$

$4f + 3l \leqslant 50$

(ii)

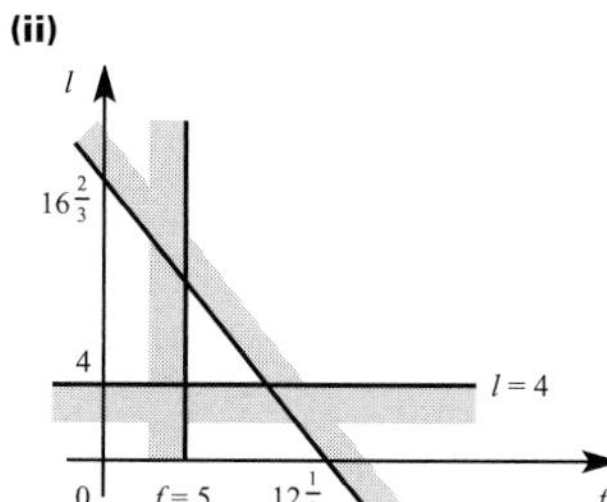

(iii) He can have at most nine fruit trees (with four others) or at most ten others (with five fruit trees).

4 **(i)** **(a)** $d \geqslant t$

(b) $d \leqslant 2t$

(c) $15d + 18t \leqslant 1800$

$\Rightarrow\ 5d + 6t \leqslant 600$

(ii)

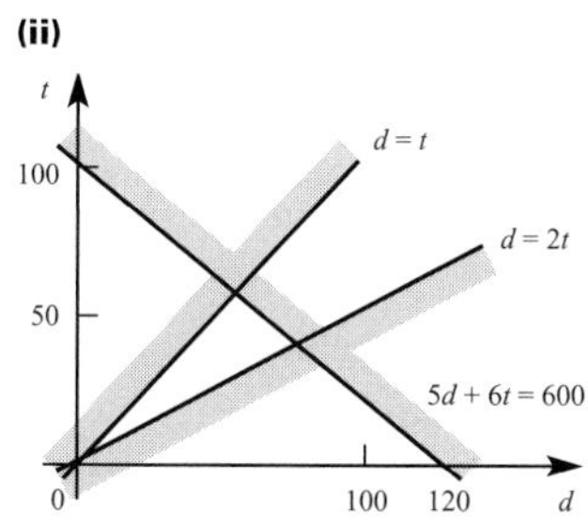

5 **(i)** **(a)** The company only has 4 fifty-seat coaches.

(b) The company has 6 thirty-seat coaches available.

(c) 240 seats are needed so $50f + 30t \geqslant 240$.

(ii)

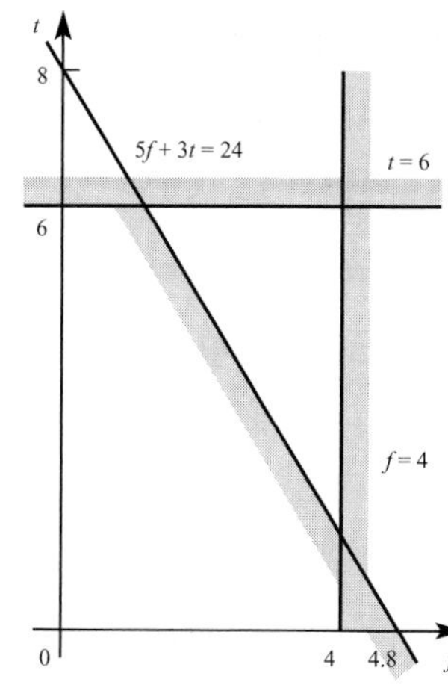

(iii) Crosses marked at (2, 6), (3, 6), (4, 6), (2, 5), (3, 5), (4, 5), (3, 4), (4, 4), (3, 3), (3, 4) and (4, 2).

(iv) 3 fifty-seat coaches plus 3 thirty-seat coaches leaves no empty seats.

(v) On a fifty-seat coach there will be only 48 seats for the children and on a thirty-seat coach there will be only 29 seats for children, so $48f + 29t \geqslant 240$.

Exercise 6C (Page 114)

1 **(i)** $b \geqslant 7,\ r \geqslant 3,\ b + r \leqslant 18$

(ii)

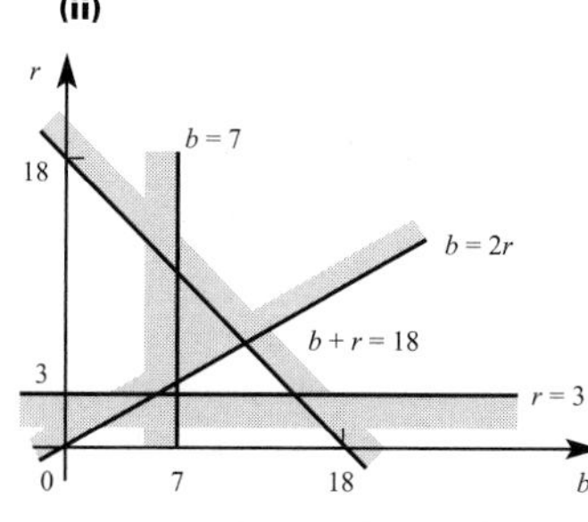

(iii) $b \geqslant 2r$

(iv) Objective function = $3b + 4r$, £60

2 **(i)** **(a)** $2x + 2y$ **(b)** $3x + 5y$

(ii) **(a)** $x + y \leqslant 50$

(b) $3x + 5y \leqslant 200$

(c) $x \geqslant 2y$

(d) $x \leqslant 4y$

(iii)

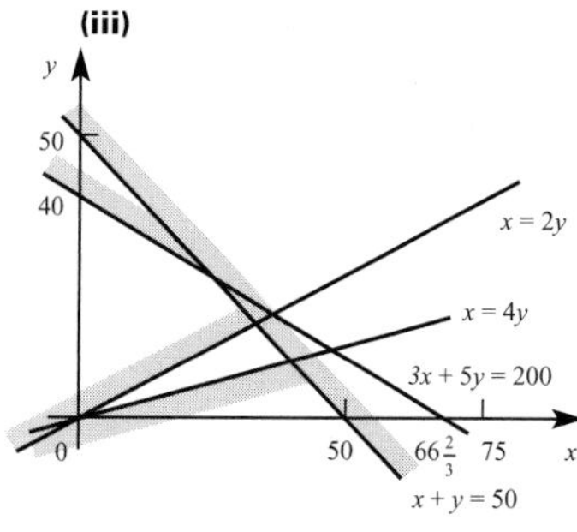

(iv) Objective function = x; $x = 40,\ y = 10$

3 **(i)** $t + c \leqslant 400,\ t \leqslant 2c,$ $4t + 3c \geqslant 1000$

(ii)

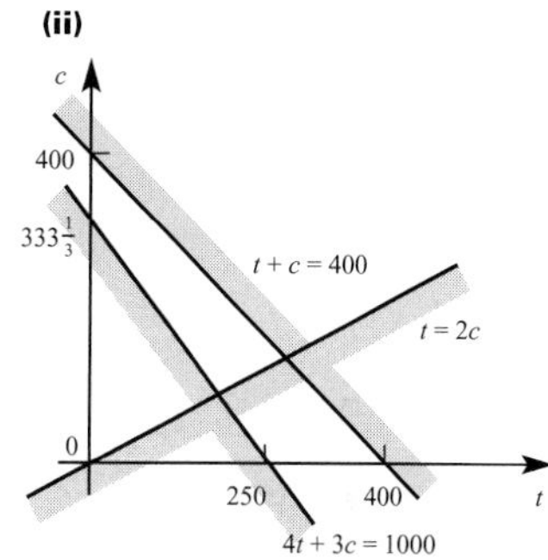

(iii) Objective function = $4t + 3c$

(a) $c = 134,\ t = 266$

(b) £1466

4 **(i)** $1.2a + 1.6b \leqslant 1000,$ $0.4a + 0.3b \leqslant 240$

(ii)

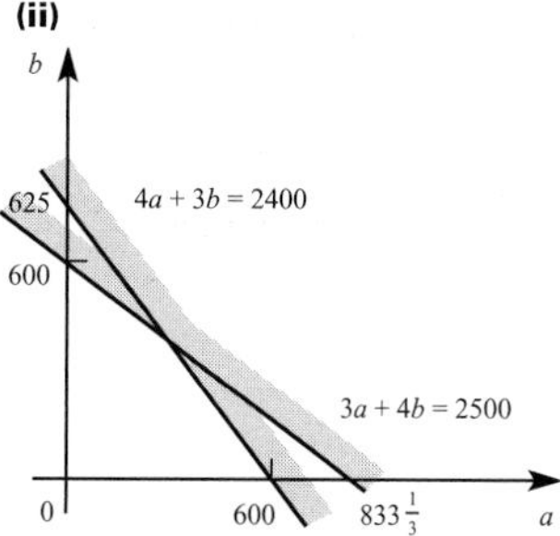

(iii) 700; $a = 300,\ b = 400$

5 **(i)** $x + y = 100$,

$0.6x + 0.84y \geqslant 75$,

$0.2x + 0.1y \geqslant 12$

(ii)

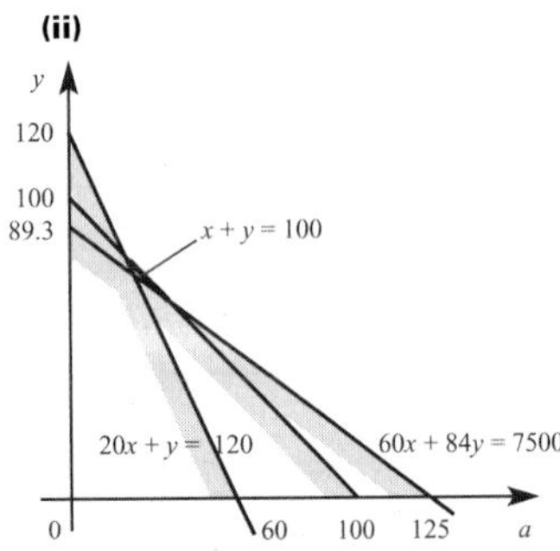

(iii) Objective function = $3x + 4.5y$; $x = 37\frac{1}{2}$, $y = 63\frac{1}{2}$

6 **(i)** $3c + 2m \leqslant 20$, $2c + 5m \leqslant 30$

(ii)

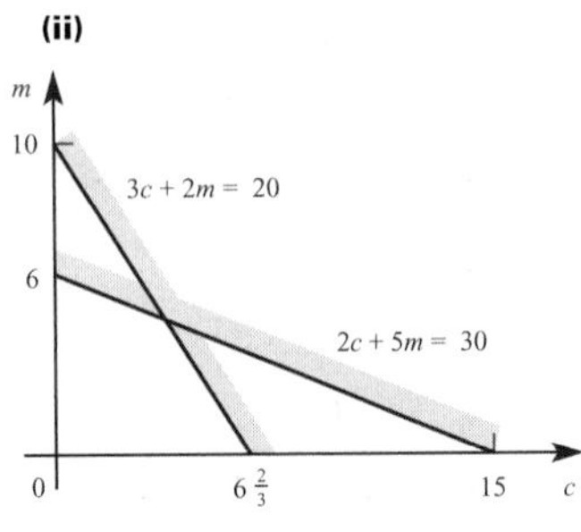

(iii) Objective function = $12c + 16m$; 112 cakes ($m = 4$, $c = 4$)

Chapter 7

❓ (Page 121)

(i) You need to know the radius of the arm and the angle it makes with the vertical.

(ii) You need to know the radius of the arm and the angle it makes with the horizontal.

❓ (Page 122)

No: since they are defined using the sides of a right-angled triangle they are restricted to $0 < \theta < 90°$.

❓ (Page 123)

You need at least 3 decimal places: $\tan^{-1} 0.714 = 35.5°$, but $\tan^{-1} 0.71 = 35.4°$.

❓ (Page 124)

The best function would be $\tan\theta$, since this does not use the value of h that you calculated earlier.

Exercise 7A (Page 125)

1 **(i)** 11.2 cm **(ii)** 7.7 cm **(iii)** 12.1 cm **(iv)** 15.1 cm **(v)** 6.8 cm **(vi)** 7.7 cm

2 **(i)** 30.6° **(ii)** 50.4° **(iii)** 55.7° **(iv)** 41.4° **(v)** 45.0° **(vi)** 64.2°

3 **(i)** 63.6° **(ii)** 14.9 cm **(iii)** 9.1 cm

4 4.5 m

5 78.2 m

6 282.7 m

7 33.7°

8 **(i)** 119 km **(ii)** 33° **(iii)** 333 km h^{-1}

❓ (Page 127)

Yes, it is possible provided that the definitions are changed to ones that do not require that the angle is in a right-angled triangle.

Activity 7.1 (Page 129)

The curve continues in the same manner repeating the wave pattern every 360° both to the right and the left.

❓ (Page 130)

The graph of $y = \sin(\theta + 90°)$ is obtained when the graph of $y = \sin\theta$ is translated 90° to the left. It then coincides with the graph of $y = \cos\theta$.

Activity 7.2 (Page 130)

(Note: In this activity other answers are possible.)

(i) See figure 7.9, points P_0 to P_3.
For $90° \leqslant \theta \leqslant 180°$, reflect the curve for $0° \leqslant \theta \leqslant 90°$ in the line $\theta = 90°$.
For $-180° \leqslant \theta \leqslant 0°$, rotate the curve for $0° \leqslant \theta \leqslant 180°$ through 180° about the origin.
For $180° \leqslant \theta \leqslant 360°$, translate the curve for $-180° \leqslant \theta \leqslant 0°$ to the right through 360°.
There is a line of symmetry at $\theta = 90°$.

(ii) See figure 7.11, points P_0 to P_3.
For $-90° \leqslant \theta \leqslant 0$, reflect the curve for $0° \leqslant \theta \leqslant 90°$ in the y axis.
For $90° \leqslant \theta \leqslant 270°$, rotate the curve for $-90° \leqslant \theta \leqslant 90°$ through 180° about the point (90°, 0).
The rest of the curve is obtained by translating the curve for $-90° \leqslant \theta \leqslant 270°$ through 360° either to the right or the left.
There is a line of symmetry at $\theta = 0°$ (the y axis).

❓ (Page 130)

Undefined means that you cannot find a value for it.

When $\theta = 90°$, $x = 0$ and $\cos\theta = 0$, so neither definition works since you cannot divide by zero.

$\tan\theta$ is also undefined for $\theta = 90° \pm$ any multiple of 180°.

❓ (Page 131)

It is a line that is very close to the shape of the curve for large values of x or y.

❓ (Page 131)

The period is 180° since it repeats itself every 180°.

For $-90 \leqslant \theta \leqslant 0°$, rotate the part of the curve for $0° \leqslant \theta \leqslant 90°$ through 180° about the origin. This gives one complete branch of the curve. Translating this branch through multiples of 180° to the right or left gives the rest of the curve.

Activity 7.3 (Page 131)

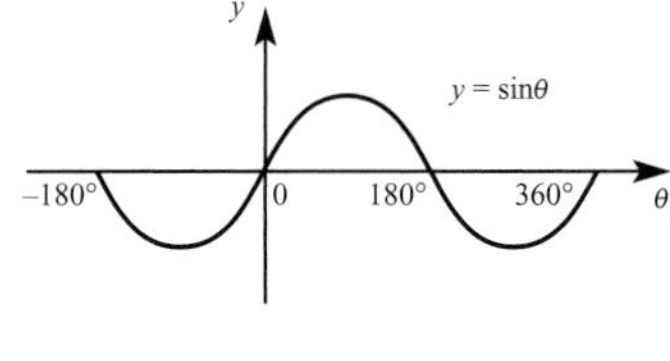

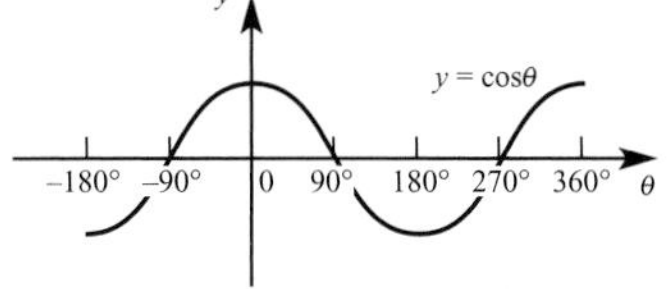

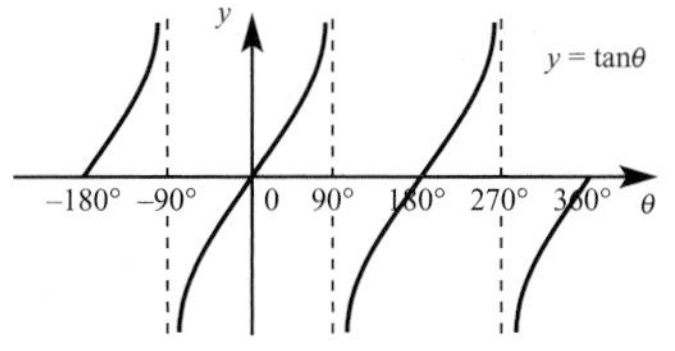

❓ (Page 131)

The equation will have infinitely many roots since the curve continues to oscillate and the line $y = 0.5$ crosses it infinitely many times.

❓ (Page 132)

$293.6° = -66.4° + 360°$

Activity 7.4 (Page 133)

$\sin\theta > 0$ $\cos\theta < 0$ $\tan\theta < 0$	$\sin\theta > 0$ $\cos\theta > 0$ $\tan\theta > 0$
$\sin\theta < 0$ $\cos\theta < 0$ $\tan\theta > 0$	$\sin\theta < 0$ $\cos\theta > 0$ $\tan\theta < 0$

Exercise 7B (Page 133)

1 **(i)** 60°, 300°
(ii) 45°, 225°
(iii) 60°, 120°
(iv) 210°, 330°
(v) 90°, 270°
(vi) 101.3°, 281.3°
(vii) 0°, 180°
(viii) 122.7°, 237.3°
(ix) 90°

2 **(i)** 303.2°
(ii) 99.8°
(iii) 32.9°
(iv) 138.6°
(v) 180°
(vi) 206.6°

3 **(i)** 65°, 295°
(ii) 60°, 120°
(iii) –135°, 45°
(iv) –130°, –50°
(v) –90°, 90°
(vi) 105°, 285°

4 **(ii)** $x = 60°$, 120° or 420°
(iv) $x = -60°$, 60°, 300° or 420°
(v) –60° occurs when $\sin x = -\frac{\sqrt{3}}{2}$ and $\cos x = 0.5$; 120° occurs when $\sin x = \frac{\sqrt{3}}{2}$ and $\cos x = -0.5$.

5 **(i)** $\theta = 48.2°$ or 311.8°
(ii) $\theta = 45.6°$ or 134.4°
(iii) $\theta = 69.4°$ or 249.4°
(iv) $\theta = 236.4°$ or 303.6°
(v) $\theta = 113.6°$ or 246.4°
(vi) $\theta = 150.9°$ or 330.9°

6 **(i)** $\theta = 11.5°$ or 168.5°
(ii) $\varphi = 31.5°$

7 **(i)** $\theta = 60°$, 120°, 240° or 300°
(ii) $\theta = 45°$, 135°, 225° or 315°
(iii) $\theta = 45°$, 135°, 225° or 315°

8 **(i)** $(2x - 1)(x + 1)$
(ii) $x = 0.5$ or –1
(iii) **(a)** $\theta = 30°$, 150° or 270°
(b) $\theta = 60°$, 180° or 300°
(c) $\theta = 26.6°$, 135°, 206.6° or 315°

Activity 7.5 (Page 134)

$\sin\theta = \dfrac{BC}{AC}$ and $\angle ABC = (90° - \theta)$

$\Rightarrow \cos(90° - \theta) = \dfrac{BC}{AC}$

Checking $\sin\theta$ and $\cos(90° - \theta)$ on your calculator for different values of θ shows that it seems to be true.

The cosine curve is symmetrical about the y axis, so $\cos(90° - \theta) = \cos(\theta - 90°)$. The graph of $y = \cos(\theta - 90°)$ is obtained from the graph of $y = \cos\theta$ by translating it 90° to the right, and this can be seen to be the same as the graph of $y = \sin\theta$.

? (Page 135)

(i) Lines of symmetry are $\theta = (90° \pm$ any multiple of $180°)$

Rotational symmetry through 180° about the points on the θ axis where $\theta = (0 \pm$ any multiple of $180°)$

The curve has a period of 360°.

(ii) Lines of symmetry are $\theta = (0 \pm$ any multiple of $180°)$

Rotational symmetry through 180° about the points on the θ axis where $\theta = (90° \pm$ any multiple of $180°)$

The curve has a period of 360°.

(iii) No line symmetry.

Rotational symmetry through 180° about the points on the θ axis where $\theta = (0 \pm$ any multiple of $180°)$

The curve has a period of 180°.

Investigation (Page 135)

(i) $\cos\theta = 0.4 \Rightarrow \theta = 66.4°$

Reflect in $\theta = 0°$ to give $\theta = -66.4°$.

Relabel 66.4° as –293.6° and –66.4° as 293.6°.

(ii)

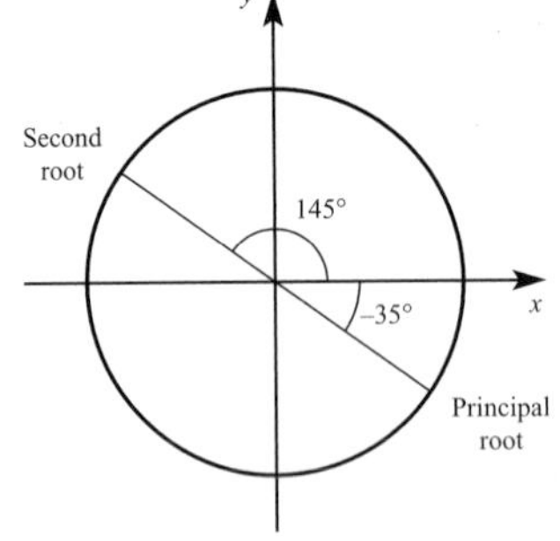

Principal root = –35°.

Second root = 145°

Relabel –35° as 325° and 145° as –215°.

? (Page 138)

An identity, such as $\tan\theta = \dfrac{\sin\theta}{\cos\theta}$, is true for all values of the variable θ, but an equation, such as $\sin\theta = 0.7$, is true for only certain values of θ.

? (Page 139)

$(2\sin\theta - 1)(\sin\theta - 2) = 0 \Rightarrow$

$\sin\theta = 0.5$ or $\sin\theta = 2$.

$\sin\theta = 2$ has no solutions, since $-1 \leqslant \sin\theta \leqslant 1$ for all values of the angle θ.

This means that the solution is just those values of θ for which $\sin\theta = 0.5$.

? (Page 139)

If the quadratic equation does not factorise, you would try to solve it using the quadratic formula.

Exercise 7C (Page 139)

1 **(i)** **(a)** $2\sin^2\theta - \sin\theta - 1 = 0$

(b) $\theta = 90°, 210°$ or $330°$

(ii) **(a)** $\cos^2\theta - \cos\theta - 2 = 0$

(b) $\theta = 180°$

(iii) **(a)** $2\cos^2\theta + \cos\theta - 1 = 0$

(b) $\theta = 60°, 180°$ or $300°$

(iv) **(a)** $\sin^2\theta - \sin\theta = 0$

(b) $\theta = 0°, 90°, 180°$ or $360°$

(v) **(a)** $2\sin^2\theta + \sin\theta - 1 = 0$

(b) $\theta = 30°, 150°, 270°$

2 **(i)** **(a)** $\cos^2\theta + 2\cos\theta - 2 = 0$

(b) $\theta = -42.9°$ or $42.9°$

(ii) **(a)** $\sin^2\theta + \sin\theta - 1 = 0$

(b) $\theta = 38.2°$ or $141.8°$

(iii) **(a)** $\cos^2\theta + 3\cos\theta - 1 = 0$

(b) $\theta = -72.4°$ or $72.4°$

3 **(i)** $\tan\theta = 2$

(ii) $\theta = -116.6°$ or $63.4°$

4 **(i)** $\theta = 153.4°$ or $333.4°$

(ii) $\theta = 0°, 30°, 180°, 330°$ or $360°$

(iii) $\theta = 14.5°$ or $165.5°$

Exercise 7D (Page 142)

Answers given to 3 significant figures.

1 **(i)** $9.85\,\text{cm}^2$

(ii) $19.5\,\text{cm}^2$

(iii) $15.2\,\text{cm}^2$

(iv) $20.5\,\text{cm}^2$

2 $127\,\text{cm}^2$

3 **(i)** $23.8\,\text{cm}^2$

(ii) $5.56\,\text{cm}$

(iii) $126\,\text{cm}^2$

4 **(i)** $308\,\text{cm}^2$

(ii) 325

(iii) There is likely to be a lot of wastage when tiles are cut for the edges, so he will need more tiles.

5 $173\,\text{cm}^2$

? (Page 144)

It is easier to solve an equation involving fractions if the unknown quantity is in the numerator.

? (Page 145)

$\frac{\sin Z}{6} = \frac{\sin 78°}{8} \Rightarrow Z = 47.2°$ or

$Z = 132.8°$, but 132.8° is too large to fit into a triangle where one of the other angles is 78°.

Exercise 7E (Page 146)

1 **(i)** 4.6 m
(ii) 11.0 cm
(iii) 5.6 cm

2 **(i)** 57.7°
(ii) 16.5°
(iii) 103.3°
(Reject 76.7° since the angle in the diagram is obtuse.)

Exercise 7F (Page 149)

1 **(i)** 6.4 cm
(ii) 8.8 cm
(iii) 13.3 cm

2 **(i)** 41.4°
(ii) 107.2°
(iii) 90°

3 9.1 cm, 12.3 cm

4 **(i)** 10 cm
(ii) 111.8°

5 55.8°

? (Page 151)

You know the lengths of all three sides of triangle ABC so there is only one possible value for ∠A. Using the cosine rule confirms that ∠A = 62°.

? (Page 152)

The cosine rule involves three sides and one angle and you want to find an angle and know three sides. The sine rule involves two angles and two sides and you do not know any angles.

Exercise 7G (Page 152)

1 12.2 cm

2 6.1 km

3 **(i)** 26.5 m
(ii) 19.4 m

4 **(i)** 57.1°, 57.1°, 122.9°, 122.9°
(ii) 14.5 cm

5 **(i)** 10.2 km
(ii) 117°

6 **(i)** 29.9 km
(ii) 12.9 km h^{-1}

7 BD = 2.1 m, EG = 2.1 m

8 4.8 km

Chapter 8

? (Page 155)

On the surface of the Earth (or just above it) the lines of latitude do not represent the shortest distance between two points (unless those points are on the equator). In general, aircraft fly long distances along the most economical route.

? (Page 157)

One possible example is a ramp used for disabled access to a building.

? (Page 158)

Possible answers

The shelves of a bookcase are *parallel*.

The side of a filing cabinet meets the floor *in a line*.

Exercise 8A (Page 161)

1 **(i)** 14.1 cm
(ii) 17.3 cm
(iii) 35.3°
(iv) 70.5°

2 **(i)** 3 cm
(ii) 72.1°
(iii) 76.0°

3 **(i)** 18.4°
(ii) 13 cm
(iii) 17.1°
(iv) Half way along

4 **(i)** 75 m
(ii) 67.5 m
(iii) 42°

5 **(i)** 33.4 m
(ii) 66.7 m
(iii) 115.6 m
(iv) 22.8°

6 **(i)** 28.3 cm
(ii) 42.4 cm
(iii) 40.6 cm

7 **(i)** 41.8°
(ii) 219 m
(iii) 186 m
(iv) 17 918 litres

8 **(i)** 1.57 m
(ii) 15.47 m
(iii) 3.05 m

9 **(i)** 346 km
(ii) 71°

10 **(i)** 5.2 cm

(ii) 5.2 cm

(iii) 54.7°

(iv) 16.9 cm

Investigation (Page 164)

(i) 4095 km

(ii) 8790 km

(iii) 7198 km

(iv) 6370 km

(v) 68.8°

(vi) 7649 km

? (Page 165)

When two points are joined by an arc of a circle, the arc length decreases as the radius of the circle increases. Arc ACB is shorter than arc ADB.

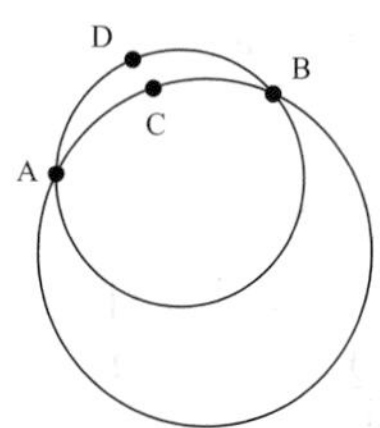

The maximum radius of any circle on the Earth is 6370 km, so the arc on a circle of this radius gives the shortest distance and hence the most economical flight path.

Chapter 9

? (Page 171)

Yes: drawing chords from P to points to the left of P will again give a sequence that eventually gives the gradient of the tangent. Activity 9.1 does this in greater detail.

Activity 9.1 (Page 171)

Taking $R_1 = (2, 4)$, $R_2 = (2.5, 6.25)$, $R_3 = (2.9, 8.41)$, $R_4 = (2.99, 8.9401)$ and $R_5 = (2.999, 8.994001)$ gives the gradient sequence 5, 5.5, 5.9, 5.99, 5.999. Again the sequence seems to converge to 6.

Activity 9.2 (Page 172)

(i) The gradient of the chord is $4 + h$. The gradient of the tangent is 4.

(ii) The gradient of the chord is $-2 + h$. The gradient of the tangent is -2.

(iii) The gradient of the chord is $-6 + h$. The gradient of the tangent is -6.

In each case the gradient of the tangent is twice the value of the x co-ordinate.

Exercise 9A (Page 173)

1 **(i)** Gradient 34.481, 32.241…, 32.024…, approaching a limit of 32.

(ii) Gradient 113.521, 108.541…, 108.054…, approaching a limit of 108.

-3.439, $-3.940…$, $-3.994…$, approaching a limit of -4.

2 The gradient of the chord is $2x + h$, with a limit of $2x$.

3 $(x + h)^4 = x^4 + 4x^3h + 6x^2h^2 + 4xh^3 + h^4$

The gradient of the chord is $4x^3 + 6x^2h + 4xh^2 + h^3$ with a limit of $4x^3$.

4

$f(x)$	$f'(x)$
x^2	$2x$
x^3	$3x^2$
x^4	$4x^3$
x^5	$5x^4$
x^6	$6x^5$
⋮	⋮
x^n	$nx^{(n-1)}$

Activity 9.3 (Page 175)

(i), (ii)

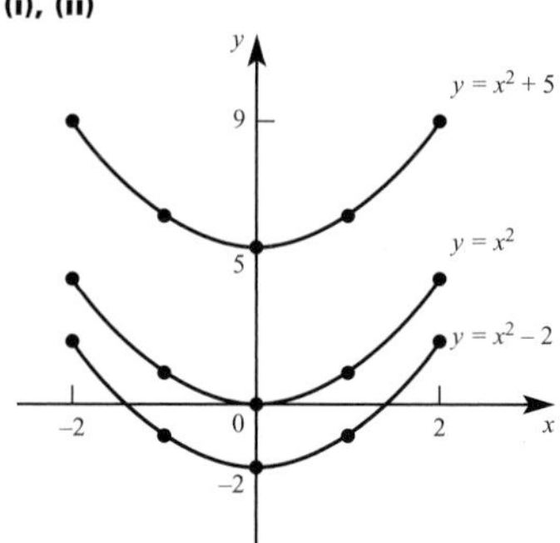

(iii) When $x = 0$, the gradients of all three curves are zero. When $x = 1$, the tangents to the three curves are parallel, so the gradients are the same. When $x = -1$, the tangents to the three curves are parallel, so the gradients are the same.

(iv) When $y = x^2 + c$, where c is a constant, $\frac{dy}{dx} = 2x$, so the value of c does not affect the value of $\frac{dy}{dx}$. This shows that the value of $\frac{dy}{dx}$ for all three curves in **(i)** and **(ii)** will be the same for any particular value of x.

Exercise 9B (Page 177)

1 **(i)** $\frac{dy}{dx} = 4x^3$

(ii) $\frac{dy}{dx} = 6x^2$

(iii) $\frac{dy}{dx} = 10x$

(iv) $\frac{dy}{dx} = 63x^8$

(v) $\frac{dy}{dx} = -18x^5$

(vi) $\frac{dy}{dx} = 0$

(vii) $\frac{dy}{dx} = 10$

(viii) $\frac{dy}{dx} = 10x^4 + 8x$

(ix) $\frac{dy}{dx} = 12x^3 + 8$

(x) $\frac{dy}{dx} = 3x^2$

(xi) $\frac{dy}{dx} = 1 - 15x^2$

(xii) $\frac{dy}{dx} = 15x^4 + 16x^3 - 6x$

(xiii) $\frac{du}{dx} = 12x^2 + 2$

(xiv) $\frac{dp}{dx} = 2$

(xv) $\frac{dz}{dx} = 5x^4 + 36x^2 + 3$

(xvi) $\frac{dv}{dt} = 15t^4$

(xvii) $\frac{dd}{dp} = \frac{3}{4}p^2$

(xviii) $\frac{dh}{dr} = 3r^2 + 84r - 5$

(xix) $\frac{dC}{dr} = 2\pi$

(xx) $\frac{dA}{dr} = 2\pi r$

Exercise 9C (Page 179)

1 **(i)** $\frac{dy}{dx} = 5 - 2x$

(ii) -1

(iii) $x + y - 9 = 0$

(iv) $x - y + 3 = 0$

2 **(i)** **(a)** $\frac{dy}{dx} = 6x - 3x^2$

(b) 0

(c) $y = 4$

(d) $x = 2$

(ii) **(a)** $(3, 0)$

(b) -9

(c) $9x + y = 27$

(iii) $y = 0$

3 **(i)** $(1, 0)$

(ii) $y = 2x - 2$

(iii) $x + 2y - 2 = 0$

(iv) $Q(0, -2)$, $R(0, \frac{1}{2})$; $1\frac{1}{4}$ units^2

4 **(i)** $f'(x) = 3x^2 - 6x + 4$

(ii) **(a)** 5

(b) $y = 4x - 3$

(c) $x + 4y - 22 = 0$

(iii) $x = -1, x = 3$

5 **(i)** $\frac{dy}{dx} = 3x^2 - 18x + 23$

(ii) -1

(iii) $x + y = 5$

(iv) $(4, -3)$

(v) $x + y = 1$

6 **(i)** $2p - q = 16$

(ii) $\frac{dy}{dx} = 3x^2 - p$

(iii) $p = 12$

(iv) $(-2, 24)$

(v) $(0, 8)$

(vi) $x - 12y + 96 = 0$

7 **(i)** $y = 3x - 5$

(ii) $-\frac{1}{3}$

(iii) $(\frac{1}{3}, -1\frac{2}{9})$

8 **(i)** $\frac{dy}{dx} = 10 - 2x$

(ii) $2x + y - 15 = 0$

(iii) $x - 2y = 0$

(iv) The normal

9 **(ii)** At $(0, 0)$ the tangent is $y = 2x$ and the normal is $x + 2y = 0$.

At $(1, 0)$ the tangent is $x + 2y - 1 = 0$ and the normal is $2x - y - 2 = 0$.

Activity 9.4 (Page 182)

(i)

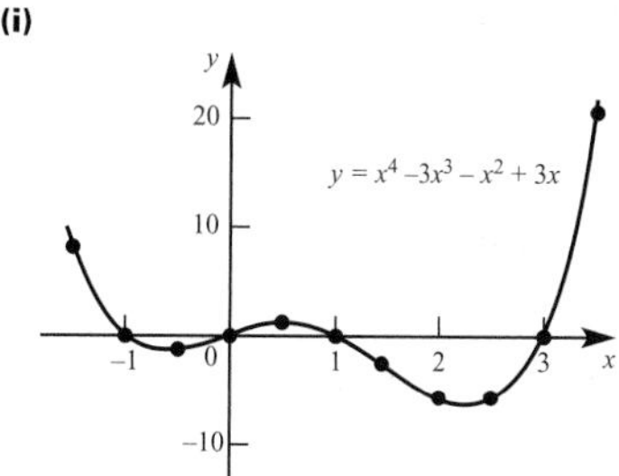

(ii) Three

(iii) Zero

(iv) Minimum near $(-0.5, -1.3)$, maximum near $(0.5, 0.9)$, minimum near $(-2.3, -6.9)$

(v) No, for example $x = -1.5$ and $x = 3.5$ give higher points.

(vi) No. $\frac{dy}{dx} = 4x^3 - 9x^2 - 2x + 3$ and this does not equal zero at the points plotted.

(vii) About -6.9

? (Page 183)

The gradient is positive both to the left and to the right of D.

Activity 9.5 (Page 183)

When $x = 0°$ the gradient is zero. It then decreases through negative values and is least when $x = 90°$. It increases to zero when $x = 180°$ and continues to increase through positive values until it is greatest when $x = 270°$. The gradient then decreases to zero when $x = 360°$.

? (Page 186)

There are no more values when $\frac{dy}{dx} = 0$, so there are no more turning points. As x increases beyond the point where $x = 2$, $\frac{dy}{dx}$ takes positive values and so the curve will cross the x axis again. To the left of $x = -2$ the gradient is always negative, giving a further point of intersection with the x axis.

? (Page 186)

(i) The curve crosses the x axis when $x^3 - 12x + 3 = 0$. This does not factorise, so the values of x cannot be found easily.

(ii) Only when the equation obtained when $y = 0$ factorises.

Exercise 9D (Page 187)

1 (i) (a) $\frac{dy}{dx} = 1 - 4x$; $x = \frac{1}{4}$

(b) Max

(c) $y = 1\frac{1}{8}$

(d)

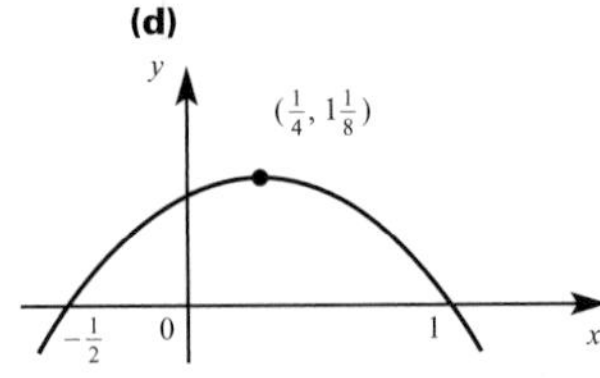

(ii) (a) $\frac{dy}{dx} = 12 + 6x - 6x^2$;

$x = -1, x = 2$

(b) Min when $x = -1$,

max when $x = 2$

(c) $x = -1, y = -7$;

$x = 2, y = 20$

(d)

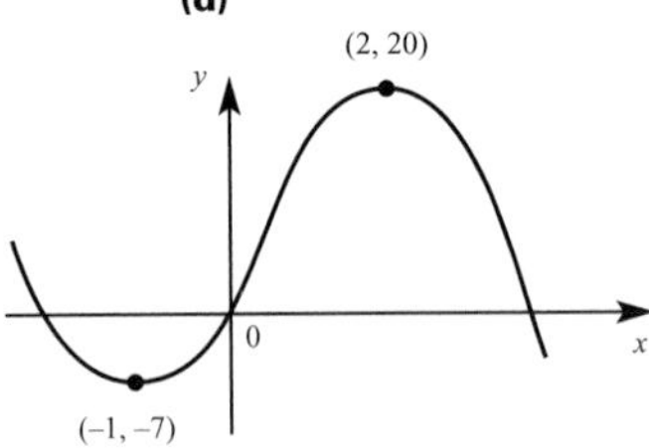

(iii) (a) $\frac{dy}{dx} = 3x^2 - 8x$; $x = 0$,

$x = 2\frac{2}{3}$

(b) Max when $x = 0$,

min when $x = 2\frac{2}{3}$

(c) $x = 0, y = 9$;

$x = 2\frac{2}{3}, y = -\frac{13}{27}$

(d)

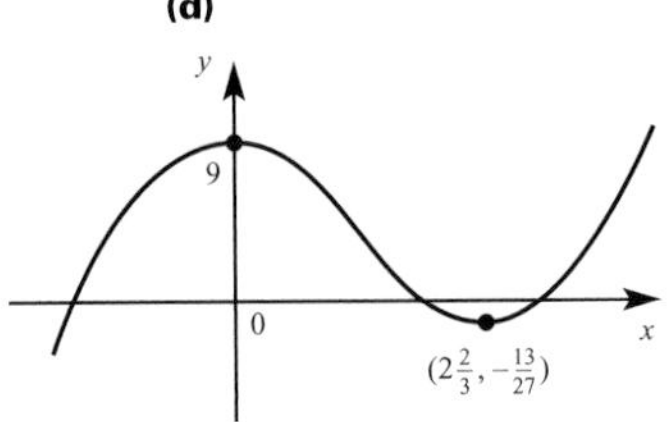

(iv) (a) $\frac{dy}{dx} = 4x^3 - 6x^2 + 2x$;

$x = 0, x = \frac{1}{2}, x = 1$

(b) Min when $x = 0$,

max when $x = \frac{1}{2}$,

min when $x = 1$

(c) $x = 0, y = 0$;

$x = \frac{1}{2}, y = \frac{1}{16}$;

$x = 1, y = 0$

(d)

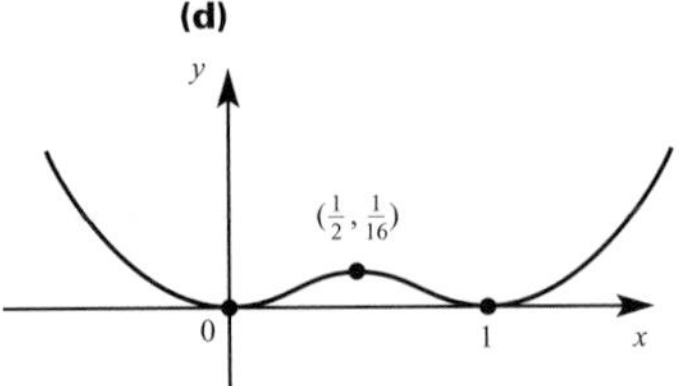

(v) (a) $\frac{dy}{dx} = 4x^3 - 16x$;

$x = -2, x = 0, x = 2$

(b) Min when $x = -2$,

max when $x = 0$,

min when $x = 2$

(c) $x = -2, y = -12$;

$x = 0, y = 4$;

$x = 2, y = -12$

(d)

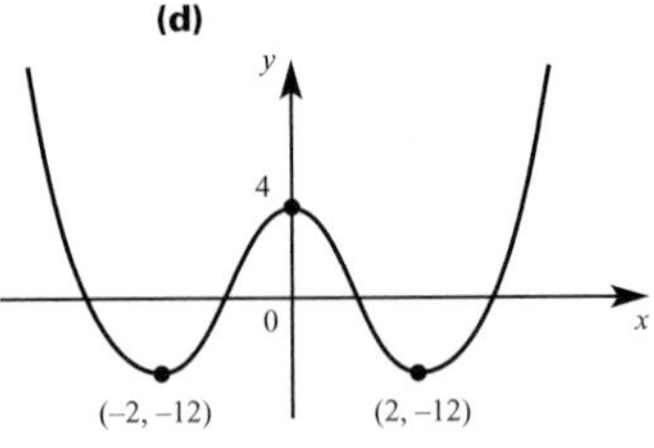

(vi) (a) $\frac{dy}{dx} = 3x^2 - 48$;

$x = -4, x = 4$

(b) Max when $x = -4$,

min when $x = 4$

(c) $x = -4, y = 128$;

$x = 4, y = -128$

(d)

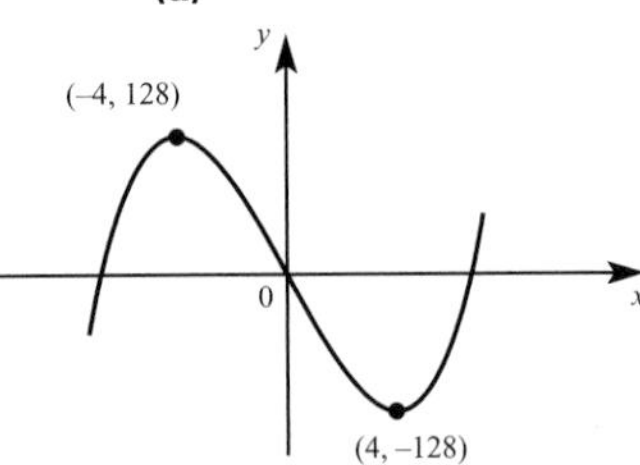

(vii) (a) $\frac{dy}{dx} = 3x^2 + 12x - 36$;

$x = -6, x = 2$

(b) Max when $x = -6$,

min when $x = 2$

(c) $x = -6, y = 241$;

$x = 2, y = -15$

(d)

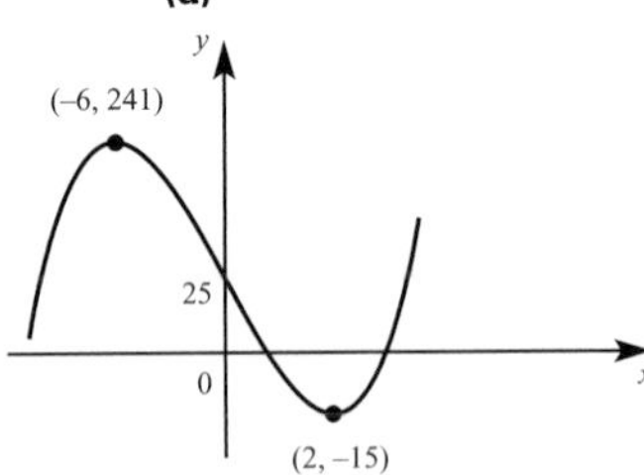

(viii) (a) $\frac{dy}{dx} = 6x^2 - 30x + 24$;

$x = 1, x = 4$

(b) Max when $x = 1$,

min when $x = 4$

(c) $x = 1, y = 19$;

$x = 4, y = -8$

(d)

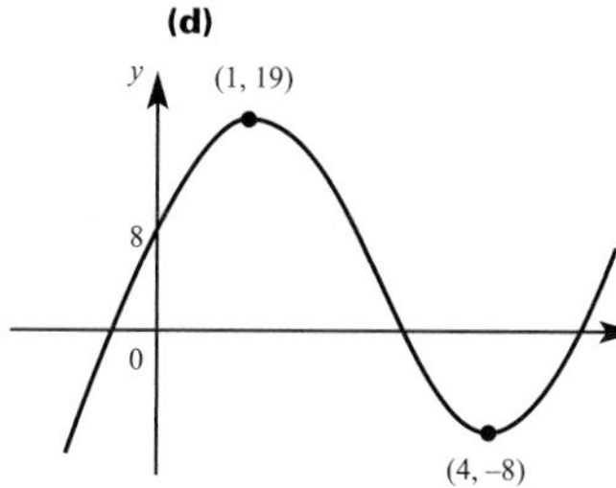

2 (i) $p = 4, q = -3$

(ii) $y = 1\frac{1}{3}, x = \frac{2}{3}$

3 (i) Min at $\left(-\frac{1}{2}, -\frac{5}{16}\right)$, max at $(0, 0)$, min at $(1, -2)$

(ii)

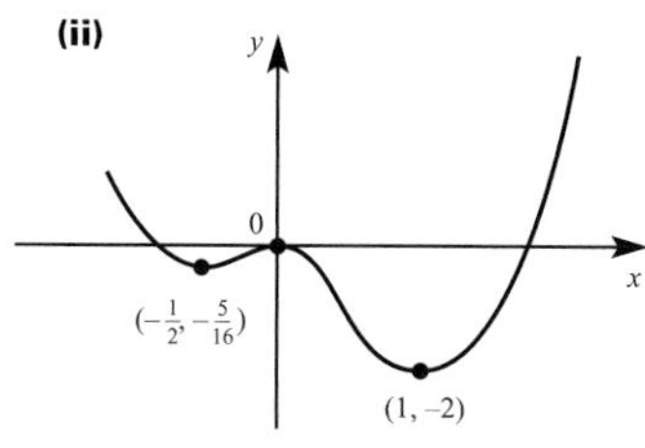

Exercise 9E (Page 189)

1 (i) **(a)** $\frac{dy}{dx} = 3x^2 + 12x + 12$; $x = -2$

(b) Point of inflection

(c) $x = -2, y = 0$

(d)

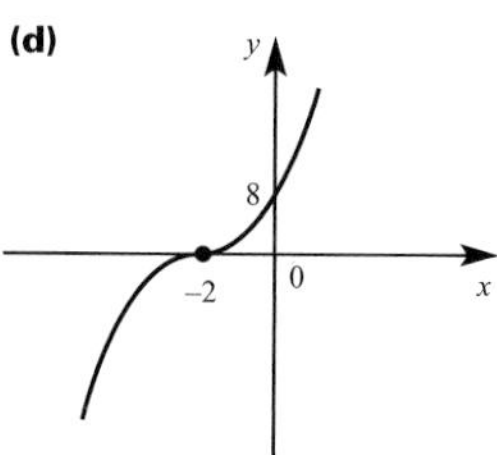

(ii) **(a)** $\frac{dy}{dx} = 12x^3 + 12x^2$; $x = -1, x = 0$

(b) Min when $x = -1$, point of inflection when $x = 0$

(c) $x = -1, y = -1$; $x = 0, y = 0$

(d)

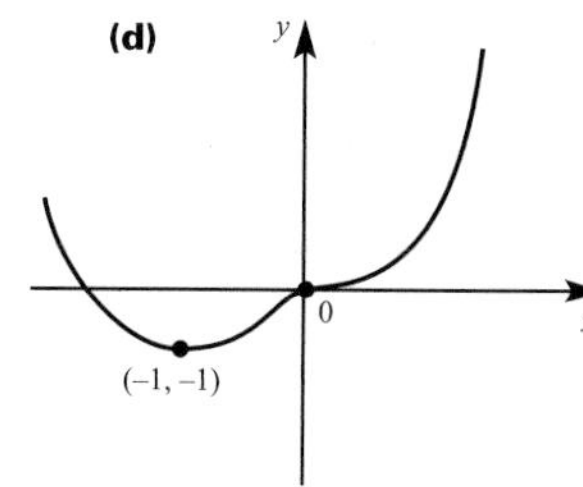

(iii) **(a)** $\frac{dy}{dx} = 12x^2 - 4x^3$; $x = 0, x = 3$

(b) Point of inflection when $x = 0$, max when $x = 3$

(c) $x = 0, y = 3$; $x = 3, y = 30$

(d)

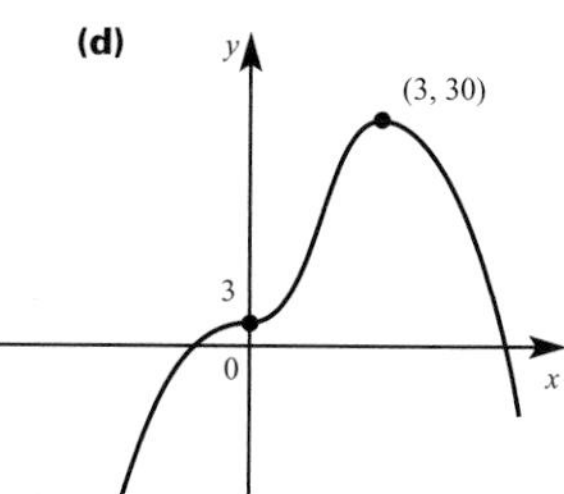

(iv) **(a)** $\frac{dy}{dx} = 15x^4 - 15x^2$; $x = -1, x = 0, x = 1$

(b) Max when $x = -1$, point of inflection when $x = 0$, min when $x = 1$

(c) $x = -1, y = 2$; $x = 0, y = 0$; $x = 1, y = -2$

(d)

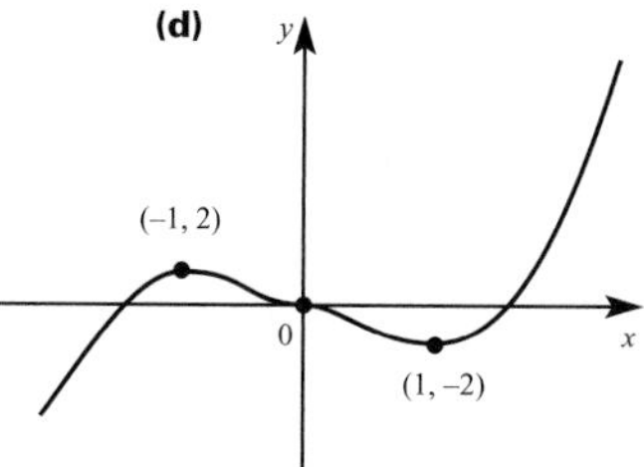

(v) **(a)** $\frac{dy}{dx} = 24x^2 - 16x^3$; $x = 0, x = 1\frac{1}{2}$

(b) Point of inflection when $x = 0$; max when $x = 1\frac{1}{2}$

(c) $x = 0, y = 0$; $x = 1\frac{1}{2}, y = 6\frac{3}{4}$

(d)

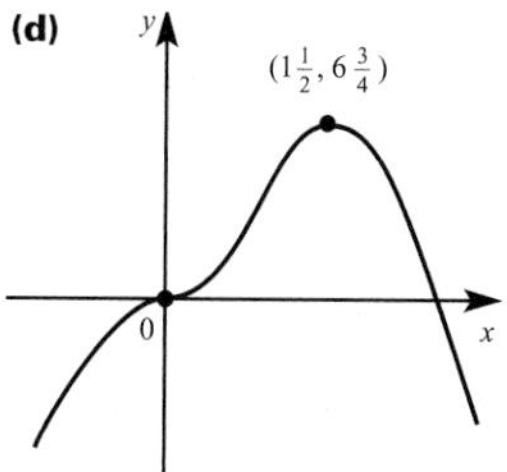

(vi) **(a)** $\frac{dy}{dx} = 3x^2 - 6x + 3$; $x = 1$

(b) Point of inflection when $x = 1$

(c) $x = 1, y = 2$

(d)

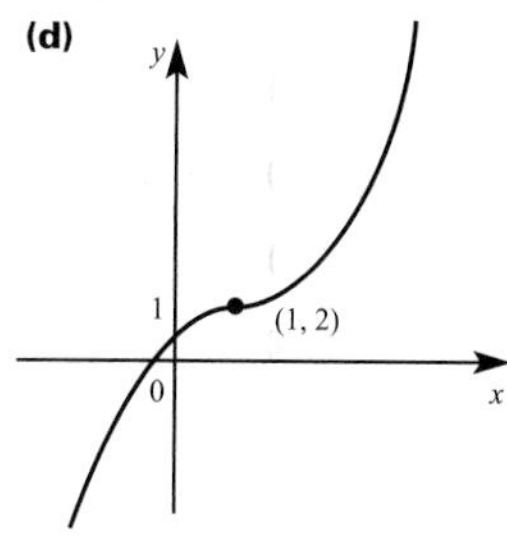

2 (i) Point of inflection at $(1, 3)$

(ii) The curve is the same shape as the curve in question 1**(vi)**, translated up one unit.

3 (i) $p + q = -3$

(ii) $4p + 3q = 0$

(iii) $p = 9, q = -12$

(iv) Minimum turning point

(v) Point of inflection at $(0, 0)$

Chapter 10

? (Page 191)

You need to know the co-ordinates of any one point on the curve.

Activity 10.1 (Page 191)

(i) **(a)** $\frac{dy}{dx} = 3x^2$

(b) $\frac{dy}{dx} = 3x^2$

(c) $\frac{dy}{dx} = 3x^2$

(ii) The answers in **(i)** are all the same. They are not affected by the value of the constant.

? (Page 192)

There are infinitely many, since $y = x^3 + c$, where c can take any constant value.

? (Page 193)

Write 1 as x^0, so that integrating x^0 gives $\frac{x^1}{1} + c$

? (Page 195)

You would need to expand the brackets to give $2x^2 - 7x - 4$.

Exercise 10A (Page 195)

1 **(i)** $y = 2x^2 + 2x + c$

(ii) $y = 2x^3 - \frac{5x^2}{2} - x + c$

(iii) $y = 3x - \frac{5x^4}{4} + c$

(iv) $y = x^3 - 2x^2 - 4x + c$

(v) $f(x) = \frac{5x^2}{2} + 3x + c$

(vi) $f(x) = \frac{x^5}{5} + \frac{x^4}{2} - \frac{x^2}{2} + 8x + c$

(vii) $f(x) = \frac{x^4}{4} + \frac{4x^3}{3} + x^2 - 8x + c$

(viii) $f(x) = \frac{x^3}{3} - 7x^2 + 49x + c$

2 **(i)** $\frac{5x^4}{4} + c$

(ii) $x^2 - 3x + c$

(iii) $\frac{3x^4}{4} - 2x^2 + 3x + c$

(iv) $9x - 3x^2 + \frac{x^3}{3} + c$

(v) $4x + c$

(vi) $\frac{2x^3}{3} - \frac{5x^2}{2} - 3x + c$

(vii) $\frac{x^3}{3} + x^2 + x + c$

(viii) $\frac{4x^3}{3} - 2x^2 + x + c$

3 **(i)** $y = x^2 - 3x + 6$

(ii) $y = 4x + \frac{3x^4}{4} - 210$

(iii) $y = \frac{5x^2}{2} - 6x - 18$

(iv) $f(x) = \frac{x^3}{3} + 3x + 9$

(v) $f(x) = \frac{x^3}{3} - \frac{x^2}{2} - 2x - 44$

(vi) $f(x) = \frac{4x^3}{3} + 2x^2 + x - 5\frac{1}{3}$

4 **(i)** $y = x^2 + 3x + c$

(ii) $y = x^2 + 3x - 11$

5 **(i)** $y = x^3 - 2x^2 + x + 9$

6 **(i)** $y = 2x^2 - x + c$

(ii) $y = 2x^2 - x + 1$

(iii) Above

7 -1

8 **(i)** $\frac{dy}{dx} = 0$ when $x = 0$ and $x = 2 \Rightarrow \frac{dy}{dx} = kx(x - 2)$ where k is a constant.

Putting $k = 1$ shows that $\frac{dy}{dx} = x^2 - 2x$ is a possibility.

(ii) $y = \frac{x^3}{3} - x^2 + 2$

? (Page 197)

'Evaluate' means 'find the value of'. It is appropriate to use it in Example 10.7 because it is possible to find the value of a definite integral whereas it is not possible to find the value of an indefinite integral.

Activity 10.2 (Page 198)

(i) $\int_1^3 x^2 dx = 8\frac{2}{3}$; $\int_3^1 x^2 dx = -8\frac{2}{3}$

(ii) $\int_{-1}^4 (x + 3)\, dx = 22\frac{1}{2}$;

$\int_4^{-1} (x + 3)\, dx = -22\frac{1}{2}$

When the limits are reversed, the answer has the same magnitude but the opposite sign.

? (Page 198)

$\int_a^b f(x)\, dx = -\int_b^a f(x)\, dx$

Exercise 10B (Page 198)

1 **(i)** 7

(ii) 255

(iii) 4

(iv) 16

(v) $20\frac{2}{3}$

(vi) 30

(vii) 591

(viii) $25\frac{1}{3}$

(ix) $-1\frac{1}{3}$

(x) 12

(xi) 27

(xii) $5\frac{1}{3}$

(xiii) 92

(xiv) $-1\frac{1}{3}$

(xv) $13\frac{1}{2}$

(xvi) 0

(xvii) 28

(xviii) $26\frac{2}{3}$

Activity 10.3 (Page 199)

(i) A(2, 5), B(4, 9)

(ii) 14 units^2

(iii) 14; this has the same magnitude as the area.

Activity 10.4 (Page 199)

(i) 5 units^2, smaller; 14 units^2, larger

(ii) **(a)** 6.875 units^2, smaller; 11.375 units^2, larger

(b) 8.555 units^2, smaller; 9.455 units^2, larger

(iii) $\int_0^3 x^2\,dx = 9$

(iv) The value of the integral is between the upper and lower values of the sums of rectangles. The upper and lower sums seem to approach the value of the integral.

Exercise 10C (Page 201)

1 **(i)** 9 units^2

(ii) 36 units^2

(iii) 2 units^2

(iv) $6\frac{2}{3}$ units^2

(v) $\frac{1}{4}$ units^2

(vi) $520\frac{5}{6}$ units^2

(vii) 36 units^2

(viii) $13\frac{1}{2}$ units^2

(ix) $8\frac{8}{15}$ units^2

(x) $21\frac{1}{3}$ units^2

Activity 10.5 (Page 202)

(i) Area A = 2 units^2; area B = 4.5 units^2

(ii) $\int_{-2}^{0} x\,dx = -2$; $\int_0^3 x\,dx = 4.5$

The areas have the same numerical value as the integral, but when the area is below the x axis the value of the integral is negative.

(iii) $\int_{-2}^{3} x\,dx = 2.5$

The areas above and below the x axis have cancelled out.

? (Page 203)

The curve has rotational symmetry of order 2 about the origin, so the areas of P and Q will be the same.

? (Page 204)

The area is 8 units^2. You should always calculate areas above and below the x axis separately.

Exercise 10D (Page 204)

1 $\frac{1}{4}$ units^2

2 **(i)** P: $\frac{5}{12}$ units^2; Q: $2\frac{2}{3}$ units^2

(ii) $3\frac{1}{12}$ units^2

3 **(i)** $(\sqrt[3]{2}, 0)$

(ii) 0.952 units^2

4 $21\frac{1}{12}$ units^2

5 **(i) and (ii)**

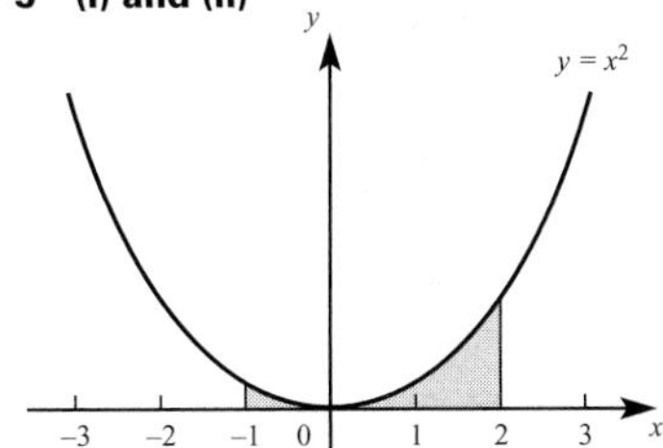

(iii) 3 units^2

6 **(i)**

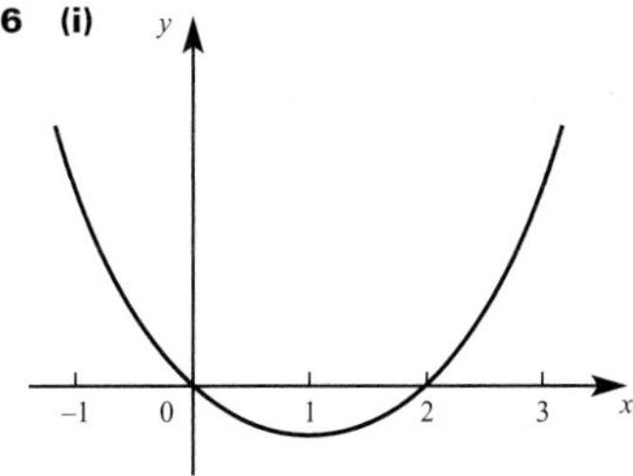

(ii) $0 < x < 2$

(iii) $1\frac{1}{3}$ units^2

7 **(i) and (ii)**

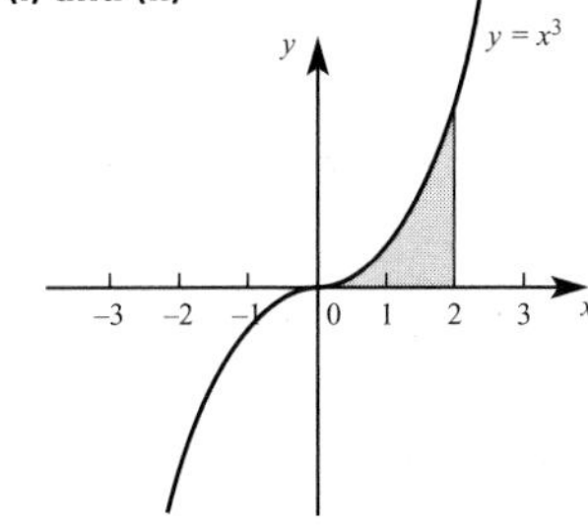

(iii) 4 units^2

(iv) 0. Since the curve has rotational symmetry about the origin, the areas above and below the x axis will cancel out.

8 **(i)**

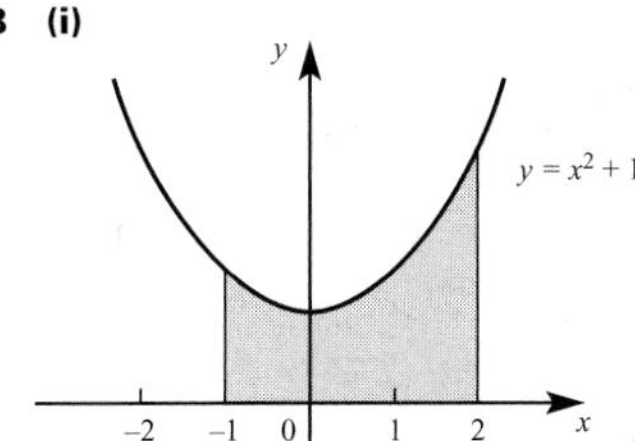

(ii) 6 units^2

9 **(i)** 18 units^2

(ii)

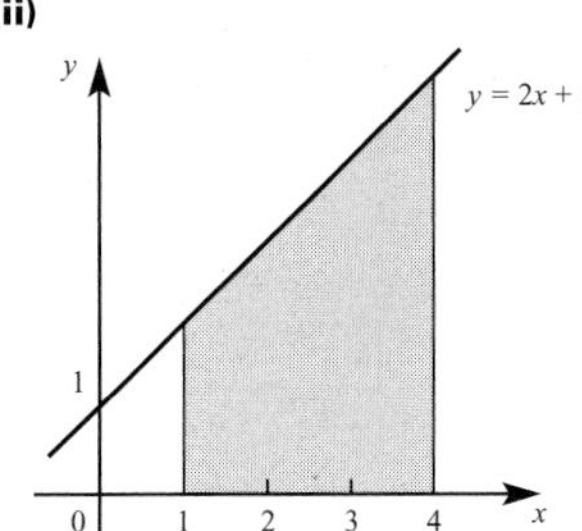

10 (i)

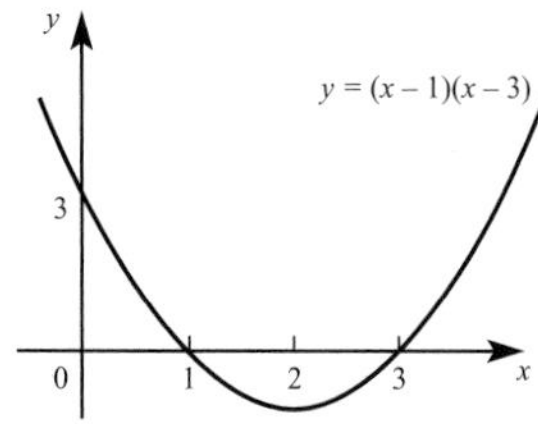

The curve crosses the x axis at (1, 0) and (3, 0).

(ii) $1\frac{1}{3}$ units^2

(iii) The curve is symmetrical about the line $x = 2$.

11 (i) A: $x = -20$; B: $x = 20$

(ii) $32\,\text{m}^2$

(iii) $128\,\text{m}^2$

(iv) $737\,280\,\text{m}^3$

12 (i) 22

(ii) 8

(iii) $154\frac{2}{3}\,\text{m}^2$

(iv) $21\frac{1}{3}\,\text{m}^2$

(v) $6.4\,\text{m}^3$

13 (i) (7.625, 0)

(ii) $6.56\,\text{m}^2$ (2 d.p.)

? (Page 209)

By collecting like terms together you reduce the number of integrations required.

? (Page 210)

Because the bottom curve is partly above and partly below the x axis.

Exercise 10E (Page 210)

1 (i) (a)

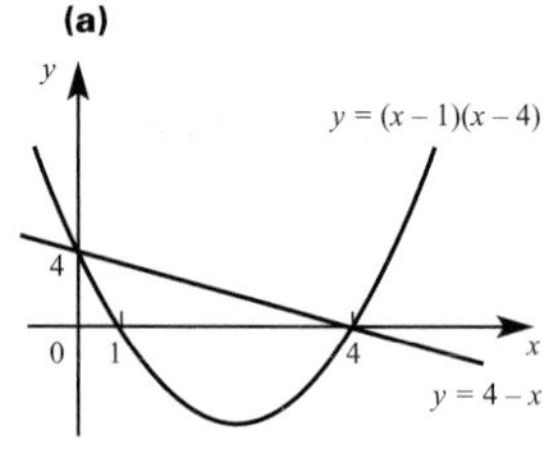

(b) $x = 0, x = 4$

(c) $10\frac{2}{3}$ units^2

(ii) (a)

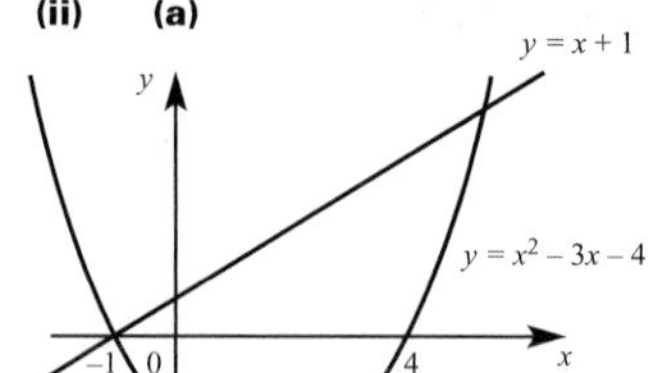

(b) $x = -1, x = 5$

(c) 36 units^2

(iii) (a)

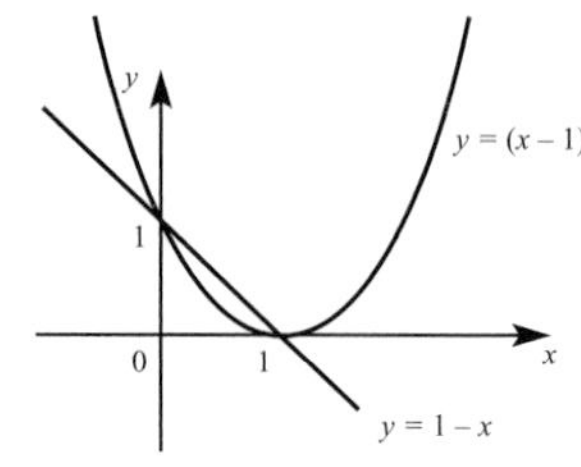

(b) $x = 0, x = 1$

(c) $\frac{1}{6}$ units^2

(iv) (a)

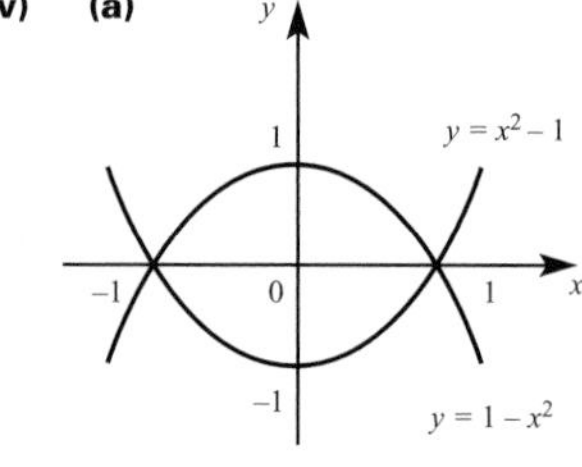

(b) $x = -1, x = 1$

(c) $2\frac{2}{3}$ units^2

(v) (a)

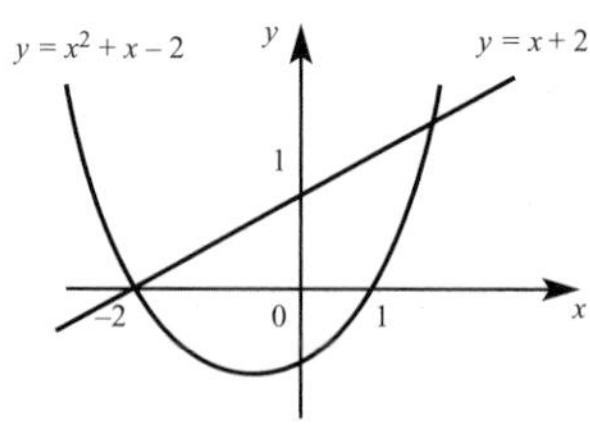

(b) $x = -2, x = 2$

(c) $10\frac{2}{3}$ units^2

(vi) (a)

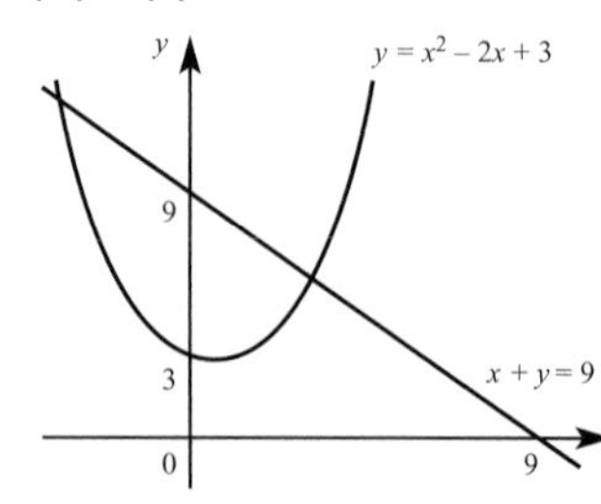

(b) $x = -2, x = 3$

(c) $21\frac{5}{6}$ units^2

(vii) (a)

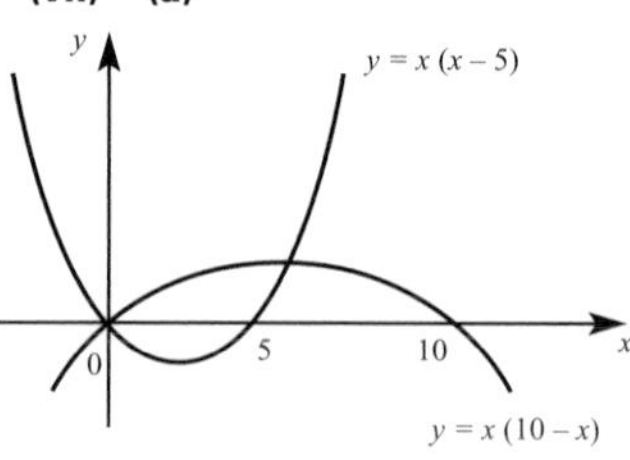

(b) $x = 0, x = 7.5$

(c) 140.625 units^2

(viii) (a)

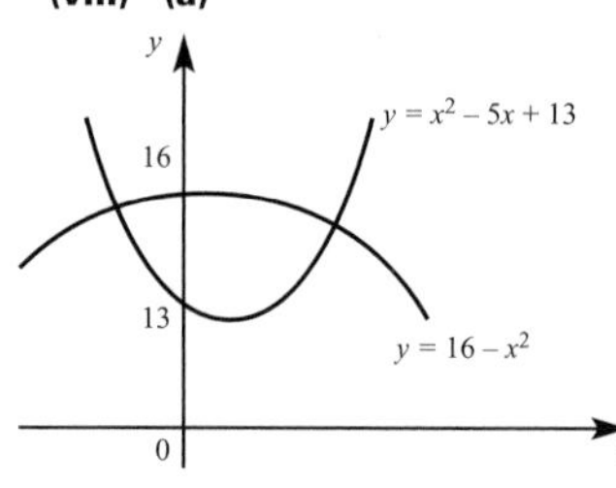

(b) $x = -\frac{1}{2}, x = 3$

(c) 12.71 units^2

(ix) (a)

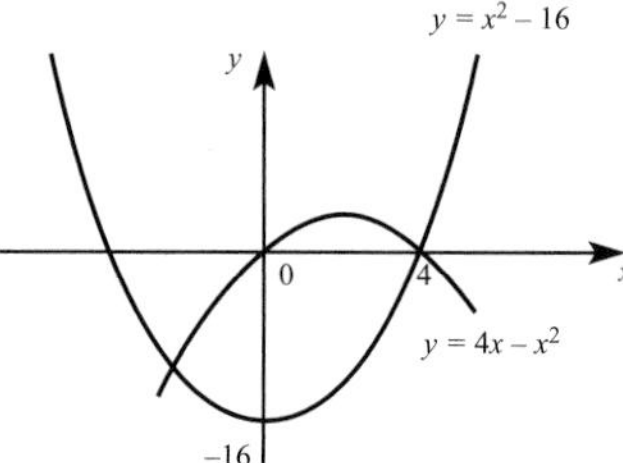

(b) $x = -2, x = 4$

(c) 72 units^2

(x) (a)

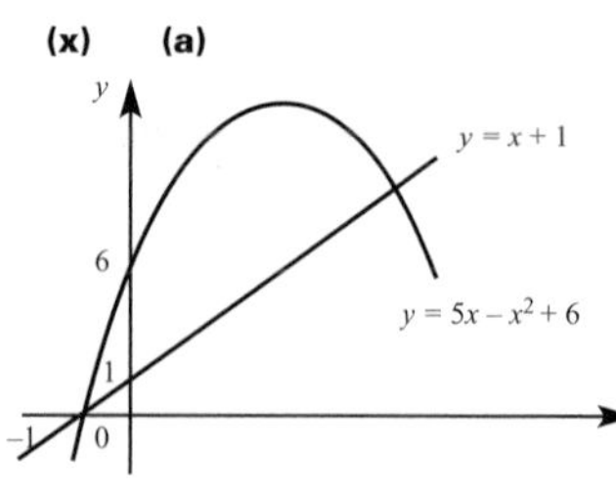

(b) $x = -1, x = 5$

(c) 36 units^2

2 (i)

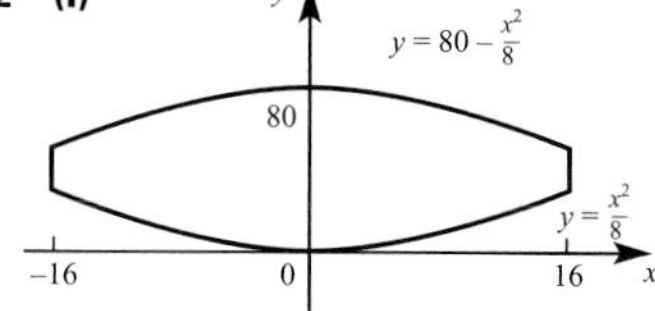

(ii) 1877 units^2

3 (i) OA: $y = x(x+3)$

OB: $y = x - \frac{1}{4}x^2$

AB: $y = x^2 - 6x + 9$

(ii) A: (1, 4), B: (2, 1)

(iii) $2\frac{5}{6}\,\text{m}^2$

4 (i) (7.00, 12.12) (2 d.p.)

(ii) $61.4\,\text{cm}^2$

(iii) $737.2\,\text{cm}^2$

Chapter 11

❓ (Page 213)

Your answer will depend on your journey. Here is a description of the start of one particular journey. The car turned left out of the drive and accelerated until a speed of 30 mph was reached. This speed was maintained for about two minutes until we slowed down as we approached some traffic lights. The lights changed to green as we got nearer, but we had to stop before turning right because of the oncoming traffic ….

❓ (Page 214)

$\frac{ds}{dt}$

❓ (Page 214)

Speeds AB: $5\,\text{ms}^{-1}$, BC: $2.5\,\text{ms}^{-1}$, CD: $10\,\text{ms}^{-1}$, DE: $5\,\text{ms}^{-1}$

Nature of road: AB: level, BC: uphill, CD: downhill, DE: level

Activity 11.1 (Page 214)

(i)

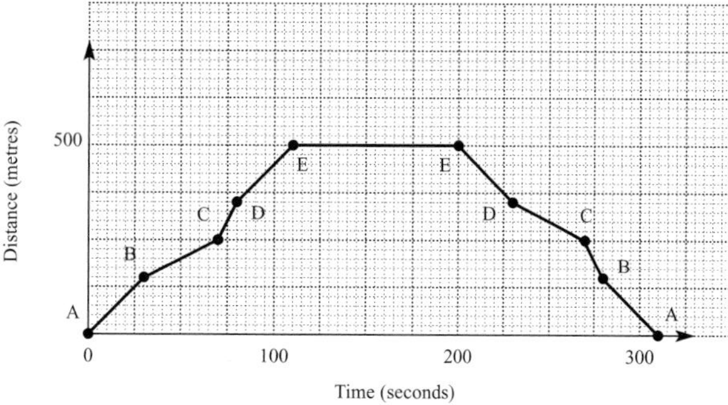

(ii)

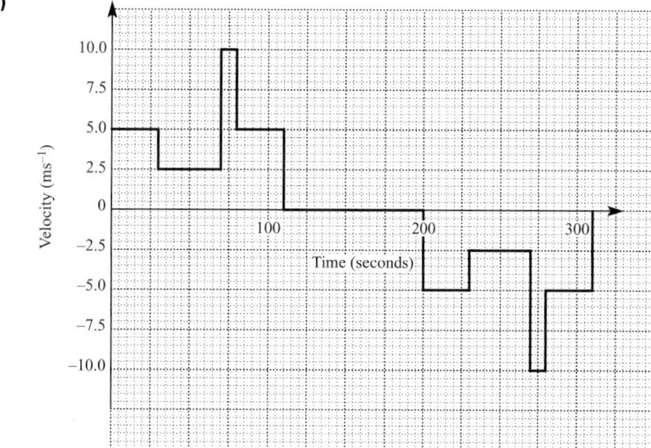

❓ (Page 215)

(i) Displacement

(ii) Acceleration

❓ (Page 216)

The ball is at the point where it is about to change direction by 180°.

❓ (Page 216)

(i) Ground level or the starting point

(ii) Upwards

❓ (Page 217)

It would be the same shape but translated down by 1 unit so that it starts at the origin and ends at a displacement of –1 m.

❓ (Page 217)

A negative acceleration is either a deceleration when the object is moving in the positive direction, or an acceleration when it is moving in the negative direction.

❓ (Page 218)

(i) The acceleration has a constant value and, since it is the gradient of the velocity–time graph, the velocity–time graph must be a straight line.

(ii) $2.5\,\text{ms}^{-1}$

❓ (Page 221)

The word 'dropped' implies that the initial velocity is zero.

❓ (Page 221)

a and s are both negative, but the time is the same.

❓ (Page 221)

$v = u + at$

Exercise 11A (Page 222)

1 (i) $s = ut + \frac{1}{2}at^2$

(ii) $v^2 = u^2 + 2as$

(iii) $v = u + at$

(iv) $s = \frac{u+v}{2} \times t$

(v) $v = u + at$

(vi) $v^2 = u^2 + 2as$

(vii) $s = ut + \frac{1}{2}at^2$

(viii) $v = u + at$

(ix) $s = \frac{u+v}{2} \times t$

(x) $v^2 = u^2 + 2as$

2 (i) 60

(ii) 12

(iii) 13

(iv) –5

(v) 2

3 $20\,\text{ms}^{-1}$

4 $3.4\,\text{ms}^{-2}$

5 5 seconds

6 $4.5\,\text{ms}^{-2}$

7 (i) $4.2\,\text{ms}^{-1}$

(ii) 0.43 seconds

8 (i)

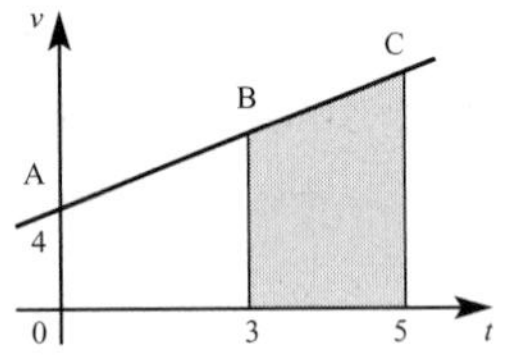

(ii) $14\,\text{ms}^{-1}$

(iii) $2\,\text{ms}^{-2}$

(iv) $10\,\text{ms}^{-1}$

9 (i) 1 second

(ii) 5 m

(iii) 2 seconds

10 (i) 1.5 seconds

(ii) 4.5 m

(iii) After 3 seconds

(iv) After 5 seconds

11 (i)

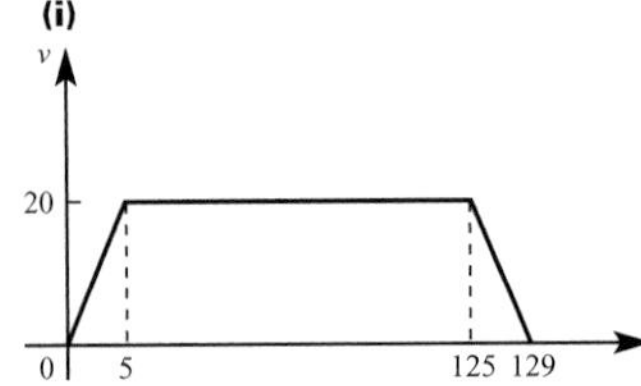

(ii) 129 seconds

(iii) 2490 m

❓ (Page 224)

Because a depends on t.

❓ (Page 225)

The particle is moving towards the origin during the first second.

Exercise 11B (Page 225)

1 (i) (a) $v = 10t - 1,\ a = 10$

(b) $s = 3,\ u = -1,\ a = 10$

(c) $t = 0.1,\ s = 2.95$

(ii) (a) $v = 3 - 3t^2,\ a = -6t$

(b) $s = 0,\ v = 3,\ a = 0$

(c) $t = 1,\ s = 2$

(iii) (a) $v = 4t^3 - 4,\ a = 12t^2$

(b) $s = -6,\ v = -4,\ a = 0$

(c) $t = 1,\ s = -9$

(iv) (a) $v = 12t^2 - 3,\ a = 24t$

(b) $s = 5,\ v = -3,\ a = 0$

(c) $t = 0.5,\ s = 4$

(v) (a) $v = -4t + 1,\ a = -4$

(b) $s = 5,\ v = 1,\ a = -4$

(c) $t = 0.25,\ s = 5.125$

2 (i) $v = 6t - 3t^2,\ a = 6 - 6t$

(ii) $t = 0$ and $t = 2$ seconds

(iii) 4 m

(iv) $v = -24\,\text{ms}^{-1}$; the body is moving towards O.

(v) $6\,\text{ms}^{-2}$

3 (i) 1 m

(ii) $v = 4 - 10t$

(iii) After 0.4 seconds

(iv) 1.8 m

(v) After 1 second

(vi)

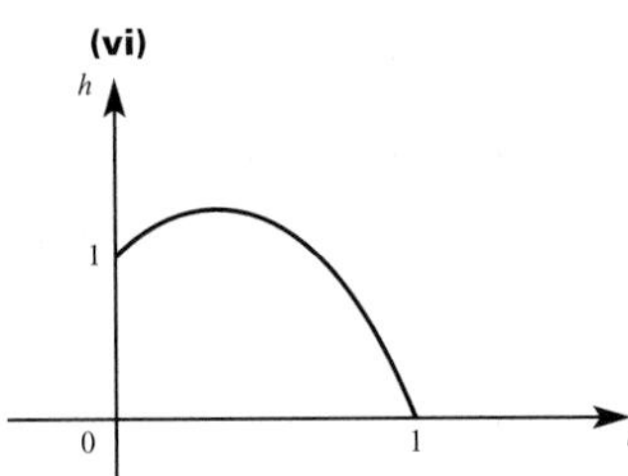

(vii) $6\,\text{ms}^{-1}$

4 (i) $v = \frac{2}{3}t^3,\ a = 2t^2$

(ii) After 6 seconds

(iii) $h = 216\,\text{m},\ v = 144\,\text{ms}^{-1}$

5 (i) After 2.5 seconds and 5 seconds

(ii) $5\,\text{ms}^{-2}$ and $-5\,\text{ms}^{-2}$

(iii) $3.125\,\text{ms}^{-1}$

(iv)

v
0
2.5
5
t
–25

❓ (Page 227)

No

❓ (Page 227)

c is the velocity when $t = 0$, i.e. the initial velocity.

❓ (Page 227)

k is the displacement when $t = 0$.

❓ (Page 228)

No

❓ (Page 230)

Method 1 is easier if the graph is already drawn, but it can only be used when the graph of v against t is a straight line.

❓ (Page 230)

(i) Method 2 must be used when the graph of v against t is not a straight line, as is the case when $v = 3t^2 + 2$.

(ii) No, $a = 6t$; the expression for acceleration is in terms of t so the acceleration varies with time.

(iii) No, you can only use the *suvat* equations when acceleration is constant.

❓ (Page 230)

Yes

Exercise 11C (Page 230)

1 (i) $v = 2t - 3t^2 + 1$,
$s = t^2 - t^3 + t$

(ii) $v = 2t^2 + 4$,
$s = \frac{2}{3}t^3 + 4t + 3$

(iii) $v = 4t^3 - 4t + 2$,
$s = t^4 - 2t^2 + 2t + 1$

(iv) $v = 2t + 2$,
$s = t^2 + 2t + 4$

(v) $v = 4t + \dfrac{t^2}{2} + 1$,
$s = 2t^2 + \dfrac{t^3}{6} + t + 3$

2 (i) $v = 3t^2 - 12t + 9$,
$s = t^3 - 6t^2 + 9t$

(ii) After 3 seconds

3 (i) $-3\,\text{ms}^{-1}$

(ii) After 2 seconds, 4 m

4 (i)

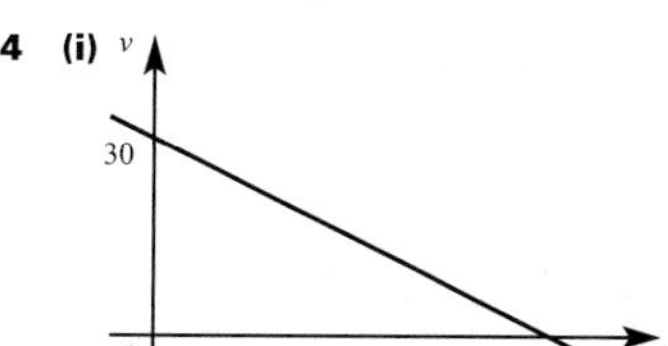

(ii) 6 seconds

(iii) 90 m

5 (i) $v = 4t + 6t^2$

(ii) 48 m

6 (i) $a = 3t^2 - 8t + 4$

(ii) After $\frac{2}{3}$ second and after 2 seconds; slowing down

(iii) 8.25

Index